Te-Ming Huang, Vojislav Kecman, Ivica Kopriva

Kernel Based Algorithms for Mining Huge Data Sets

Studies in Computational Intelligence, Volume 17

Editor-in-chief
Prof. Janusz Kacprzyk
Systems Research Institute
Polish Academy of Sciences
ul. Newelska 6
01-447 Warsaw
Poland
E-mail: kacprzyk@ibspan.waw.pl

Further volumes of this series can be found on our homepage: springer.com

Vol. 3. Bożena Kostek
Perception-Based Data Processing in Acoustics, 2005
ISBN 3-540-25729-2

Vol. 4. Saman K. Halgamuge, Lipo Wang (Eds.)
Classification and Clustering for Knowledge Discovery, 2005
ISBN 3-540-26073-0

Vol. 5. Da Ruan, Guoqing Chen, Etienne E. Kerre, Geert Wets (Eds.)
Intelligent Data Mining, 2005
ISBN 3-540-26256-3

Vol. 6. Tsau Young Lin, Setsuo Ohsuga, Churn-Jung Liau, Xiaohua Hu, Shusaku Tsumoto (Eds.)
Foundations of Data Mining and Knowledge Discovery, 2005
ISBN 3-540-26257-1

Vol. 7. Bruno Apolloni, Ashish Ghosh, Ferda Alpaslan, Lakhmi C. Jain, Srikanta Patnaik (Eds.)
Machine Learning and Robot Perception, 2005
ISBN 3-540-26549-X

Vol. 8. Srikanta Patnaik, Lakhmi C. Jain, Spyros G. Tzafestas, Germano Resconi, Amit Konar (Eds.)
Innovations in Robot Mobility and Control, 2006
ISBN 3-540-26892-8

Vol. 9. Tsau Young Lin, Setsuo Ohsuga, Churn-Jung Liau, Xiaohua Hu (Eds.)
Foundations and Novel Approaches in Data Mining, 2005
ISBN 3-540-28315-3

Vol. 10. Andrzej P. Wierzbicki, Yoshiteru Nakamori
Creative Space, 2005
ISBN 3-540-28458-3

Vol. 11. Antoni Ligêza
Logical Foundations for Rule-Based Systems, 2006
ISBN 3-540-29117-2

Vol. 13. Nadia Nedjah, Ajith Abraham, Luiza de Macedo Mourelle (Eds.)
Genetic Systems Programming, 2006
ISBN 3-540-29849-5

Vol. 14. Spiros Sirmakessis (Ed.)
Adaptive and Personalized Semantic Web, 2006
ISBN 3-540-30605-6

Vol. 15. Lei Zhi Chen, Sing Kiong Nguang, Xiao Dong Chen
Modelling and Optimization of Biotechnological Processes, 2006
ISBN 3-540-30634-X

Vol. 16. Yaochu Jin (Ed.)
Multi-Objective Machine Learning, 2006
ISBN 3-540-30676-5

Vol. 17. Te-Ming Huang, Vojislav Kecman, Ivica Kopriva
Kernel Based Algorithms for Mining Huge Data Sets, 2006
ISBN 3-540-31681-7

Te-Ming Huang
Vojislav Kecman
Ivica Kopriva

Kernel Based Algorithms for Mining Huge Data Sets

Supervised, Semi-supervised, and Unsupervised Learning

Te-Ming Huang
Vojislav Kecman

Faculty of Engineering
The University of Auckland
Private Bag 92019
1030 Auckland, New Zealand
E-mail: huangtm@learning-from-data.com
v.kecman@auckland.ac.nz

Ivica Kopriva

Department of Electrical and
Computer Engineering
22nd St. NW 801
20052 Washington D.C., USA
E-mail: ikopriva@gmail.com

ISSN print edition: 1860-949X
ISSN electronic edition: 1860-9503
ISBN 978-3-642-06856-0
e-ISBN 978-3-540-31689-3

Springer is a part of Springer Science+Business Media
springer.com

Softcover reprint of the hardcover 1st edition 2006

To Our Parents

Jun-Hwa Huang & Wen-Chuan Wang,
Danica & Mane Kecman,
Štefanija & Antun Kopriva,

and to Our Teachers

Preface

This is a book about (machine) learning from (experimental) data. Many books devoted to this broad field have been published recently. One even feels tempted to begin the previous sentence with an adjective *extremely*. Thus, there is an urgent need to introduce both the motives for and the content of the present volume in order to highlight its distinguishing features.

Before doing that, few words about the very broad meaning of data are in order. Today, we are surrounded by an ocean of all kind of experimental data (i.e., examples, samples, measurements, records, patterns, pictures, tunes, observations,..., etc) produced by various sensors, cameras, microphones, pieces of software and/or other human made devices. The amount of data produced is enormous and ever increasing. The first obvious consequence of such a fact is - humans can't handle such massive quantity of data which are usually appearing in the numeric shape as the huge (rectangular or square) matrices. Typically, the number of their rows (n) tells about the number of data pairs collected, and the number of columns (m) represent the dimensionality of data. Thus, faced with the Giga- and Terabyte sized data files one has to develop new approaches, algorithms and procedures. Few techniques for coping with huge data size problems are presented here. This, possibly, explains the appearance of a wording *'huge data sets'* in the title of the book.

Another direct consequence is that (instead of attempting to dive into the sea of hundreds of thousands or millions of high-dimensional data pairs) we are developing other 'machines' or 'devices' for analyzing, recognizing and/or learning from, such huge data sets. The so-called 'learning machine' is predominantly a piece of software that implements both the learning algorithm and the function (network, model) which parameters has to be determined by the learning part of the software. Today, it turns out that some models used for solving machine learning tasks are either originally based on using kernels (e.g., support vector machines), or their newest extensions are obtained by an introduction of the kernel functions within the existing standard techniques. Many classic data mining algorithms are extended to the applications in the high-dimensional feature space. The list is long as well as the fast growing one, and just the most recent extensions are mentioned here. They are - kernel principal component analysis, kernel independent component analysis, kernel least squares, kernel discriminant analysis, kernel k-means clustering, kernel self-organizing feature map, kernel Mahalanobis distance, kernel subspace classification methods and kernel functions based dimensionality reduction. What the kernels are, as well as why and how they became so popular in the learning from data sets tasks, will be shown shortly. As for now, their wide use as well as their efficiency in a numeric part of the algorithms (achieved by avoiding the calculation of the scalar products between extremely high dimensional feature vectors), explains their appearance in the title of the book.

Next, it is worth of clarifying the fact that many authors tend to label similar (or even same) models, approaches and algorithms by different names. One is just destine to cope with concepts of data mining, knowledge discovery, neural networks, Bayesian networks, machine learning, pattern recognition, classification, regression, statistical learning, decision trees, decision making etc. All of them usually have a lot in common, and they often use the same set of techniques for adjusting, tuning, training or learning the parameters defining the models. The common object for all of them is a training data set. All the various approaches mentioned start with a set of data pairs $(\mathbf{x}_i, y_i)$ where $\mathbf{x}_i$ represent the input variables (causes, observations, records) and y_i denote the measured outputs (responses, labels, meanings). However, even with the very commencing point in machine learning (namely, with the training data set collected), the real life has been tossing the coin in providing us either with

- a set of genuine training data pairs $(\mathbf{x}_i, y_i)$ where for each input $\mathbf{x}_i$ there is a corresponding output y_i or with,
- the partially labeled data containing both the pairs $(\mathbf{x}_i, y_i)$ and the sole inputs $\mathbf{x}_i$ without associated known outputs y_i or, in the worst case scenario, with
- the set of sole inputs (observations or records) $\mathbf{x}_i$ without any information about the possible desired output values (labels, meaning) y_i.

It is a genuine challenge indeed to try to solve such differently posed machine learning problems by the unique approach and methodology. In fact, this is exactly what did not happen in the real life because the development in the field followed a natural path by inventing different tools for unlike tasks. The answer to the challenge was a, more or less, independent (although with some overlapping and mutual impact) development of three large and distinct sub-areas in machine learning - *supervised, semi-supervised* and *unsupervised* learning. This is where both the subtitle and the structure of the book are originated from. Here, all three approaches are introduced and presented in details which should enable the reader not only to acquire various techniques but also to equip him/herself with all the basic knowledge and requisites for further development in all three fields on his/her own.

The presentation in the book follows the order mentioned above. It starts with seemingly most powerful *supervised learning* approach in solving classification (pattern recognition) problems and regression (function approximation) tasks at the moment, namely with support vector machines (SVMs). Then, it continues with two most popular and promising *semi-supervised approaches* (with graph based semi-supervised learning algorithms; with the Gaussian random fields model (GRFM) and with the consistency method (CM)). Both the original setting of methods and their improved versions will be introduced. This makes the volume to be the first book on semi-supervised learning at all. The book's final part focuses on the two most appealing and widely used *unsupervised methods* labeled as principal component analysis (PCA) and independent component analysis (ICA). Two algorithms are the working horses in unsupervised learning today and their presentation, as well as a pointing to their major characteristics, capacities and differences, is given the highest care here.

The models and algorithms for all three parts of machine learning mentioned are given in the way that equips the reader for their straight implementation. This is achieved not only by their sole presentation but also through the applications of the models and algorithms to some low dimensional (and thus, easy to understand, visualize and follow) examples. The equations and models provided will be able to handle much bigger problems (the ones having much more data of much higher dimensionality) in the same way as they did the ones we can follow and 'see' in the examples provided. In the authors' experience and opinion, the approach adopted here is the most accessible, pleasant and useful way to master the material containing many new (and potentially difficult) concepts.

The structure of the book is shown in Fig. 0.1.

The basic motivations and presentation of three different approaches in solving three unlike learning from data tasks are given in Chap. 1. It is a kind of both the background and the stage for a book to evolve.

Chapter 2 introduces the constructive part of the SVMs without going into all the theoretical foundations of statistical learning theory which can be found in many other books. This may be particularly appreciated by and useful for the applications oriented readers who do not need to know all the theory back to its roots and motives. The basic quadratic programming (QP) based learning algorithms for both classification and regression problems are presented here. The ideas are introduced in a gentle way starting with the learning algorithm for classifying linearly separable data sets, through the classification tasks having overlapped classes but still a linear separation boundary, beyond the linearity assumptions to the nonlinear separation boundary, and finally to the linear and nonlinear regression problems. The appropriate examples follow each model derived, just enabling in this way an easier grasping of concepts introduced. The material provided here will be used and further developed in two specific directions in Chaps. 3 and 4.

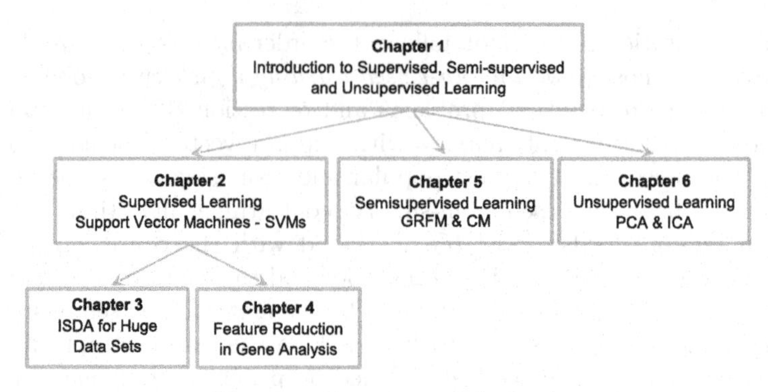

Fig. 0.1. Structure of the book

Chapter 3 resolves the crucial problem of the QP based learning coming from the fact that the learning stage of SVMs scales with the number of training data pairs. Thus, when having more than few thousands data pairs, the size of the original Hessian matrix appearing in the cost function of the QP problem setting goes beyond the capacities of contemporary computers. The fact that memory chips are increasing is not helping due to the much faster increase in the size of data files produced. Thus, there is a need for developing an iterative learning algorithm that does not require a calculation of the complete Hessian matrix. The Iterative Single Data Algorithm (ISDA) that in each iteration step needs a single data point only is introduced here. Its performance seems to be superior to other known iterative approaches.

Chapter 4 shows how SVMs can be used as a feature reduction tools by coupling with the idea of recursive feature elimination. The Recursive Feature Elimination with Support Vector Machines (RFE-SVMs) developed in [61] is the first approach that utilizes the idea of margin as a measure of relevancy for feature selection. In this chapter, an improved RFE-SVM is also proposed and it is applied to the challenging problem of DNA microarray analysis. DNA microarray is a powerful tool which allows biologists to measure thousands of genes' expression in a single experiment. This technology opens up the possibility of finding out the causal relationship between genes and certain phenomenon in the body, e.g. which set of genes is responsible for certain disease or illness. However, the high cost of the technology and the limited number of samples available make the learning from DNA microarray data a very difficult task. This is due to the fact that the training data set normally consists of a few dozens of samples, but the number of genes (i.e., the dimensionality of the problem) can be as high as several thousands. The results of applying the improved RFE-SVM to two DNA microarray data sets show that the performance of RFE-SVM seems to be superior to other known approaches such as the nearest shrunken centroid developed in [137].

Chapter 5 presents two very promising semi-supervised learning techniques, namely, GRFM and CM. Both methods are based on the theory of graphical models and they explore the manifold structure of the data set which leads to their global convergence. An in depth analysis of both approaches when facing with unbalanced labeled suggests that the performance of both approaches can deteriorate very significantly when labeled data is unbalanced (i.e., when the number of labeled data in each class is different). As a result, a novel normalization step is introduced to both algorithms improving the performance of the algorithms very significantly when faced with an unbalance in labeled data. This chapter also presents the comparisons of CM and GRFM with the various variants of transductive SVMs (TSVMs) and the results suggest that the graph-based approaches seem to have better performance in multi-class problems

Chapter 6 introduces two basic methodologies for learning from unlabeled data within the unsupervised learning approach: the Principal Component Analysis (PCA) and the Independent Component Analysis (ICA). Unsupervised learning is related to the principle of redundancy reduction which is implemented in mathematical form through minimization of the statistical dependence between observed data pairs. It is demonstrated that PCA, which decorrelates data pairs, is optimal for Gaussian sources and suboptimal for non-Gaussian ones. It is also pointed to the necessity of using ICA for non-Gaussian sources as well as that there is no reason for using it in the case of Gaussian ones. PCA algorithm known as whitening or sphering transform is derived. Batch and adaptive ICA algorithms are derived through the minimization of the mutual information which is an exact measure of statistical (in)dependence between data pairs. Both PCA and ICA derived unsupervised learning algorithms are implemented in MATLAB code, which illustrates their use on computer generated examples.

As it is both the need and the habit today, the book is accompanied with an Internet site

`www.learning-from-data.com`

The site contains the software and other material used in the book and it may be helpful for readers to make occasional visits and download the newest version of software and/or data files.

Auckland, New Zealand,
Washington, D.C., USA
October 2005

Te-Ming Huang
Vojislav Kecman
Ivica Kopriva

Contents

1

Introduction

1.1 An Overview of Machine Learning

The amount of data produced by sensors has increased explosively as a result of the advances in sensor technologies that allow engineers and scientists to quantify many processes in fine details. Because of the sheer amount and complexity of the information available, engineers and scientists now rely heavily on computers to process and analyze data. This is why machine learning has become an emerging topic of research that has been employed by an increasing number of disciplines to automate complex decision-making and problem-solving tasks. This is because the goal of machine learning is to extract knowledge from experimental data and use computers for complex decision-making, i.e. decision rules are extracted automatically from data by utilizing the speed and the robustness of the machines. As one example, the DNA microarray technology allows biologists and medical experts to measure the expressiveness of thousands of genes of a tissue sample in a single experiment. They can then identify cancerous genes in a cancer study. However, the information that is generated from the DNA microarray experiments and many other measuring devices cannot be processed or analyzed manually because of its large size and high complexity. In the case of the cancer study, the machine learning algorithm has become a valuable tool to identify the cancerous genes from the thousands of possible genes. Machine-learning techniques can be divided into three major groups based on the types of problems they can solve, namely, the supervised, semi-supervised and unsupervised learning.

The supervised learning algorithm attempts to learn the input-output relationship (dependency or function) $f(x)$ by using a training data set $\{\mathcal{X} = [\mathbf{x}_i, y_i], i = 1, \ldots, n\}$ consisting of n pairs $(\mathbf{x}_1, y_1), (\mathbf{x}_2, y_2), \ldots (\mathbf{x}_n, y_n)$, where the inputs $\mathbf{x}$ are m-dimensional vectors $\mathbf{x} \in \Re^m$ and the labels (or system responses) y are discrete (e.g., Boolean) for classification problems and continuous values ($y \in \Re$) for regression tasks. Support Vector Machines (SVMs) and Artificial Neural Network (ANN) are two of the most popular techniques in this area.

T.-M. Huang et al.: *Kernel Based Algorithms for Mining Huge Data Sets*, Studies in Computational Intelligence (SCI) **17**, 1–9 (2006)
www.springerlink.com

There are two types of supervised learning problems, namely, classification (pattern recognition) and the regression (function approximation) ones. In the classification problem, the training data set consists of examples from different classes. The simplest classification problem is a binary one that consists of training examples from two different classes (+1 or -1 class). The outputs $y_i \in \{1, -1\}$ represent the class belonging (i.e. labels) of the corresponding input vectors $\mathbf{x}_i$ in the classification. The input vectors $\mathbf{x}_i$ consist of measurements or features that are used for differentiating examples of different classes. The learning task in classification problems is to construct classifiers that can classify previously unseen examples $\mathbf{x}_j$. In other words, machines have to learn from the training examples first, and then they should make complex decisions based on what they have learned. In the case of multi-class problems, several binary classifiers are built and used for predicting the labels of the unseen data, i.e. an N-class problem is generally broken down into N binary classification problems. The classification problems can be found in many different areas, including, object recognition, handwritten recognition, text classification, disease analysis and DNA microarray studies. The term "supervised" comes from the fact that the labels of the training data act as teachers who educate the learning algorithms.

In the regression problem, the task is to find the mapping between input $\mathbf{x} \in \Re^m$ and output $y \in \Re$. The output y in regression is a continuous value instead of a discrete one in the classification. Similarly, the learning task in regression is to find the underlying function between some m-dimensional input vectors $\mathbf{x}_i \in \Re^m$ and scalar outputs $y_i \in \Re$. The regression problems can also be found in many disciplines, including time-series analysis, control system, navigation and interest rates analysis in finance.

There are two phases when applying supervised learning algorithms for problem-solving as shown in Fig. 1.1. The first phase is the so-called learning phase where the learning algorithms design a mathematical model of a

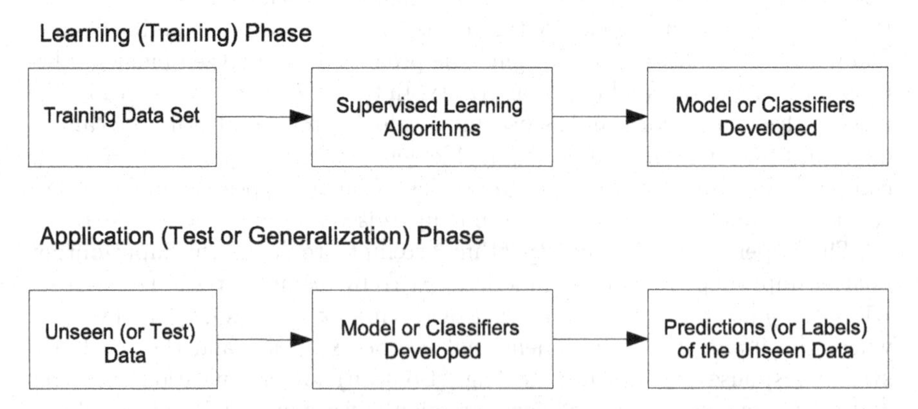

Fig. 1.1. Two Phases of Supervised Learning Algorithms.

dependency, function or mapping (in an regression) or classifiers (in a classification i.e., pattern recognition) based on the training data given. This can be a time-consuming procedure if the size of the training data set is huge. One of the mainstream research fields in learning from empirical data is to design algorithms that can be applied to large-scale problems efficiently, which is also the core of this book. The second phase is the test and/or application phase. In this phase, the models developed by the learning algorithms are used to predict the outputs y_i of the data which are unseen by the learning algorithms in the learning phase. Before an actual application, the test phase is always carried out for checking the accuracy of the models developed in the first phase.

Another large group of standard learning algorithms are those dubbed as unsupervised algorithms when there are only raw data $\mathbf{x}_i \in \Re^m$ without the corresponding labels y_i (i.e., there is a 'no-teacher' in a shape of labels). The most popular, representative, algorithms belonging to this group are various clustering techniques and (principal or independent) component analysis routines. These two algorithms will be introduced and compared in Chap. 6.

Between the two ends of the spectrum are the semi-supervised learning problems. These problems are characterized by the presence of (usually) a small percentage of labeled data and a large percentage of unlabeled ones. The cause of an appearance of the unlabeled data points is usually an expensive, difficult and slow process of obtaining labeled data. Thus, labeling brings additional costs and often it is not feasible. Typical areas where this happens are speech processing (due to the slow transcription), text categorization (due to huge number of documents and slow reading by people), web categorization, and, finally, a bioinformatics area where it is usually both expensive and slow to label a huge number of data produced. As a result, the goal of a semi-supervised learning algorithm is to predict the labels of the unlabeled data by taking the entire data set into account. In other words, the training data set consists of both labeled and unlabeled data (more details will be found in Chap. 5). At the time of writing this book, the semi-supervised learning techniques are still at the early stage of their developments and they are only applicable for solving classification problems. This is because they are designed to group the unlabeled data $\mathbf{x}_i$, but not to approximate the underlying function $f(\mathbf{x})$. This volume seems to be the first one (in the line of many books coming) on semi-supervised learning. The presentation here is focused on the widely used and the most popular graph-based (a.k.a. manifold) approaches only.

1.2 Challenges in Machine Learning

Like most areas in science and engineering, machine learning requires developments in both theoretical and practical (engineering) aspects. An activity on the theoretical side is concentrated on inventing new theories as the foundations for constructing novel learning algorithms. On the other hand, by extending existing theories and inventing new techniques, researchers who work in the engineering aspects of the field try to improve the existing learning algorithms and apply them to the novel and challenging real-world problems. This book is focused on the practical aspects of SVMs, graph-based semi-supervised learning algorithms and two basic unsupervised learning methods. More specifically, it aims at making these learning techniques more practical for the implementation to the real-world tasks. As a result, the primary goal of this book is aimed at developing novel algorithms and software that can solve large-scale SVMs, graph-based semi-supervised and unsupervised learning problems. Once an efficient software implementation has been obtained, the goal will be to apply these learning techniques to real-world problems and to improve their performance. Next four sections outline the original contributions of the book in solving the mentioned tasks.

1.2.1 Solving Large-Scale SVMs

As mentioned previously, machine learning techniques allow engineers and scientists to use the power of computers to process and analyze large amounts of information. However, the amount of information generated by sensors can easily go beyond the processing power of the latest computers available. As a result, one of the mainstream research fields in learning from empirical data is to design learning algorithms that can be used in solving large-scale problems efficiently. The book is primarily aimed at developing efficient algorithms for implementing SVMs. SVMs are the latest supervised learning techniques from statistical learning theory and they have been shown to deliver state-of-the-art performance in many real-world applications [153]. The challenge of applying SVMs on huge data sets comes from the fact that the amount of computer memory required for solving the quadratic programming (QP) problem associated with SVMs increases drastically with the size of the training data set n (more details can be found in Chap. 3). As a result, the book aims at providing a better solution for solving large-scale SVMs using iterative algorithms. The novel contributions presented in this book are as follows:

1. The development of Iterative Single Data Algorithm (ISDA) with the explicit bias term b. Such a version of ISDA has been shown to perform better (faster) than the standard SVMs learning algorithms achieving at the same time the same accuracy. These contributions are presented in Sect. 3.3 and 3.4.
2. An efficient software implementation of the ISDA is developed. The ISDA software has been shown to be significantly faster than the well-known SVMs learning software LIBSVM [27]. These contributions are presented in Sect. 3.5.

1.2.2 Feature Reduction with Support Vector Machines

Recently, more and more instances have occurred in which the learning problems are characterized by the presence of a small number of the high-dimensional training data points, i.e. n is small and m is large. This often occurs in the bioinformatics area where obtaining training data is an expensive and time-consuming process. As mentioned previously, recent advances in the DNA microarray technology allow biologists to measure several thousands of genes' expressions in a single experiment. However, there are three basic reasons why it is not possible to collect many DNA microarrays and why we have to work with sparse data sets. First, for a given type of cancer it is not simple to have thousands of patients in a given time frame. Second, for many cancer studies, each tissue sample used in an experiment needs to be obtained by surgically removing cancerous tissues and this is an expensive and time consuming procedure. Finally, obtaining the DNA microarrays is still expensive technology. As a result, it is not possible to have a relatively large quantity of training examples available. Generally, most of the microarray studies have a few dozen of samples, but the dimensionality of the feature spaces (i.e. space of input vector $\mathbf{x}$) can be as high as several thousand. In such cases, it is difficult to produce a classifier that can generalize well on the unseen data, because the amount of training data available is insufficient to cover the high dimensional feature space. It is like trying to identify objects in a big dark room with only a few lights turned on. The fact that n is much smaller than m makes this problem one of the most challenging tasks in the areas of machine learning, statistics and bioinformatics.

The problem of having high-dimensional feature space led to the idea of selecting the most relevant set of genes or features first, and only then the classifier is constructed from these selected and "'important"' features by the learning algorithms. More precisely, the classifier is constructed over a reduced space (and, in the comparative example above, this corresponds to an object identification in a smaller room with the same number of lights). As a result such a classifier is more likely to generalize well on the unseen data. In the book, a feature reduction technique based on SVMs (dubbed Recursive Feature Elimination with Support Vector Machines (RFE-SVMs)) developed in [61], is implemented and improved. In particular, the focus is on gene selection for cancer diagnosis using RFE-SVMs. RFE-SVM is included in the book because it is the most natural way to harvest the discriminative power of SVMs for microarray analysis. At the same time, it is also a natural extension of the work on solving SVMs efficiently. The original contributions presented in the book in this particular area are as follows:

1. The effect of the penalty parameter C which was neglected in most of the studies is explored in order to develop an improved RFE-SVMs for feature reduction. The simulation results suggest that the performance improvement can be as high as 35% on the popular colon cancer data-set

[8]. Furthermore, the improved RFE-SVM outperforms several other techniques including the well-known nearest shrunken centroid method [137] developed at the Stanford University. These contributions are contained in Sects. 4.4, 4.5 and 4.6.
2. An investigation of the effect of different data preprocessing procedures on the RFE-SVMs was carried out. The results suggest that the performance of the algorithms can be affected by different procedures. They are presented in Sect. 4.5.2
3. The book also tries to determine whether gene selection algorithms such as RFE-SVMs can help biologists to find the right set of genes causing a certain disease. A comparison of the genes' ranking from different algorithms shows a great deal of consensus among all nine different algorithms tested in the book. This indicates that machine learning techniques may help narrowing down the scope of searching for the set of 'optimal' genes. This contribution is presented in Sect. 4.7.

1.2.3 Graph-Based Semi-supervised Learning Algorithms

As mentioned previously, semi-supervised learning (SSL) is the latest development in the field of machine learning. It is driven by the fact that in many real-world problems the cost of labeling data can be quite high and there is an abundance of unlabeled data. The original goal of this book was to develop large-scale solvers for SVMs and apply SVMs to real-world problems only. However, it was found that some of the techniques developed in SVMs can be extended naturally to the graph-based semi-supervised learning, because the optimization problems associated with both learning techniques are identical (more details shortly).

In the book, two very popular graph-based semi-supervised learning algorithms, namely, the Gaussian random fields model (GRFM) introduced in [160] and [159], and the consistency method (CM) for semi-supervised learning proposed in [155] were improved. The original contributions to the field of SSL presented in this book are as follows:

1. An introduction of the novel normalization step into both CM and GRFM. This additional step improves the performance of both algorithms significantly in the cases where labeled data are unbalanced. The labeled data are regarded as unbalanced when each class has a different number of labeled data in the training set. This contribution is presented in Sect. 5.3 and 5.4.
2. The world first large-scale graph-based semi-supervised learning software SemiL is developed as part of this book. The software is based on a Conjugate Gradient (CG) method which can take box-constraints into account and it is used as a backbone for all the simulation results in Chap. 5. Furthermore, SemiL has become a very popular tool in this area at the time of writing this book, with approximately 100 downloads per month. The details of this contribution are given in Sect. 5.6.

Both CM and GRFM are also applied to five benchmarking data sets in order to compare them with Low Density Separation (LDS) method developed in [29]. The detailed comparison shows the strength and the weakness of different semi-supervised learning approaches. It is presented in Sect. 5.5. Although SVMs and graph-based semi-supervised learning algorithms are totally different in terms of their theoretical foundations, the same Quadratic Programming (QP) problem needs to be solved for both of them in order to learn from the training data. In SVMs, when positive-definite kernels are used without bias term, the QP problem has the following form:

$$\max L_d(\boldsymbol{\alpha}) = -0.5\boldsymbol{\alpha}^T\mathbf{H}\boldsymbol{\alpha} + \mathbf{p}^T\boldsymbol{\alpha}, \tag{1.1a}$$

$$\text{s.t.} \quad 0 \leq \alpha_i \leq C, i = 1, \ldots, k, \tag{1.1b}$$

where, in the classification $k = n$ (n is the size of the data set) and the Hessian matrix $\mathbf{H}$ is an $n \times n$ symmetric positive definite matrix, while in regression $k = 2n$ and $\mathbf{H}$ is a $2n \times 2n$ symmetric semi-positive definite one, α_i are the Lagrange multipliers in SVMs, in classification $\mathbf{p}$ is a unit $n \times 1$ vector, and C is the penalty parameter in SVMs. The task is to find the optimal $\boldsymbol{\alpha}$ that gives the maximum of L_d (more details can be found in Chap. 2 and 3). Similarly, in graph-based semi-supervised learning, the following optimization problem which is in the same form as (1.1) needs to be solved (see Sect. 5.2.2)

$$\max \quad \mathbf{Q}(\mathbf{f}) = -\frac{1}{2}\mathbf{f}^T\mathcal{L}\mathbf{f} + \mathbf{y}^T\mathbf{f} \tag{1.2a}$$

$$\text{s.t.} \quad -C \leq f_i \leq C \quad i = 1 \ldots n \tag{1.2b}$$

where $\mathcal{L}$ is the normalized Laplacian matrix, $\mathbf{f}$ is the output of graph-based semi-supervised learning algorithm, C is the parameter that restricts the size of the output $\mathbf{f}$, $\mathbf{y}$ is a $n \times 1$ vector that contains the information about the labeled data and n is the size of the data set.

The Conjugate Gradient (CG) method for box constraints implemented in SemiL (in Sect. 5.6.3) was originally intended and developed to solve large-scale SVMs. Because the $\mathbf{H}$ matrix in the case of SVMs is extremely dense, it was found that CG is not as efficient as ISDA for solving SVMs. However, it is ideal for the graph-based semi-supervised learning algorithms, because matrix $\mathcal{L}$ can be a sparse one in the graph-based semi-supervised learning. This is why the main contributions of the book is across the two major subfields of machine learning. The algorithms developed for solving the SVMs learning problem are the ones successfully implemented in this part of the book, too.

1.2.4 Unsupervised Learning Based on Principle of Redundancy Reduction

SVMs as the latest supervised learning technique from the statistical learning theory as well as any other supervised learning method require labeled data in order to train the learning machine. As already mentioned, in many real world problems the cost of labeling data can be quite high. This presented motivation for most recent development of the semi-supervised learning where only small amount of data is assumed to be labeled. However, there exist classification problems where accurate labeling of the data is sometime even impossible. One such application is classification of remotely sensed multispectral and hyperspectral images [46, 47]. Recall that typical family RGB color image (photo) contains three spectral bands. In other words we can say that family photo is a three-spectral image. A typical hyperspectral image would contain more than one hundred spectral bands. As remote sensing and its applications receive lots of interests recently, many algorithms in remotely sensed image analysis have been proposed [152]. While they have achieved a certain level of success, most of them are supervised methods, i.e., the information of the objects to be detected and classified is assumed to be known a priori. If such information is unknown, the task will be much more challenging. Since the area covered by a single pixel is very large, the reflectance of a pixel can be considered as the mixture of all the materials resident in the area covered by the pixel. Therefore, we have to deal with mixed pixels instead of pure pixels as in conventional digital image processing. Linear spectral unmixing analysis is a popular approach used to uncover material distribution in an image scene [127, 2, 125, 3]. Formally, the problem is stated as:

$$\mathbf{r} = \mathbf{M}\boldsymbol{\alpha} + \mathbf{n} \tag{1.3}$$

where $\mathbf{r}$ is a reflectance column pixel vector with dimension L in a hyperspectral image with L spectral bands. An element r_i in the $\mathbf{r}$ is the reflectance collected in the ith wavelength band. $\mathbf{M}$ denotes a matrix containing p independent material spectral signatures (referred to as endmembers in linear mixture model), i.e., $\mathbf{M} = [\mathbf{m}_1, \mathbf{m}_2, \ldots, \mathbf{m}_p]$, $\boldsymbol{\alpha}$ represents the unknown abundance column vector of size $p \times 1$ associated with $\mathbf{M}$, which is to be estimated and $\mathbf{n}$ is the noise term. The ith item α_i in $\boldsymbol{\alpha}$ represents the abundance fraction of $\mathbf{m}_i$ in pixel $\mathbf{r}$. When $\mathbf{M}$ is known, the estimation of $\boldsymbol{\alpha}$ can be accomplished by least squares approach. In practice, it may be difficult to have *prior* information about the image scene and endmember signatures. Moreover, in-field spectral signatures may be different from those in spectral libraries due to atmospheric and environmental effects. So an unsupervised classification approach is preferred. However, when $\mathbf{M}$ is also unknown, i.e., in unsupervised analysis, the task is much more challenging since both $\mathbf{M}$ and $\boldsymbol{\alpha}$ need to be estimated [47]. Under stated conditions the problem represented by linear mixture model (1.3) can be interpreted as a linear instantaneous blind source separation (BSS) problem [76] mathematically described as:

$$\mathbf{x} = \mathbf{A}\mathbf{s} + \mathbf{n} \tag{1.4}$$

where $\mathbf{x}$ represents data vector, $\mathbf{A}$ is unknown mixing matrix, $\mathbf{s}$ is vector of source signals or classes to be found by an unsupervised method and $\mathbf{n}$ is again additive noise term. The BSS problem is solved by the independent component analysis (ICA) algorithms [76]. The advantages offered by interpreting linear mixture model (1.3) as an BSS problem (1.4) in remote sensing image classification are: 1) no prior knowledge of the endmembers in the mixing process is required; 2) the spectral variability of the endmembers can be accommodated by the unknown mixing matrix $\mathbf{M}$ since the source signals are considered as scalar and random quantities; and 3) higher order statistics can be exploited for better feature extraction and pattern classification. The last advantage is consequence of the non-Gaussian nature of the classes what is assumed by each ICA method.

As noted in [67] any meaningful data are not really random but are generated by physical processes. When physical processes are independent generated source signals i.e. classes are not related too. It means they are statistically independent. Statistical independence implies that there is no redundancy between the classes. If redundancy between the classes or sources is interpreted as the amount of information which one can infer about one class having information about another one then mutual information can be used as a redundancy measure between the sources or classes. This represents mathematical implementation of the redundancy reduction principle, which was suggested in [14] as a coding strategy in neurons. The reason is that, as shown in [41], the mutual information expressed in a form of the Kullback-Leibler divergence:

$$I(s_1, s_2, ..., s_N) = D\left(p(\mathbf{s}) \left\| \prod_{n=1}^{N} p_n(s_n) \right.\right) = \int p(\mathbf{s}) \log \frac{p(\mathbf{s})}{\prod\limits_{n=1}^{N} p_n(s_n)} \mathbf{ds} \quad (1.5)$$

is a non-negative convex function with the global minimum equal to zero for $p(\mathbf{s}) = \prod\limits_{n=1}^{N} p_n(s_n)$ i..e. when classes s_n are statistically independent. Indeed, as it is shown in Chap. 6, it is possible to derive computationally efficient and completely unsupervised ICA algorithm through the minimization of the mutual information between the sources. PCA and ICA are unsupervised classification methods built upon uncorrelatedness and independence assumptions respectively. They provide very powerful tool for solving BSS problems, which have found applications in many fields such as brain mapping [93, 98], wireless communications [121], nuclear magnetic resonance spectroscopy [105] and already mentioned unsupervised classification of the multispectral remotely sensed images [46, 47]. That is why PCA and ICA as two representative groups of unsupervised learning methods are covered in this book.

2

Support Vector Machines in Classification and Regression – An Introduction

This is an introductory chapter on the supervised (machine) learning from empirical data (i.e., examples, samples, measurements, records, patterns or observations) by applying support support vector machines (SVMs) a.k.a. kernel machines[1]. The parts on the semi-supervised and unsupervised learning are given later and being entirely different tasks they use entirely different math and approaches. This will be shown shortly. Thus, the book introduces the problems gradually in an order of loosing the information about the desired output label. After the supervised algorithms, the semi-supervised ones will be presented followed by the unsupervised learning methods in Chap. 6. The basic aim of this chapter is to give, as far as possible, a condensed (but systematic) presentation of a novel learning paradigm embodied in SVMs. Our focus will be on the constructive part of the SVMs' learning algorithms for both the classification (pattern recognition) and regression (function approximation) problems. Consequently, we will not go into all the subtleties and details of the statistical learning theory (SLT) and structural risk minimization (SRM) which are theoretical foundations for the learning algorithms presented below. The approach here seems more appropriate for the application oriented readers. The theoretically minded and interested reader may find an extensive presentation of both the SLT and SRM in [146, 144, 143, 32, 42, 81, 123]. Instead of diving into a theory, a quadratic programming based learning, leading to parsimonious SVMs, will be presented in a gentle way - starting with linear separable problems, through the classification tasks having overlapped classes but still a linear separation boundary, beyond the linearity assumptions to the nonlinear separation boundary, and finally to the linear and nonlinear regression problems. Here, the adjective 'parsimonious' denotes a SVM with a small number of support vectors ('hidden layer neurons'). The scarcity of the model results from a sophisticated, QP based, learning that matches the

[1] This introduction strictly follows and partly extends the School of Engineering of The University of Auckland Report 616. The right to use the material from this report is received with gratitude.

T.-M. Huang et al.: *Kernel Based Algorithms for Mining Huge Data Sets*, Studies in Computational Intelligence (SCI) **17**, 11–60 (2006)
www.springerlink.com

model capacity to data complexity ensuring a good generalization, i.e., a good performance of SVM on the future, previously, during the training unseen, data.

Same as the neural networks (or similarly to them), SVMs possess the well-known ability of being universal approximators of any multivariate function to any desired degree of accuracy. Consequently, they are of particular interest for modeling the unknown, or partially known, highly nonlinear, complex systems, plants or processes. Also, at the very beginning, and just to be sure what the whole chapter is about, we should state clearly when there is no need for an application of SVMs' model-building techniques. In short, whenever there exists an analytical closed-form model (or it is possible to devise one) there is no need to resort to learning from empirical data by SVMs (or by any other type of a learning machine)

2.1 Basics of Learning from Data

SVMs have been developed in the reverse order to the development of neural networks (NNs). SVMs evolved from the sound theory to the implementation and experiments, while the NNs followed more heuristic path, from applications and extensive experimentation to the theory. It is interesting to note that the very strong theoretical background of SVMs did not make them widely appreciated at the beginning. The publication of the first papers by Vapnik and Chervonenkis [145] went largely unnoticed till 1992. This was due to a widespread belief in the statistical and/or machine learning community that, despite being theoretically appealing, SVMs are neither suitable nor relevant for practical applications. They were taken seriously only when excellent results on practical learning benchmarks were achieved (in numeral recognition, computer vision and text categorization). Today, SVMs show better results than (or comparable outcomes to) NNs and other statistical models, on the most popular benchmark problems.

The learning problem setting for SVMs is as follows: there is some unknown and nonlinear dependency (mapping, function) $y = f(\mathbf{x})$ between some high-dimensional input vector $\mathbf{x}$ and the scalar output y (or the vector output $\mathbf{y}$ as in the case of multiclass SVMs). There is no information about the underlying joint probability functions here. Thus, one must perform a distribution-free learning. The only information available is a training data set $\{\mathcal{X} = [\mathbf{x}(i), y(i)] \in \Re^m \times \Re,\ i = 1, \ldots, n\}$, where n stands for the number of the training data pairs and is therefore equal to the size of the training data set $\mathcal{X}$. Often, y_i is denoted as d_i (i.e., t_i), where $d(t)$ stands for a desired (target) value. Hence, SVMs belong to the supervised learning techniques.

Note that this problem is similar to the classic statistical inference. However, there are several very important differences between the approaches and assumptions in training SVMs and the ones in classic statistics and/or NNs modeling. Classic statistical inference is based on the following three fundamental assumptions:

1. Data can be modeled by a set of linear in parameter functions; this is a foundation of a parametric paradigm in learning from experimental data.
2. In the most of real-life problems, a stochastic component of data is the normal probability distribution law, that is, the underlying joint probability distribution is a Gaussian distribution.
3. Because of the second assumption, the induction paradigm for parameter estimation is the maximum likelihood method, which is reduced to the minimization of the sum-of-errors-squares cost function in most engineering applications.

All three assumptions on which the classic statistical paradigm relied turned out to be inappropriate for many contemporary real-life problems [143] because of the following facts:

1. Modern problems are high-dimensional, and if the underlying mapping is not very smooth the linear paradigm needs an exponentially increasing number of terms with an increasing dimensionality of the input space (an increasing number of independent variables). This is known as 'the curse of dimensionality'.
2. The underlying real-life data generation laws may typically be very far from the normal distribution and a model-builder must consider this difference in order to construct an effective learning algorithm.
3. From the first two points it follows that the maximum likelihood estimator (and consequently the sum-of-error-squares cost function) should be replaced by a new induction paradigm that is uniformly better, in order to model non-Gaussian distributions.

In addition to the three basic objectives above, the novel SVMs' problem setting and inductive principle have been developed for standard contemporary data sets which are typically high-dimensional and sparse (meaning, the data sets contain small number of the training data pairs).

SVMs are the so-called 'nonparametric' models. 'Nonparametric' does not mean that the SVMs' models do not have parameters at all. On the contrary, their 'learning' (selection, identification, estimation, training or tuning) is the crucial issue here. However, unlike in classic statistical inference, the parameters are not predefined and their number depends on the training data used. In other words, parameters that define the capacity of the model are data-driven in such a way as to match the model capacity to data complexity. This is a basic paradigm of the structural risk minimization (SRM) introduced by Vapnik and Chervonenkis and their coworkers that led to the new learning algorithm. Namely, there are two basic constructive approaches possible in designing a model that will have a good generalization property [144, 143]:

1. choose an appropriate structure of the model (order of polynomials, number of HL neurons, number of rules in the fuzzy logic model) and, keeping

the estimation error (a.k.a. confidence interval, a.k.a. variance of the model) fixed in this way, minimize the training error (i.e., empirical risk), or
2. keep the value of the training error (a.k.a. an approximation error, a.k.a. an empirical risk) fixed (equal to zero or equal to some acceptable level), and minimize the confidence interval.

Classic NNs implement the first approach (or some of its sophisticated variants) and SVMs implement the second strategy. In both cases the resulting model should resolve the trade-off between under-fitting and over-fitting the training data. The final model structure (its order) should ideally *match the learning machines capacity with training data complexity.* This important difference in two learning approaches comes from the minimization of different cost (error, loss) functionals. Table 2.1 tabulates the basic risk functionals applied in developing the three contemporary statistical models. In Table 2.1, d_i stands for desired values, $\mathbf{w}$ is the weight vector subject to training, λ is a regularization parameter, $\mathbf{P}$ is a smoothness operator, L_ε is a SVMs' loss function, h is a VC dimension and Ω is a function bounding the capacity of the learning machine. In classification problems L_ε is typically 0-1 loss function, and in regression problems L_ε is the so-called Vapnik's ε-insensitivity loss (error) function

Table 2.1. Basic Models and Their Error (Risk) Functionals

Model	Risk functional
Multilayer perceptron (NN)	$R = \sum_{i=1}^{n} \underbrace{(d_i - f(\mathbf{x}_i, \mathbf{w}))^2}_{Closeness\ to\ data}$
Regularization Network(Radial Basis Functions Network)	$R = \sum_{i=1}^{n} \underbrace{(d_i - f(\mathbf{x}_i, \mathbf{w}))^2}_{Closeness\ to\ data} + \lambda \underbrace{\lVert \mathbf{P}f \rVert^2}_{Smoothness}$
Support Vector Machine	$R = \sum_{i=1}^{n} \underbrace{L_\varepsilon}_{Closeness\ to\ data} + \underbrace{\Omega(n, h)}_{Capacity\ of\ a\ machine}$

Closeness to data = training error, a.k.a. empirical risk

$$L_\varepsilon = |y - f(\mathbf{x}, \mathbf{w})|_\varepsilon = \begin{cases} 0, & \text{if } |y - f(\mathbf{x}, \mathbf{w})| \le \varepsilon \\ |y - f(\mathbf{x}, \mathbf{w})| - \varepsilon, & \text{otherwise.} \end{cases} \tag{2.1}$$

where ε is a radius of a tube within which the regression function must lie, after the successful learning. (Note that for $\varepsilon = 0$, the interpolation of training data will be performed). It is interesting to note that [58] has shown that under some constraints the SV machine can also be derived from the framework of regularization theory rather than SLT and SRM. Thus, *unlike the classic* *adaptation algorithms (that work in the L_2 norm), SV machines represent novel learning techniques which perform SRM.* In this way, the SV machine creates a model with minimized VC dimension and when the VC dimension of the model is low, the expected probability of error is low as well. This means good performance on previously unseen data, i.e. a good generalization. This property is of particular interest because the model that generalizes well is a good model and not the model that performs well on training data pairs. Too good a performance on training data is also known as an extremely undesirable overfitting.

As it will be shown below, in the 'simplest' pattern recognition tasks, support vector machines use a linear separating hyperplane to create a *classifier with a maximal margin.* In order to do that, the learning problem for the SV machine will be cast as a *constrained nonlinear optimization* problem. In this setting the cost function will be quadratic and the constraints linear (i.e., one will have to solve a classic *quadratic programming problem*).

In cases when given classes cannot be linearly separated in the original input space, the SV machine first (non-linearly) transforms the original input space into a higher dimensional feature space. This transformation can be achieved by using various nonlinear mappings; polynomial, sigmoid as in multilayer perceptrons, RBF mappings having as the basis functions radially symmetric functions such as Gaussians, or multiquadrics or different spline functions. After this nonlinear transformation step, the task of a SV machine in finding the linear optimal separating hyperplane in this feature space is 'relatively trivial'. Namely, the optimization problem to solve in a feature space will be of the same kind as the calculation of a maximal margin separating hyperplane in original input space for linearly separable classes. How, after the specific nonlinear transformation, nonlinearly separable problems in input space can become linearly separable problems in a feature space will be shown later.

In a probabilistic setting, there are three basic components in all supervised learning from data tasks: a *generator* of random inputs $\mathbf{x}$, a *system* whose *training responses* y (i.e., d) are used for training the learning machine, and a *learning machine* which, by using inputs $\mathbf{x}_i$ and system's responses y_i, should learn (estimate, model) the unknown dependency between these two sets of variables (namely, $\mathbf{x}_i$ and y_i) defined by the weight vector $\mathbf{w}$ (Fig.2.1). The figure shows the most common learning setting that some readers may have already seen in various other fields - notably in statistics, NNs, control system identification and/or in signal processing. During the (successful) training phase a learning machine should be able to find the relationship between an input space X and an output space Y , by using data $\mathcal{X}$ in regression tasks (or to find a function that separates data within the input space, in classification ones). The result of a learning process is an 'approximating function' $f_a(\mathbf{x}, \mathbf{w})$, which in statistical literature is also known as, a *hypothesis* $f_a(\mathbf{x}, \mathbf{w})$. This function approximates the underlying (or true) dependency between the input and output in the case of regression, and the decision boundary, i.e.,

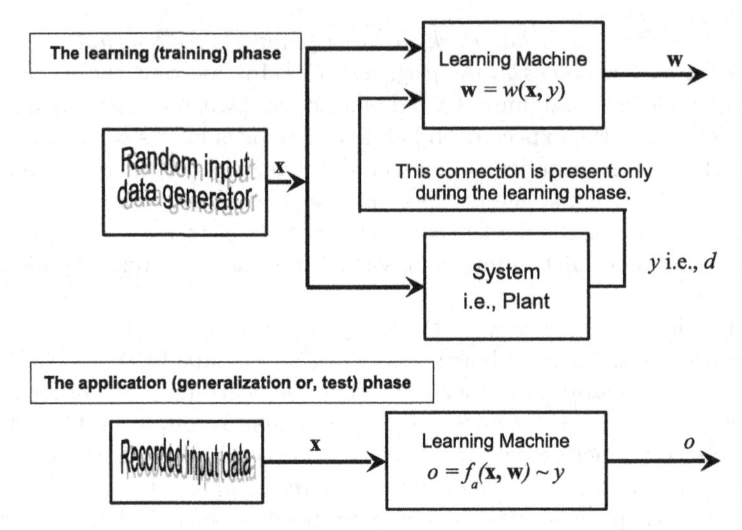

Fig. 2.1. A model of a learning machine (top) $\mathbf{w} = w(\mathbf{x}, y)$ that during *the training phase* (by observing inputs $\mathbf{x}_i$ to, and outputs y_i from, the system) estimates (learns, adjusts, trains, tunes) its parameters (weights) $\mathbf{w}$, and in this way learns mapping $y = f(\mathbf{x}, \mathbf{w})$ performed by the system. The use of $f_a(\mathbf{x}, \mathbf{w}) \sim y$ denotes that *we will rarely* try to *interpolate* training data pairs. We would rather seek an *approximating* function that can generalize well. After the training, at the *generalization* or *test phase*, the output from a machine $o = f_a(\mathbf{x}, \mathbf{w})$ is expected to be 'a good' estimate of a system's true response y.

separation function, in a classification. The chosen hypothesis $f_a(\mathbf{x}, \mathbf{w})$ belongs to a *hypothesis space of functions* $H (f_a \in H)$, and it is a function that minimizes some *risk functional* $R(\mathbf{w})$.

It may be practical to remind the reader that under the general name 'approximating function' we understand any mathematical structure that maps inputs $\mathbf{x}$ into outputs y. Hence, an 'approximating function' may be: a multilayer perceptron NN, RBF network, SV machine, fuzzy model, Fourier truncated series or polynomial approximating function. Here we discuss SVMs. A set of parameters $\mathbf{w}$ is the very subject of learning and generally these parameters are called weights. These parameters may have different geometrical and/or physical meanings. Depending upon the hypothesis space of functions H we are working with the parameters $\mathbf{w}$ are usually:

- the hidden and the output layer weights in multilayer perceptrons,
- the rules and the parameters (for the positions and shapes) of fuzzy subsets,
- the coefficients of a polynomial or Fourier series,
- the centers and (co)variances of Gaussian basis functions as well as the output layer weights of this RBF network,
- the support vector weights in SVMs.

There is another important class of functions in learning from examples tasks. A learning machine tries to capture an unknown *target function* $f_o(\mathbf{x})$ that is believed to belong to some target space T, or to a class T, that is also called a *concept class*. Note that we rarely know the target space T and that our learning machine generally does not belong to the same class of functions as an unknown target function $f_o(\mathbf{x})$. Typical examples of target spaces are continuous functions with s continuous derivatives in m variables; Sobolev spaces (comprising square integrable functions in m variables with s square integrable derivatives), band-limited functions, functions with integrable Fourier transforms, Boolean functions, etc. In the following, we will assume that the target space T is a space of differentiable functions. The basic problem we are facing stems from the fact that we know very little about the possible underlying function between the input and the output variables. All we have at our disposal is a training data set of labeled examples drawn by independently sampling $a(X \times Y)$ space according to some unknown probability distribution.

The learning-from-data problem is ill-posed. (This will be shown on Figs. 2.2 and 2.3 for a regression and classification examples respectively). The basic source of the ill-posedness of the problem is due to the infinite number of possible solutions to the learning problem. At this point, just for the sake of illustration, it is useful to remember that all functions that interpolate data points will result in a zero value for training error (empirical risk) as shown (in the case of regression) in Fig. 2.2. The figure shows a simple example of three-out-of-infinitely-many different interpolating functions of training data pairs sampled from a noiseless function $y = \sin(x)$.

In Fig. 2.2, each interpolant results in a training error equal to zero, but at the same time, each one is a very bad model of the true underlying dependency between x and y, because all three functions perform very poorly outside the training inputs. In other words, none of these three particular interpolants can generalize well. However, not only interpolating functions can mislead. There are many other approximating functions (learning machines) that will minimize the empirical risk (approximation or training error) but not necessarily the generalization error (true, expected or guaranteed risk). This follows from the fact that a learning machine is trained by using some particular sample of the true underlying function and consequently it always produces biased approximating functions. These approximants depend necessarily on the specific training data pairs (i.e., the training sample) used.

Figure 2.3 shows an extremely simple classification example where the classes (represented by the empty training circles and squares) are linearly separable. However, in addition to a linear separation (dashed line) the learning was also performed by using a model of a high capacity (say, the one with Gaussian basis functions, or the one created by a high order polynomial, over the 2-dimensional input space) that produced a perfect separation boundary (empirical risk equals zero) too. However, such a model is overfitting the data and it will definitely perform very badly on, during the training unseen, test examples. Filled circles and squares in the right hand graph are all wrongly

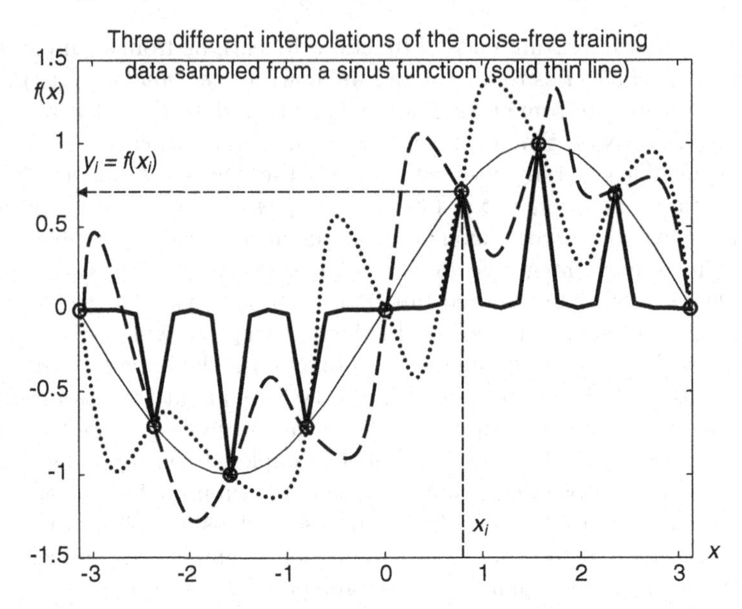

Fig. 2.2. Three-out-of-infinitely-many interpolating functions resulting in a training error equal to 0. However, a thick solid, dashed and dotted lines are bad models of a true function $y = \sin(x)$ (thin solid line).

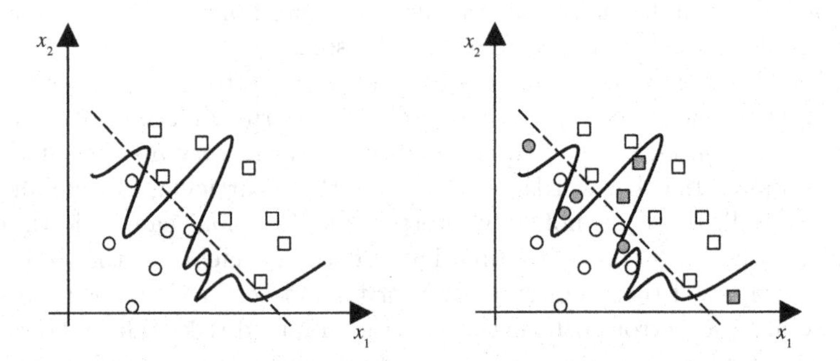

Fig. 2.3. Overfitting in the case of linearly separable classification problem. *Left*: The perfect classification of the training data (empty circles and squares) by both low order linear model (dashed line) and high order nonlinear one (solid wiggly curve). *Right*: Wrong classification of all the test data shown (filled circles and squares) by a high capacity model, but correct one by the simple linear separation boundary.

classified by the nonlinear model. Note that a simple linear separation boundary correctly classifies both the training and the test data.

A solution to this problem proposed in the framework of the SLT is restricting the hypothesis space H of approximating functions to a set smaller than that of the target function T while simultaneously controlling the flexibility (complexity) of these approximating functions. This is ensured by an introduction of a novel induction principle of the SRM and its algorithmic realization through the SV machine. The Structural Risk Minimization principle [141] tries to minimize an expected risk (the cost function) R comprising two terms as given in Table 2.1 for the SVMs $R = \Omega(n, h) + \sum_{i=1}^{n} L_\varepsilon = \Omega(n, h) + R_{emp}$ and it is based on the fact that for the classification learning problem with a probability of at least $1 - \eta$ the bound

$$R(\mathbf{w}_m) \leqslant \Omega(\frac{h}{n}, \frac{\ln(\eta)}{n}) + R_{emp}(\mathbf{w}_m) \tag{2.2a}$$

holds. The first term on the right hand side is named a VC confidence (confidence term or confidence interval) that is defined as

$$\Omega(\frac{h}{n}, \frac{\ln(\eta)}{n}) = \sqrt{\frac{h\left[\ln(\frac{2n}{h}) + 1\right] - \ln(\frac{\eta}{4})}{n}}. \tag{2.2b}$$

The parameter h is called the VC (Vapnik-Chervonenkis) dimension of a set of functions. It describes the capacity of a set of functions implemented in a learning machine. For a binary classification h is the maximal number of points which can be separated (shattered) into two classes in all possible 2^h ways by using the functions of the learning machine.

A SV (learning) machine can be thought of as

- a set of functions implemented in a SVM,
- an induction principle and,
- an algorithmic procedure for implementing the induction principle on the given set of functions.

The notation for risks given above by using $R(\mathbf{w}_m)$ denotes that an expected risk is calculated over a set of functions $f_{an}(\mathbf{x}, \mathbf{w}_m)$ of increasing complexity. Different bounds can also be formulated in terms of other concepts such as *growth function* or *annealed VC entropy*. Bounds also differ for regression tasks. More detail can be found in ([144], as well as in [32]). However, the general characteristics of the dependence of the confidence interval on the number of training data n and on the VC dimension h is similar and given in Fig 2.4.

Equations (2.2) show that when the number of training data increases, i.e., for $n \to \infty$(with other parameters fixed), an expected (true) risk $R(\mathbf{w}_n)$ is very close to empirical risk $R_{emp}(\mathbf{w}_n)$ because $\Omega \to 0$. On the other hand, when the probability $1 - \eta$ (also called a confidence level which should not be confused with the confidence term Ω) approaches 1, the generalization bound grows large, because in the case when $\eta \to 0$ (meaning that the confidence level $1 - \eta \to 1$), the value of $\Omega \to \infty$. This has an obvious intuitive interpretation

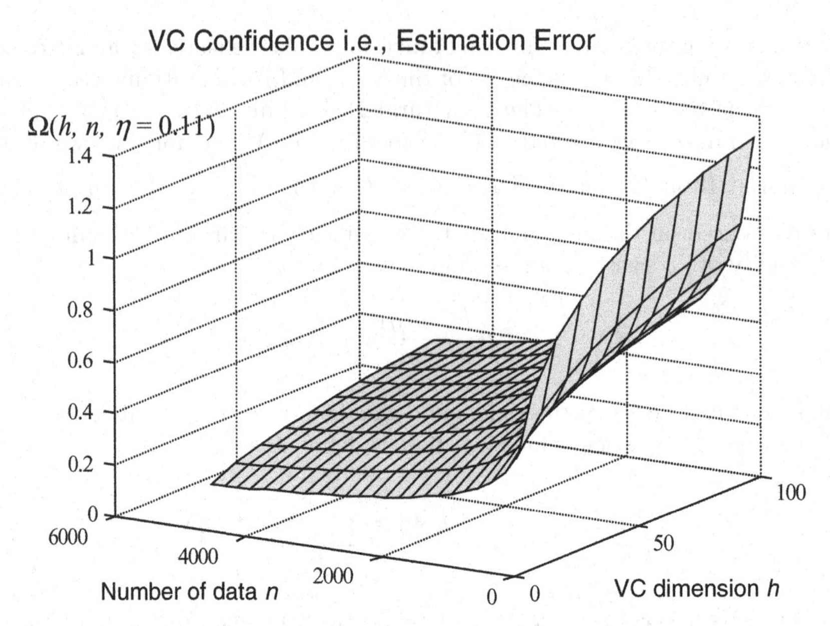

Fig. 2.4. The dependency of VC confidence interval $\Omega(h, n, \eta)$ on the number of training data n and the VC dimension $h(h < n)$ for a fixed confidence level $1 - \eta = 1 - 0.11 = 0.89$.

[32] in that any learning machine (model, estimates) obtained from a finite number of training data cannot have an arbitrarily high confidence level. There is always a trade-off between the accuracy provided by bounds and the degree of confidence (in these bounds). Fig 2.4 also shows that the VC confidence interval increases with an increase in a VC dimension h for a fixed number of the training data pairs n.

The SRM is a novel inductive principle for learning from finite training data sets. It proved to be very useful when dealing with small samples. The basic idea of the SRM is to choose (from a large number of possibly candidate learning machines), a model of the right capacity to describe the given training data pairs. As mentioned, this can be done by restricting the hypothesis space H of approximating functions and simultaneously by controlling their flexibility (complexity). Thus, learning machines will be those parameterized models that, by increasing the number of parameters (typically called weights w_i here), form a nested structure in the following sense

$$H_1 \subset H_2 \subset H_3 \subset \ldots H_{n-1} \tag{2.3}$$

In such a nested set of functions, every function always contains a previous, less complex, function. Typically, H_n may be: a set of polynomials in one variable of degree n; fuzzy logic model having n rules; multilayer perceptrons, or RBF network having n HL neurons, SVM structured over n support vectors. The goal of learning is one of a *subset selection* that matches training data complexity with approximating model capacity. In other words, a learning algorithm chooses an optimal polynomial degree or, an optimal number of HL neurons or, an optimal number of FL model rules, for a polynomial model or NN or FL model respectively. For learning machines linear in parameters, this complexity (expressed by the VC dimension) is given by the number of weights, i.e., by the number of 'free parameters'. For approximating models nonlinear in parameters, the calculation of the VC dimension is often not an easy task. Nevertheless, even for these networks, by using simulation experiments, one can find a model of appropriate complexity.

2.2 Support Vector Machines in Classification and Regression

Below, we focus on the algorithm for implementing the SRM induction principle on the given set of functions. It implements the strategy mentioned previously - it keeps the training error fixed and minimizes the confidence interval. We first consider a 'simple' example of linear decision rules (i.e., the separating functions will be hyperplanes) for binary classification (dichotomization) of linearly separable data. In such a problem, we are able to perfectly classify data pairs, meaning that an empirical risk can be set to zero. It is the easiest classification problem and yet an excellent introduction of all relevant and important ideas underlying the SLT, SRM and SVM.

Our presentation will gradually increase in complexity. It will begin with a *Linear Maximal Margin Classifier for Linearly Separable Data* where there is no sample overlapping. Afterwards, we will allow some degree of overlapping of training data pairs. However, we will still try to separate classes by using linear hyperplanes. This will lead to the *Linear Soft Margin Classifier for Overlapping Classes.* In problems when linear decision hyperplanes are no longer feasible, the mapping of an input space into the so-called feature space (that 'corresponds' to the HL in NN models) will take place resulting in the *Nonlinear Classifier.* Finally, in the subsection on *Regression by SV Machines* we introduce same approaches and techniques for solving regression (i.e., function approximation) problems.

2.2.1 Linear Maximal Margin Classifier for Linearly Separable Data

Consider the problem of binary classification or dichotomization. Training data are given as

$$(\mathbf{x}_1, y), (\mathbf{x}_2, y), \ldots, (\mathbf{x}_n, y_n), \;\; \mathbf{x} \in \Re^m, \quad y \in \{+1, -1\}. \tag{2.4}$$

For reasons of visualization only, we will consider the case of a two-dimensional input space, i.e., ($\mathbf{x} \in \Re^2$). Data are linearly separable and there are many different hyperplanes that can perform separation (Fig. 2.5). (Actually, for $\mathbf{x} \in \Re^2$, the separation is performed by 'planes' $w_1x_1 + w_2x_2 + b = d$. In other words, the decision boundary, i.e., the separation line in input space is defined by the equation $w_1x_1 + w_2x_2 + b = 0$.). How to find 'the best' one? The difficult part is that all we have at our disposal are sparse training data. Thus, we want to find the optimal separating function without knowing the underlying probability distribution $P(\mathbf{x}, y)$. There are many functions that can solve given pattern recognition (or functional approximation) tasks. In such a problem setting, the SLT (developed in the early 1960s by Vapnik and Chervonenkis [145]) shows that it is crucial to restrict the class of functions implemented by a learning machine to one with a complexity that is suitable for the amount of available training data.

In the case of a classification of linearly separable data, this idea is transformed into the following approach - among all the hyperplanes that minimize the training error (i.e., empirical risk) find the one with the largest margin. This is an intuitively acceptable approach. Just by looking at Fig 2.5 we will find that the dashed separation line shown in the *right graph* seems to promise *probably* good classification while facing previously unseen data (meaning, in the generalization, i.e. test, phase). Or, at least, it seems to probably be better in generalization than the dashed decision boundary having smaller margin shown in the left graph. This can also be expressed as that a classifier with smaller margin will have higher expected risk. By using given training examples, during the learning stage, our machine finds parameters $\mathbf{w} = [w_1\ w_2\ \ldots w_m]^T$ and b of a discriminant or decision function $d(\mathbf{x}, \mathbf{w}, b)$ given as

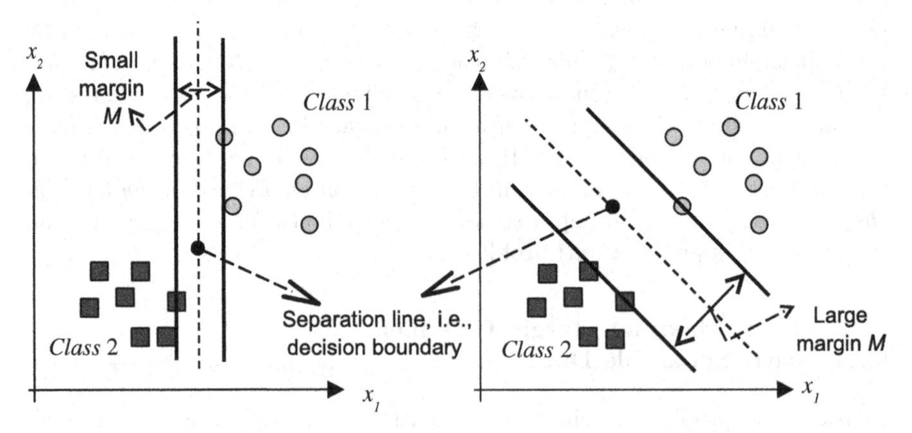

Fig. 2.5. Two-out-of-many separating lines: a good one with a large margin (right) and a less acceptable separating line with a small margin, (left).

$$d(\mathbf{x}, \mathbf{w}, b) = \mathbf{w}^T \mathbf{x} + b = \sum_{i=1}^{m} w_i x_i + b, \tag{2.5}$$

where $\mathbf{x}, \mathbf{w} \in \Re^m$, and the scalar b is called a *bias*.(Note that the dashed separation lines in Fig. 2.5 represent the line that follows from $d(\mathbf{x}, \mathbf{w}, b) = 0$). After the successful training stage, by using the weights obtained, the learning machine, given previously unseen pattern $\mathbf{x}_p$, produces output o according to an *indicator function* given as

$$i_F = o = \operatorname{sign}(d(\mathbf{x}_p, \mathbf{w}, b)). \tag{2.6}$$

where o is the standard notation for the *output* from the learning machine. In other words, *the decision rule* is:

- if $d(\mathbf{x}_p, \mathbf{w}, b) > 0$, the pattern $\mathbf{x}_p$ belongs to a class 1 (i.e., $o = y_p = +1$),
- and if $d(\mathbf{x}_p, \mathbf{w}, b) < 0$ the pattern $\mathbf{x}_p$, belongs to a class 2 (i.e., $o = y_p = -1$).

The *indicator function* i_F given by (2.6) is a step-wise (i.e., a stairs-wise) function (see Figs. 2.6 and 2.7). At the same time, the decision (or discriminant) function $d(\mathbf{x}, \mathbf{w}, b)$ is a hyperplane. Note also that both a decision hyperplane d and the indicator function i_F live in an $n+1$-dimensional space or they lie 'over' a training pattern's n-dimensional input space. There is one more mathematical object in classification problems called a separation boundary that lives in the same n-dimensional space of input vectors $\mathbf{x}$. Separation boundary separates vectors $\mathbf{x}$ into two classes. Here, in cases of linearly separable data, the boundary is also a (separating) hyperplane but of a lower order than $d(\mathbf{x}, \mathbf{w}, b)$. The decision (separation) *boundary* is an intersection of a decision *function* $d(\mathbf{x}, \mathbf{w}, b)$ and a space of input features. It is given by

$$d(\mathbf{x}, \mathbf{w}, b) = 0. \tag{2.7}$$

All these functions and relationships can be followed, for two-dimensional inputs $\mathbf{x}$, in Fig. 2.6. In this particular case, the decision boundary i.e., separating (hyper)plane is actually a separating line in a $x_1 - x_2$ plane and, a decision function $d(\mathbf{x}, \mathbf{w}, b)$ is a plane over the 2-dimensional space of features, i.e., over a $x_1 - x_2$ plane. In the case of 1-dimensional training patterns x (i.e., for 1-dimensional inputs x to the learning machine), decision function $d(\mathbf{x}, \mathbf{w}, b)$ is a straight line in an $x - y$ plane. An intersection of this line with an x-axis defines a point that is a separation boundary between two classes. This can be followed in Fig. 2.7. Before attempting to find an optimal separating hyperplane having the largest margin, we introduce the concept of the *canonical hyperplane*. We depict this concept with the help of the 1-dimensional example shown in Fig. 2.7. Not quite incidentally, the decision plane $d(\mathbf{x}, \mathbf{w}, b)$ shown in Fig. 2.6 is also a *canonical* plane. Namely, the values of d and of i_F are the same and both are equal to $|1|$ for the support vectors depicted by stars. At the same time, for all other training patterns $|d| > |i_F|$.

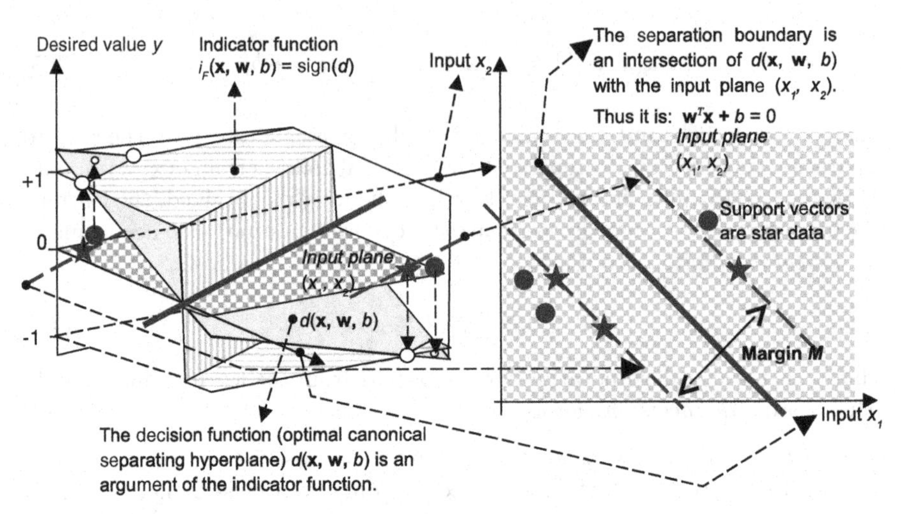

Fig. 2.6. The definition of a decision (discriminant) *function* or hyperplane $d(\mathbf{x}, \mathbf{w}, b)$, a decision boundary $d(\mathbf{x}, \mathbf{w}, b) = 0$ and an indicator function $i_F = \text{sign}(d(\mathbf{x}, \mathbf{w}, b))$ whose value represents a learning machine's output o.

In order to present a notion of this new concept of the canonical plane, first note that there are many hyperplanes that can correctly separate data. In Fig. 2.7 three different decision functions $d(\mathbf{x}, \mathbf{w}, b)$ are shown. There are infinitely many more. In fact, given $d(\mathbf{x}, \mathbf{w}, b)$, all functions $d(\mathbf{x}, k\mathbf{w}, kb)$, where k is a positive scalar, are correct decision functions too. Because parameters $(\mathbf{w}, b)$ describe the same separation hyperplane as parameters $(k\mathbf{w}, kb)$ there is a need to introduce the notion of a *canonical hyperplane*:

A hyperplane is in the canonical form with respect to the training data x_i, $i = 1, \ldots, n$, if

$$\underbrace{\min}_{\mathbf{x}_i \in \mathcal{X}} \left| \mathbf{w}^T \mathbf{x}_i + b \right| = 1. \tag{2.8}$$

The solid line $d(\mathbf{x}, \mathbf{w}, b) = -2x + 5$ in Fig. 2.7 fulfills (2.8) because its minimal *absolute value for the given six training patterns* belonging to two classes is 1. It achieves this value for two patterns, chosen as support vectors, namely for $x_3 = 2$, and $x_4 = 3$. For all other patterns, $|d| > 1$. Note an interesting detail regarding the notion of a canonical hyperplane that is easily checked. There are many different hyperplanes (planes and straight lines for 2-D and 1-D problems in Figs. 2.6 and 2.7 respectively) that have the same separation boundary (solid line and a dot in Figs. 2.6 (right) and 2.7 respectively). At the same time there are far fewer hyperplanes that can be defined as canonical ones fulfilling (2.8). In Fig. 2.7, i.e., for a 1-dimensional input vector x, the canonical hyperplane is unique. This is not the case for training patterns of higher dimension. Depending upon the configuration of class' elements, various canonical hyperplanes are possible.

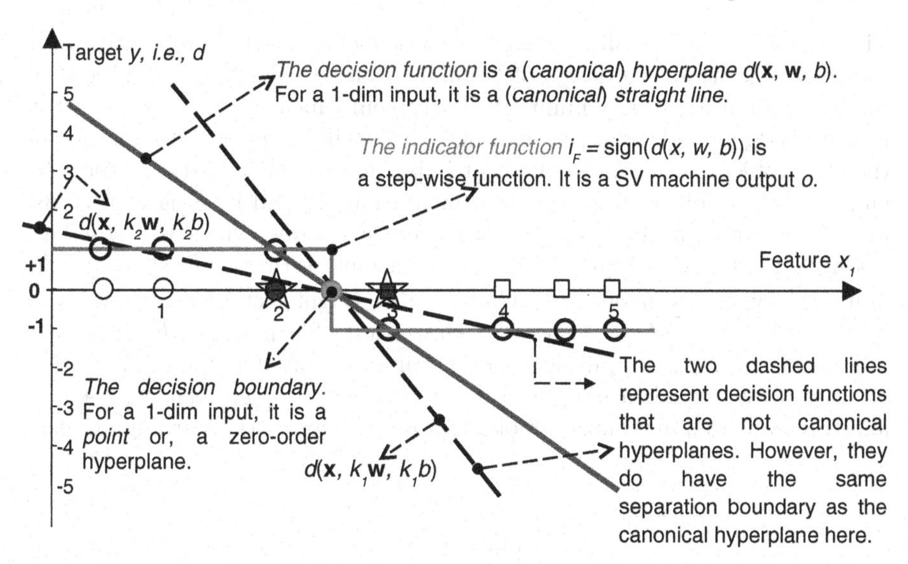

Fig. 2.7. SV classification for 1-dimensional inputs by the linear decision function. Graphical presentation of a canonical hyperplane. For 1-dimensional inputs, it is actually a canonical straight line (depicted as a thick straight solid line) that passes through points (+2, +1) and (+3, -1) defined as the support vectors (stars). The two dashed lines are the two other decision hyperplanes (i.e., straight lines). The training input patterns $\{x_1 = 0.5, x_2 = 1, x_3 = 2\} \in Class1$ have a desired or target value (label) $y_1 = +1$. The inputs $\{x_4 = 3, x_5 = 4, x_6 = 4.5, x_7 = 5\} \in Class2$ have the label $y_2 = -1$.

Therefore, there is a need to define an *optimal* separation canonical hyperplane (OCSH) as a canonical hyperplane having a *maximal margin.* This search for a separating, maximal margin, canonical hyperplane is the ultimate learning goal in statistical learning theory underlying SV machines. Carefully note the adjectives used in the previous sentence. The hyperplane obtained from a limited training data must have a *maximal margin* because it will *probably* better classify new data. It must be in *canonical* form because this will ease the quest for significant patterns, here called support vectors. The canonical form of the hyperplane will also simplify the calculations. Finally, the resulting hyperplane must ultimately *separate* training patterns.

We avoid the derivation of an expression for the calculation of a distance (margin M) between the closest members from two classes for its simplicity here. Instead, the curious reader can find a derivation of (2.9) in the Appendix A. There are other ways to get (2.9) which can be found in other books or monographs on SVMs. The margin M can be derived by both the geometric and algebraic argument and is given as

$$M = \frac{2}{\|\mathbf{w}\|}. \tag{2.9}$$

This important result will have a great consequence for the constructive (i.e., learning) algorithm in a design of a maximal margin classifier. It will lead to solving a quadratic programming (QP) problem which will be shown shortly. Hence, the 'good old' gradient learning in NNs will be replaced by solution of the QP problem here. This is the next important difference between the NNs and SVMs and follows from the implementation of SRM in designing SVMs, instead of a minimization of the sum of error squares, which is a standard cost function for NNs. Equation (2.9) is a very interesting result showing that minimization of a norm of a hyperplane normal weight vector $\|\mathbf{w}\| = \sqrt{\mathbf{w}^T\mathbf{w}} = \sqrt{w_1^2 + w_1^2 + \ldots + w_m^2}$ leads to a maximization of a margin M. Because a minimization of $\sqrt{f}$ is equivalent to the minimization of f, the minimization of a norm $\|\mathbf{w}\|$ equals a minimization of $\mathbf{w}^T\mathbf{w} = \sum_{i=1}^{m} w_1^2 + w_1^2 + \ldots + w_m^2$, and this leads to a maximization of a margin M. Hence, the learning problem is

$$\min \frac{1}{2}\mathbf{w}^T\mathbf{w} \tag{2.10a}$$

subject to constraints introduced and given in (2.10b) below. (A multiplication of $\mathbf{w}^T\mathbf{w}$ by 0.5 is for numerical convenience only, and it doesn't change the solution). Note that in the case of linearly separable classes empirical error equals zero ($R_{emp} = 0$ in (2.2a)) and minimization of $\mathbf{w}^T\mathbf{w}$ corresponds to a minimization of a confidence term Ω. The OCSH, i.e., a separating hyperplane with the largest margin defined by $M = 2/\|\mathbf{w}\|$, specifies *support vectors*, i.e., training data points closest to it, which satisfy $y_j[\mathbf{w}^T\mathbf{x}_j + b] \equiv 1,\ j = 1,\ N_{SV}$. For all the other (non-SVs data points) the OCSH satisfies inequalities $y_i[\mathbf{w}^T\mathbf{x}_i + b] > 1$. In other words, for all the data, OCSH should satisfy the following constraints

$$y_i(\mathbf{w}^T\mathbf{x}_i + b) \geq 1 \quad i = 1, \ldots, n. \tag{2.10b}$$

where n denotes a number of training data points, and N_{SV} stands for a number of SVs. The last equation can be easily checked visually in Figs. 2.6 and 2.7 for 2-dimensional and 1-dimensional input vectors $\mathbf{x}$ respectively. Thus, in order to find the OCSH having a maximal margin, a learning machine should minimize $\|\mathbf{w}\|^2$ subject to the inequality constraints (2.10b). This is a *classic quadratic optimization problem with inequality constraints.* Such an optimization problem is solved by the *saddle point* of the Lagrange functional (Lagrangian) [2].

$$L_p(\mathbf{w}, b, \boldsymbol{\alpha}) = \frac{1}{2}\mathbf{w}^T\mathbf{w} - \sum_{i=1}^{n} \alpha_i\{y_i[\mathbf{w}^T\mathbf{x}_i + b] - 1\}. \tag{2.11}$$

[2] In forming the Lagrangian, for constraints of the form $f_i > 0$, the inequality constraints equations are multiplied by *nonnegative* Lagrange multipliers (i.e., $\alpha_i \geq 0$) and *subtracted* from the objective function.

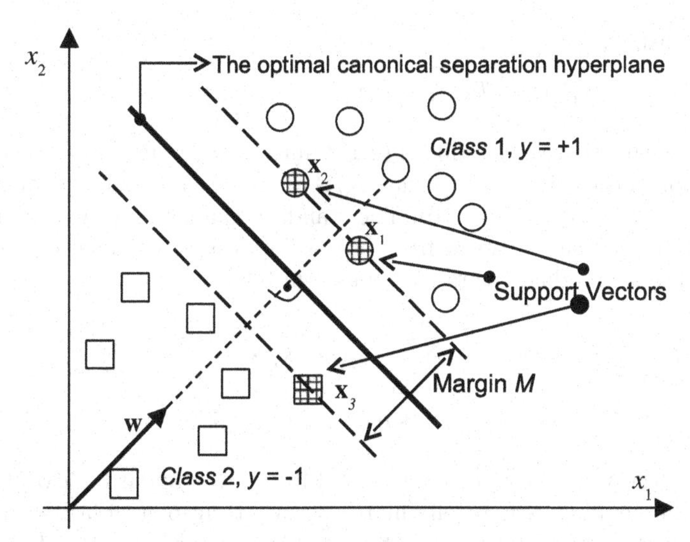

Fig. 2.8. The optimal canonical separation hyperplane with the largest margin intersects halfway between the two classes. The points closest to it (satisfying $y_j\left|\mathbf{w}^T\mathbf{x}_j+b\right| = 1, j = 1, N_{SV}$) are *support vectors* and the OCSH satisfies $y_i(\mathbf{w}^T\mathbf{x}_i+b) \geq 1,\ i=1,n$ (where n denotes the number of training data and N_{SV} stands for the number of SV). Three support vectors ($\mathbf{x}_1$ and $\mathbf{x}_2$ from class 1, and $\mathbf{x}_3$ from class 2) are the textured training data.

where the α_i are Lagrange multipliers. The search for an optimal *saddle point* $(\mathbf{w}_o, b_o, \boldsymbol{\alpha}_o)$ is necessary because Lagrangian L_p must be *minimized* with respect to $\mathbf{w}$ and b, and has to be *maximized* with respect to nonnegative α_i (i.e., $\alpha_i \geq 0$ should be found). This problem can be solved either in a *primal space* (which is the space of parameters $\mathbf{w}$ and b) or in a *dual space* (which is the space of Lagrange multipliers α_i). The second approach gives insightful results and we will consider the solution in a dual space below. In order to do that, we use Karush-Kuhn-Tucker (KKT) conditions for the optimum of a constrained function. In our case, both the objective function (2.11) and constraints (2.10b) are *convex* and KKT conditions are *necessary* and *sufficient* conditions for a maximum of (2.11). These conditions are: at the saddle point $(\mathbf{w}_o, b_o, \boldsymbol{\alpha}_o)$, derivatives of Lagrangian L_p with respect to primal variables should vanish which leads to,

$$\frac{\partial L_p}{\partial \mathbf{w}_o} = 0 \quad \mathbf{w}_o = \sum_{i=1}^{n} \alpha_i y_i \mathbf{x}_i\,, \tag{2.12a}$$

$$\frac{\partial L_p}{\partial b} = 0 \quad \sum_{i=1}^{n} \alpha_i y_i = 0. \tag{2.12b}$$

and the KKT complementarity conditions below (stating that at the solution point the products between dual variables and constraints equals zero) must also be satisfied,

$$\alpha_i\{y_i[\mathbf{w}^T\mathbf{x}_i + b] - 1\} = 0, \quad i = 1, \ldots, n \tag{2.13}$$

This also means that the condition (2.13) must also be satisfied to ensure that the solution of the primal Lagrangian L_p is the same as the solution of the original optimization problem (2.10). The standard change to a dual Lagrangian problem is to first substitute $\mathbf{w}$ from (2.12a) into the primal Lagrangian (2.11) and this leads to a dual Lagrangian problem below

$$L_d(\alpha) = \sum_{i=1}^{n} \alpha_i - \frac{1}{2} \sum_{i,j=1}^{n} y_i y_j \alpha_i \alpha_j \mathbf{x}_i^T \mathbf{x}_j - \sum_{i=1}^{n} \alpha_i y_i b, \tag{2.14a}$$

$$\text{s.t. } \alpha_i \geq 0, \quad i = 1, \ldots, n \tag{2.14b}$$

subject to the inequality constraints (2.14b). In a standard SVMs formulation, (2.12b) is also used to eliminate the last term of (2.14a), so one only needs to maximize L_d with respect to α_i. As a result, the dual function to be maximized is (2.15a) with inequality constraints (2.15b) and an equality constraint (2.15c).

$$\max L_d(\alpha) = \sum_{i=1}^{n} \alpha_i - \frac{1}{2} \sum_{i,j=1}^{n} y_i y_j \alpha_i \alpha_j \mathbf{x}_i^T \mathbf{x}_j \tag{2.15a}$$

$$\text{s.t. } \alpha_i \geq 0, \quad i = 1, \ldots, n \quad \text{and} \tag{2.15b}$$

$$\sum_{i=1}^{n} \alpha_i y_i = 0. \tag{2.15c}$$

Note that the dual Lagrangian $L_d(\boldsymbol{\alpha})$ is expressed in terms of training data and depends *only* on the *scalar products* of input patterns ($\mathbf{x}_i^T \mathbf{x}_j$). The dependency of $L_d(\boldsymbol{\alpha})$ on a scalar product of inputs will be very handy later when analyzing nonlinear decision boundaries and for general nonlinear regression. Note also that the number of unknown variables equals the number of training data n. After learning, the number of free parameters is equal to the number of SVs but it does not depend on the dimensionality of input space. Such *a standard quadratic optimization problem* can be expressed in a *matrix notation* and formulated as follows:

$$\max \; L_d(\boldsymbol{\alpha}) = -0.5\boldsymbol{\alpha}^T \mathbf{H} \boldsymbol{\alpha} + \mathbf{p}^T \boldsymbol{\alpha}, \tag{2.16a}$$

$$\text{s.t. } \; \mathbf{y}^T \boldsymbol{\alpha} = 0, \tag{2.16b}$$

$$\alpha_i \geq 0, \quad i = 1, \ldots, n, \tag{2.16c}$$

where $\boldsymbol{\alpha} = [\alpha_1, \alpha_2, \ldots, \alpha_n]^T$, $\mathbf{H}$ denotes a symmetric Hessian matrix (with elements $H_{ij} = y_i y_j \mathbf{x}_i^T \mathbf{x}_j$), and $\mathbf{p}$ is an $n \times 1$ unit vector $\mathbf{p} = \mathbf{1} = [1 \; 1 \ldots 1]^T$. (Note that maximization of (2.16a) equals a minimization of $L_d(\boldsymbol{\alpha}) = 0.5\boldsymbol{\alpha}^T \mathbf{H} \boldsymbol{\alpha} - \mathbf{p}^T \boldsymbol{\alpha}$, subject to the same constraints).The Hessian matrix has a size of n by n and it is always a dense matrix. This means that the learning of SVMs scales with the size of training data. This is the main reason for the development of a fast iterative learning algorithm which does not need to store the complete Hessian matrix in Chap. 3.1 . Solution $\boldsymbol{\alpha}_o$ of the dual optimization (2.15) determines the parameters of the optimal hyperplane $\mathbf{w}_o$ and b_o according to (2.12a) and (2.13) as follows,

$$\mathbf{w}_o = \sum_{i=1}^{n} \alpha_{oi} y_i \mathbf{x}_i \tag{2.17a}$$

$$b_o = \frac{1}{N_{SV}} \sum\nolimits_{s=1}^{N_{SV}} \left(\frac{1}{y_s} - \mathbf{x}_s^T \mathbf{w}_o\right) = \frac{1}{N_{SV}} \sum\nolimits_{s=1}^{N_{SV}} \left(y_s - \mathbf{x}_s^T \mathbf{w}_o\right) \quad s = 1, \ldots, N_{sv} \tag{2.17b}$$

In deriving (2.17b) the fact that y can be either +1 or -1, and $1/y = y$ is used. N_{SV} denotes the number of support vectors. There are two important observations about the calculation of $\mathbf{w}_o$. First, an optimal weight vector $\mathbf{w}_o$, is obtained in (2.17a) as a linear combination of the training data points and second, $\mathbf{w}_o$ (same as the bias term b_o) is calculated by using only the selected data points called support vectors (SVs) . It is because they have nonzero α_{oi} and they are the data which support forming the decision function. Thus, the data having $\alpha_{oi} = 0$ are called non-SVs here. The fact that the summation in (2.17a) goes over all training data (i.e., from 1 to n) is irrelevant because the Lagrange multipliers α_{oi} for all non-SVs are equal to zero. Furthermore, if now all the non-SVs are removed from the training data set and only the SVs are used for training, then the same solution (i.e., the same values of $\mathbf{w_o}$ and b_o) will be obtained as the ones obtained by using the complete training data set. This is a very pleasing property of SVMs, because the solutions of good models are generally sparse (only 10-20% of the complete data set are SVs). For linearly separable training data, all support vectors lie on the margin and they are generally just a small portion of all training data (typically, $N_{SV} << n$). Figs. 2.5, 2.7 and 2.8 show the geometry of standard results for non-overlapping classes. Also, in order to satisfy the KKT complementarity conditions (2.13), the output of the decision function $(\mathbf{w}^T \mathbf{x}_s + b)$ at SVs must have a magnitude of 1. This fact is used in the derivation of (2.17b). Equation (2.17b) tries to find out the optimal bias term b_o by taking the average deviation between the desired output y_s (+1 or -1) and $\mathbf{x}_s \mathbf{w}_o$ over all SVs. After calculating $\mathbf{w}_o$ and b_o, a decision hyperplane and an indicator function are obtained as follows

$$d(\mathbf{x}) = \sum\nolimits_{i=1}^{n} w_{oi} x_i + b_o = \sum\nolimits_{i=1}^{n} y_i \alpha_i \mathbf{x}_i^T \mathbf{x} + b_o, \quad i_F = o = \mathrm{sign}(d(\mathbf{x})). \tag{2.18}$$

Before presenting a derivation of an OCSH for both overlapping classes and classes having nonlinear decision boundaries, we will comment only on whether and how SV based linear classifiers actually implement the SRM principle. The more detailed presentation of this important property can be found in [81, 123]. First, it can be shown that an increase in margin reduces the number of points that can be shattered i.e., the increase in margin reduces the VC dimension, and this leads to the decrease of the SVM capacity. In short, by minimizing $\|\mathbf{w}\|$ (i.e., maximizing the margin) the SV machine training actually minimizes the VC dimension and consequently a generalization error (expected risk) at the same time. This is achieved by imposing a structure on the set of canonical hyperplanes and then, during the training, by choosing the one with a minimal VC dimension. A structure on the set of canonical hyperplanes is introduced by considering various hyperplanes having different $\|\mathbf{w}\|$. In other words, we analyze sets S_A such that $\|\mathbf{w}\| \leq A$. Then, if $A_1 \leq A_2 \leq \ldots \leq A_m$, we introduced a nested set $S_{A1} \subset S_{A2} \subset S_{A3} \ldots \subset S_{Am}$. Thus, if we impose the constraint $\|\mathbf{w}\| \leq A$, then the canonical hyperplane cannot be closer than $1/A$ to any of the training points $\mathbf{x}_i$. Vapnik in [144] states that the VC dimension h of a set of canonical hyperplanes in $\Re^m$ such that $\|w\| \leq A$ is

$$H \leq min[R^2, A^2, m] + 1, \tag{2.19}$$

where all the training data points (vectors) are enclosed by a sphere of the smallest radius R. Therefore, a small $\|\mathbf{w}\|$ results in a small h, and minimization of $\|\mathbf{w}\|$ is an implementation of the SRM principle. In other words, a minimization of the canonical hyperplane weight norm $\|\mathbf{w}\|$ minimizes the VC dimension according to (2.19). See also Fig. 2.4 that shows how the estimation error, meaning the expected risk (because the empirical risk, due to the linear separability, equals zero) decreases with a decrease of a VC dimension. Finally, there is an interesting, simple and powerful result [144] connecting the generalization ability of learning machines and the number of support vectors. Once the support vectors have been found, we can calculate the bound on the expected probability of committing an error on a test example as follows

$$E_n\left[P(\text{error})\right] \leq \frac{E\left[\text{number of support vectors}\right]}{n}, \tag{2.20}$$

where E_n denotes expectation over all training data sets of size n. Note how easy it is to estimate this bound that is independent of the dimensionality of the input space. Therefore, an SV machine having a small number of support vectors will have good generalization ability even in a very high-dimensional space.

Example below shows the SVM's learning of the weights for a simple separable data problem in both the primal and the dual domain. The small number and low dimensionality of data pairs is used in order to show the optimization steps analytically and graphically. The same reasoning will be in the case of high dimensional and large training data sets but for them, one has to rely on computers and the insight in solution steps is necessarily lost.

Example 2.1. Consider a design of SVM classifier for 3 data shown in Fig. 2.9 below. First we solve the problem in the primal domain: From the constraints (2.10b) it follows

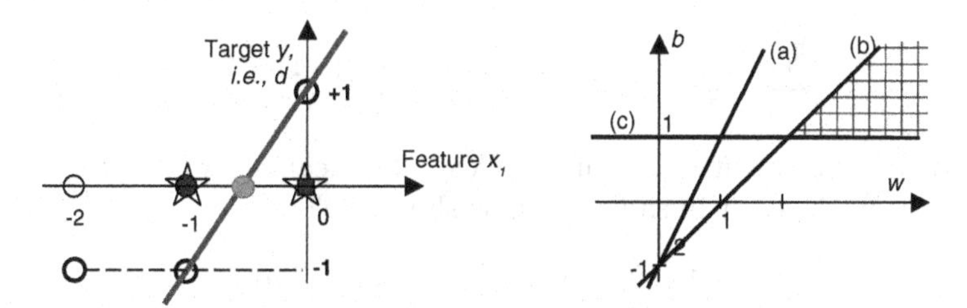

Fig. 2.9. *Left:* Solving SVM classifier for 3 data shown. SVs are star data. *Right:* Solution space $w-b$.

$$\begin{aligned} &2w - 1 \geq b && (a) \\ &w - 1 \geq b && (b) \\ &b \geq 1 && (c) \end{aligned} \tag{2.21}$$

The three straight lines corresponding to the equalities above are shown in Fig. 2.9 right. The textured area is a feasible domain for the weight w and bias b. Note that the area is not defined by the inequality (a), thus pointing to the fact that the point -2 is not a support vector. Points -1 and 0 define the textured area and they will be the supporting data for our decision function. The task is to minimize (2.10a), and this will be achieved by taking the value $w = 2$. Then, from (b), it follows that $b = 1$. Note that (a) must not be used for the calculation of the bias term b.

Because both the cost function (2.10a) and the constraints (2.10b) are convex, the primal and the dual solution must produce same w and b. Dual solution follows from maximizing (2.15a) subject to (2.15b) and (2.15c) as follows

$$L_d = \alpha_1 + \alpha_2 + \alpha_3 - \frac{1}{2}[\alpha_1\ \alpha_2\ \alpha_3] \begin{bmatrix} 4 & 2 & 0 \\ 2 & 1 & 0 \\ 0 & 0 & 0 \end{bmatrix} \begin{bmatrix} \alpha_1 \\ \alpha_2 \\ \alpha_3 \end{bmatrix},$$

$$\text{s.t.} \qquad -\alpha_1 - \alpha_2 + \alpha_3 = 0,$$
$$\alpha_1 \geqslant 0, \ \ \alpha_2 \geqslant 0, \ \ \alpha_3 \geqslant 0,$$

The dual Lagrangian is obtained in terms of α_1 and α_2 after expressing α_3 from the equality constraint and it is given as $L_d = 2\alpha_1 + 2\alpha_2 - 0.5(4\alpha_1^2 + 4\alpha_1\alpha_2 + \alpha_2^2)$. L_d will have maximum for $\alpha_1 = 0$, and it follows that we have to find the maximum of $L_d = 2\alpha_2 - 0.5\alpha_2^2$ which will be at $\alpha_2 = 2$. Note that the Hessian matrix **H** is extremely bad conditioned and, if the QP problem is to be solved by computer, **H** should be regularized first. From the equality constraint it follows that $\alpha_3 = 2$ too. Now, we can calculate the weight vector w and the bias b from (2.17a) and (2.17b) as follows,

$$w = \sum_{i=1}^{3} \alpha_i y_i \mathbf{x}_i = 0(-1)(-2) + 2(-1)(-1) + 2(1)0 = 2$$

The bias can be calculated by using SVs only, meaning from either point -1 or point 0. Both result in same value as shown below

$$b = -1 - 2(-1) = 1, \text{ or } b = 1 - 2(0) = 1$$

2.2.2 Linear Soft Margin Classifier for Overlapping Classes

The learning procedure presented above is valid for linearly separable data, meaning for training data sets without overlapping. Such problems are rare in practice. At the same time, there are many instances when linear separating hyperplanes can be good solutions even when data are overlapped (e.g., normally distributed classes having the same covariance matrices have a linear separation boundary). However, quadratic programming solutions as given above cannot be used in the case of overlapping because the constraints $y_i[\mathbf{w}^T\mathbf{x}_i + b] \geq 1, i = 1, n$ given by (2.10b) cannot be satisfied. In the case of an overlapping (see Fig. 2.10), the overlapped data points cannot be correctly classified and for any misclassified training data point $\mathbf{x}_i$, the corresponding α_i will tend to infinity. This particular data point (by increasing the corresponding α_i value) attempts to exert a stronger influence on the decision boundary in order to be classified correctly. When the α_i value reaches the maximal bound, it can no longer increase its effect, and the corresponding point will stay misclassified. In such a situation, the algorithm introduced above chooses all training data points as support vectors. To find a classifier with a maximal margin, the algorithm presented in the Sect. 2.2.1, must be changed allowing some data to be unclassified. Better to say, we must leave some data on the 'wrong' side of a decision boundary. In practice, we allow a *soft* margin and all data inside this margin (whether on the correct side of the separating line or on the wrong one) are neglected. The width of a soft margin can be controlled by a corresponding penalty parameter C (introduced below) that determines the trade-off between the training error and VC dimension of the model.

The question now is how to measure the degree of misclassification and how to incorporate such a measure into the hard margin learning algorithm given by (2.10). The simplest method would be to form the following learning problem

$$\min \quad \frac{1}{2}\mathbf{w}^T\mathbf{w} + C(\text{number of misclassified data}) \tag{2.22}$$

where C is a penalty parameter, trading off the margin size (defined by $\|\mathbf{w}\|$, i.e., by $\mathbf{w}^T\mathbf{w}$) for the number of misclassified data points. Large C leads to small number of misclassifications, bigger $\mathbf{w}^T\mathbf{w}$ and consequently to the smaller margin and vice versa. Obviously taking $C = \infty$ requires that the number of misclassified data is zero and, in the case of an overlapping this is not possible. Hence, the problem may be feasible only for some value $C < \infty$.

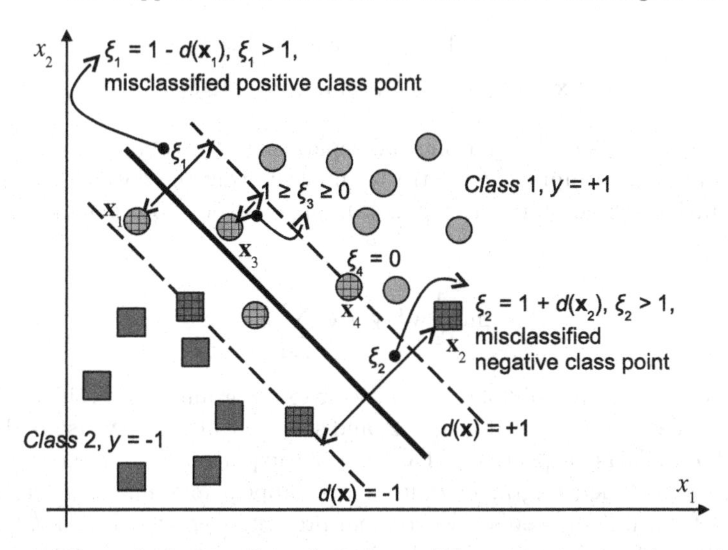

Fig. 2.10. The soft decision boundary for a dichotomization problem with data overlapping. Separation line (solid), margins (dashed) and support vectors (textured training data points).). 4 SVs in positive class (circles) and 3 SVs in negative class (squares). 2 misclassifications for positive class and 1 misclassification for negative class.

However, the serious problem with (2.22) is that the error's counting can't be accommodated within the handy (meaning reliable, well understood and well developed) quadratic programming approach. Also, the counting only can't distinguish between huge (or disastrous) errors and close misses! The possible solution is to measure the distances ξ_i of the points crossing the margin from the corresponding margin and trade their sum for the margin size as given below

$$\min \quad \frac{1}{2}\mathbf{w}^T\mathbf{w} + C(\text{sum of distances of the wrong side points}), \tag{2.23}$$

In fact this is exactly how the problem of the data overlapping was solved in [39, 40] - by generalizing the optimal 'hard' margin algorithm. They introduced the nonnegative *slack variables* $\xi_i (i = 1,\ n)$ in the statement of the optimization problem for the overlapped data points. Now, instead of fulfilling (2.10a) and (2.10b), the separating hyperplane must satisfy

$$\min \frac{1}{2}\mathbf{w}^T\mathbf{w} + C\sum_{i=1}^{n}\xi_i \tag{2.24a}$$

$$\text{s.t. } y_i(\mathbf{w}^T\mathbf{x}_i + b) \geq 1 - \xi_i, \quad i = 1,\dots,n, \quad \text{and} \tag{2.24b}$$

$$\xi_i \geq 0. \tag{2.24c}$$

i.e., subject to

$$\mathbf{w}^T\mathbf{x}_i + b \geq 1 - \xi_i, \quad \text{for } y_i = +1, \ \xi_i \geq 0 \tag{2.24d}$$

$$\mathbf{w}^T\mathbf{x}_i + b \leq -1 + \xi_i, \quad \text{for } y_i = -1, \ \xi_i \geq 0 \tag{2.24e}$$

Hence, for such a generalized optimal separating hyperplane, the functional to be minimized comprises an extra term accounting the cost of overlapping errors. In fact the cost function (2.24a) can be even more general as given below

$$\min \frac{1}{2}\mathbf{w}^T\mathbf{w} + C\sum_{i=1}^{n} \xi_i^k \tag{2.24f}$$

subject to same constraints. This is a convex programming problem that is usually solved only for $k = 1$ or $k = 2$, and such soft margin SVMs are dubbed *L1* and *L2 SVMs* respectively. By choosing exponent $k = 1$, neither slack variables ξ_i nor their Lagrange multipliers β_i appear in a dual Lagrangian L_d. Same as for a linearly separable problem presented previously, for *L1 SVMs* ($k = 1$) here, the solution to a quadratic programming problem (2.24), is given by the saddle point of the primal Lagrangian $L_p(\mathbf{w}, b, \boldsymbol{\xi}, \boldsymbol{\alpha}, \boldsymbol{\beta})$ shown below

$$L_p(\mathbf{w}, b, \boldsymbol{\xi}, \boldsymbol{\alpha}, \boldsymbol{\beta}) = \frac{1}{2}\mathbf{w}^T\mathbf{w} + C(\sum_{i=1}^{n} \xi_i) - \sum_{i=1}^{n} \alpha_i\{y_i[\mathbf{w}^T\mathbf{x}_i + b] - 1 + \xi_i\} - \sum_{i=1}^{n} \beta_i\xi_i. \tag{2.25}$$

where α_i and β_i are the Lagrange multipliers. Again, we should find an *optimal* saddle point $(\mathbf{w}_o, b_o, \boldsymbol{\xi}_o, \boldsymbol{\alpha}_o, \boldsymbol{\beta}_o)$ because the Lagrangian L_p has to be *minimized* with respect to $\mathbf{w}$, b and ξ_i and *maximized* with respect to nonnegative α_i and β_i. As before, this problem can be solved in either a primal space or dual space (which is the space of Lagrange multipliers α_i and β_i.). Again, we consider a solution in a dual space as given below by using - standard conditions for an optimum of a constrained function

$$\frac{\partial L}{\partial \mathbf{w}_o} = 0, \text{ i.e., } \quad \mathbf{w}_o = \sum_{i=1}^{n} \alpha_i y_i \mathbf{x}_i \tag{2.26a}$$

$$\frac{\partial L}{\partial b_o} = 0, \text{ i.e., } \quad \sum_{i=1}^{n} \alpha_i y_i = 0 \tag{2.26b}$$

$$\frac{\partial L}{\partial \xi_{io}} = 0, \text{ i.e., } \quad \alpha_i + \beta_i = C \tag{2.26c}$$

and the KKT complementarity conditions below,

$$\alpha_i\{y_i[\mathbf{w}^T\mathbf{x}_i + b] - 1 + \xi_i\} = 0, \quad i = 1, \ldots, n, \tag{2.26d}$$

$$\beta_i\xi_i = (C - \alpha_i)\xi_i = 0, \quad i = 1, \ldots, n. \tag{2.26e}$$

At the optimal solution, due to the KKT conditions (2.26d) and (2.26e), the last two terms in the primal Lagrangian L_p given by (2.25) vanish and the *dual variables Lagrangian* $L_d(\boldsymbol{\alpha})$, for *L1 SVM*, is not a function of β_i . In fact, it is same as the hard margin classifier's L_d given before and repeated here for the soft margin one,

$$\max L_d(\boldsymbol{\alpha}) = \sum_{i=1}^{n} \alpha_i - \frac{1}{2} \sum_{i,j=1}^{n} y_i y_j \alpha_i \alpha_j \mathbf{x}_i^T \mathbf{x}_j \tag{2.27a}$$

In order to find the optimal hyperplane, a dual Lagrangian $L_d(\boldsymbol{\alpha})$ has to be *maximized* with respect to nonnegative and (unlike before) smaller than or equal to C, α_i. In other words with

$$0 \leq \alpha_i \leq C, \quad i = 1, \ldots, n \tag{2.27b}$$

and under the constraint (2.26b), i.e., under

$$\sum_{i=1}^{n} \alpha_i y_i = 0. \tag{2.27c}$$

Thus, the final quadratic optimization problem is practically same as for the separable case the only difference being in the modified bounds of the Lagrange multipliers α_i. The penalty parameter C, which is now the upper bound on α_i, is determined by the user. The selection of a 'good' or 'proper' C is always done experimentally by using some cross-validation technique. Note that in the previous linearly separable case, without data over-lapping, this upper bound $C = \infty$. We can also readily change to the matrix notation of the problem above as in (2.16). Most important of all is that the learning problem is expressed only in terms of unknown Lagrange multipliers α_i, and known inputs and outputs. Furthermore, optimization does not solely depend upon inputs $\mathbf{x}_i$ which can be of a very high (inclusive of an infinite) dimension, but it depends upon a scalar product of input vectors $\mathbf{x}_i$. It is this property we will use in the next section where we design SV machines that can create nonlinear separation boundaries. Finally, expressions for both a *decision function* $d(\mathbf{x})$ and an indicator function $i_F = \text{sign}(d(\mathbf{x}))$ for a soft margin classifier are same as for linearly separable classes and are also given by (2.18).

From (2.26d) and (2.26e) follows that there are only three possible solutions for α_i (see Fig. 2.10)

1. α_i, $\xi_i = 0$, $\rightarrow$ data point $\mathbf{x}_i$ is correctly classified,
2. $C > \alpha_i > 0$, $\rightarrow$ then, the two complementarity conditions must result in $y_i[\mathbf{w}^T \mathbf{x}_i + b] - 1 + \xi_i = 0$, and $\xi_i = 0$. Thus, $y_i[\mathbf{w}^T \mathbf{x}_i + b] = 1$ and $\mathbf{x}_i$ is a support vector. The support vectors with $C \geq \alpha_i \geq 0$ are called *unbounded* or free support vectors. They lie on the two margins,

3. $\alpha_i = C$, $\rightarrow$ then, $y_i[\mathbf{w}^T\mathbf{x}_i + b] - 1 + \xi_i = 0$, and $\xi_i \geq 0$, and $\mathbf{x}_i$ is a support vector. The support vectors with $\alpha_i = C$ are called *bounded support vectors*. They lie on the 'wrong' side of the margin. For $1 > \xi_i \geq 0$, $\mathbf{x}_i$ is still correctly classified, and if $\xi_i \geq 1$, $\mathbf{x}_i$ is misclassified.

After the learning, the parameter $\mathbf{w}_o$ of the optimal hyperplane is calculated using the same expression (2.17a) as in the linearly separable case. For computing the optimal bias term b_o, the same philosophy as in (2.17b) is used, but the bounded support vectors BSV must not be included because they are not supposed to be on the margin, i.e. the ξ_i term for BSV should be greater than 0. Therefore, the formulation for working out the optimal bias b_o does not include BSV and it is given as follow,

$$b_o = \frac{1}{N_{FSV}} \sum_{s=1}^{N_{FSV}} (y_s - \mathbf{x}_s^T \mathbf{w}_o), \quad s = 1, \ldots, N_{FSV} \tag{2.28}$$

where N_{FSV} is the number of free support vectors. The same indicator function (2.18) is used for the soft margin SVMs as in the hard margin ones.

For *L2 SVM* the second term in the cost function (2.24f) is quadratic, i.e., $C\sum_{i=1}^{n}\xi_i^2$, and this leads to changes in a dual optimization problem which is now,

$$L_d(\boldsymbol{\alpha}) = \sum_{i=1}^{n} \alpha_i - \frac{1}{2} \sum_{i,j=1}^{n} y_i y_j \alpha_i \alpha_j \left(\mathbf{x}_i^T \mathbf{x}_j + \frac{\delta_{ij}}{C} \right) \tag{2.29a}$$

subject to

$$\alpha_i \geq 0, \quad i = 1, n \tag{2.29b}$$

$$\sum_{i=1}^{n} \alpha_i y_i = 0 \tag{2.29c}$$

where, $\delta_{ij} = 1$ for $i = j$, and it is zero otherwise. Note the change in Hessian matrix elements given by second terms in (2.29a), as well as that there is no upper bound on α_i. The detailed analysis and comparisons of the *L1* and *L2 SVMs* is presented in [1]. Derivation of (2.29) is given in the Appendix A. We use the most popular *L1 SVMs* here, because they usually produce more sparse solutions, i.e., they create a decision function by using less SVs than the L2 SVMs.

2.2.3 The Nonlinear SVMs Classifier

The linear classifiers presented in two previous sections are very limited. Mostly, classes are not only overlapped but the genuine separation functions are nonlinear hypersurfaces. A nice and strong characteristic of the approach presented above is that it can be easily (and in a relatively straightforward manner) extended to create nonlinear decision boundaries. The motivation for such an extension is that an SV machine that can create a nonlinear decision hypersurface will be able to classify nonlinearly separable data. This will be achieved by considering a linear classifier in the so-called feature space that will be introduced shortly. A very simple example of a need for designing nonlinear models is given in Fig. 2.11 where the true separation boundary is quadratic. It is obvious that no errorless linear separating hyperplane can be found now. The best linear separation function shown as a dashed straight line would make six misclassifications (textured data points; 4 in the negative class and 2 in the positive one). Yet, if we use the nonlinear separation boundary we are able to separate two classes without any error. Generally, for n-dimensional input patterns, instead of a nonlinear curve, an SV machine will create a nonlinear separating hypersurface.

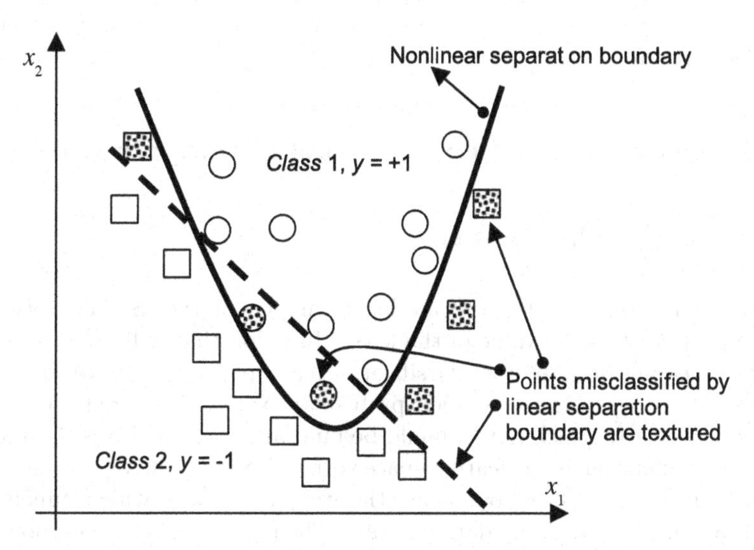

Fig. 2.11. A nonlinear SVM without data overlapping. A true separation is a quadratic curve. The nonlinear separation line (solid), the linear one (dashed) and data points misclassified by the linear separation line (the textured training data points) are shown. There are 4 misclassified negative data and 2 misclassified positive ones. SVs are not shown.

The basic idea of designing nonlinear SVMs is to map the input vectors $\mathbf{x}_i \in \Re^m$ into vectors $\boldsymbol{\Phi}(\mathbf{x}_i) \in \Re^s$ of a high dimensional feature space $\mathcal{S}$ (where $\boldsymbol{\Phi}$ represents mapping: $\Re^m \to \Re^s$) and to solve a linear classification problem in this feature space:

$$\mathbf{x} \in \Re^m \to \boldsymbol{\Phi}(\mathbf{x}) = [\phi_1(\mathbf{x})\ \phi_2(\mathbf{x}), \ldots, \phi_s(\mathbf{x})]^T \in \Re^s. \tag{2.30}$$

A mapping $\boldsymbol{\Phi}$ is chosen in advance, i.e., it is a fixed function. Note that an input space (**x**-space) is spanned by components x_i of an input vector **x** and a feature space $\mathcal{S}$ ($\boldsymbol{\Phi}$-space) is spanned by components $\phi_i(\mathbf{x})$ of a vector $\boldsymbol{\Phi}(\mathbf{x})$. By performing such a mapping, we hope that in a $\boldsymbol{\Phi}$-space, our learning algorithm will be able to linearly separate images of **x** by applying the linear SVM formulation presented above. (In fact, it can be shown that for a whole class of mappings the linear separation in a feature space is always possible. Such mappings will correspond to the positive definite kernels that will be shown shortly). We also expect this approach to again lead to solving a quadratic optimization problem with similar constraints in a $\boldsymbol{\Phi}$-space. The solution for an indicator function $i_F(\mathbf{x}) = \text{sign}(\mathbf{w}^T\boldsymbol{\Phi}(\mathbf{x}) + b) = \text{sign}\left(\sum_{i=1}^{n} y_i\alpha_i\boldsymbol{\Phi}^T(\mathbf{x}_i)\boldsymbol{\Phi}(\mathbf{x}) + b\right)$, which is a linear classifier in a feature space, will create a nonlinear separating hypersurface in the original input space given by (2.31) below. (Compare this solution with (2.18) and note the appearances of scalar products in both the original X-space and in the feature space $\mathcal{S}$).

The equation for an $i_F(\mathbf{x})$ just given above can be rewritten in a 'neural networks' form as follows

$$
\begin{aligned}
i_F(\mathbf{x}) = i_F(d(\mathbf{x})) &= \text{sign}(\mathbf{w}^T\boldsymbol{\Phi}(\mathbf{x}) + b) = \text{sign}(\sum_{i=1}^{n} y_i\alpha_i K(\mathbf{x}_i, \mathbf{x}) + b) \\
&= \text{sign}(\sum_{i=1}^{n} v_i K(\mathbf{x}_i, \mathbf{x}) + b).
\end{aligned}
\tag{2.31}
$$

where v_i corresponds to the output layer weights of the 'SVM's network' and $K(\mathbf{x}_i, \mathbf{x})$ denotes the value of the kernel function that will be introduced shortly. (v_i equals $y_i\alpha_i$ in the classification case presented above and it is equal to $(\alpha_i - \alpha_i*)$ in the regression problems). Note the difference between the weight vector **w** which norm should be minimized and which is the vector of the same dimension as the feature space vector $\boldsymbol{\Phi}(\mathbf{x})$ and the weightings $v_i = \alpha_i y_i$ that are scalar values composing the weight vector **v** which dimension equals the number of training data points n. The $(n - N_{SVs})$ of v_i components are equal to zero, and only N_{SVs} entries of **v** are nonzero elements.

A simple example below (Fig. 2.12) should exemplify the idea of a nonlinear mapping to (usually) higher dimensional space and how it happens that the data become linearly separable in the $\mathcal{S}$-space.

Example 2.2. Consider solving the simplest nonlinear 1-D classification problem in Fig. 2.12 given the three input and output (desired) values as follows: $\mathbf{x} = [-1\ 0\ 1]^T$ and $\mathbf{y} = [-1\ 1\ -1]^T$. The following mapping is chosen to form the feature space here: $\boldsymbol{\Phi}(x) = [x^2\ \sqrt{2}x\ 1]^T = [\phi_1(x)\ \phi_2(x)\ \phi_3(x)]^T$. The mapping produces the following three points in the feature space.

$$
\begin{array}{lll}
x_1 = -1 & y_1 = -1 & \boldsymbol{\Phi}(x_1) = [\,1\ -\sqrt{2}\ 1]^T \\
x_2 = 0 & y_2 = +1 \rightarrow & \boldsymbol{\Phi}(x_2) = [\,0\quad 0\quad 1]^T \\
x_3 = 1 & y_3 = -1 & \boldsymbol{\Phi}(x_3) = [\,1\ \ \sqrt{2}\ \ 1]^T
\end{array}
\tag{2.32}
$$

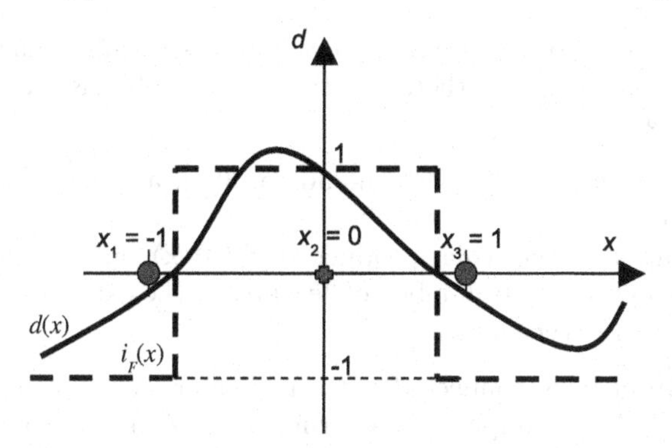

Fig. 2.12. A simple nonlinear 1-D classification problem. A linear classifier cannot be used to separate the data points successfully. One possible solution is given by the decision function $d(x)$ (solid curve). Its corresponding indicator function $\text{sign}(d(x))$ is also given as a dash line.

These three points are shown in Fig. 2.13 and they are now linearly separable in the 3-D feature space. The figure also shows that the separating boundary from the optimal separating (hyper)plane is perpendicular to the x^2 direction and it has the biggest margin. Note that the decision hyperplane cannot be visualized in Fig. 2.13, because it exists in the space which is for one dimension higher (namely, in a 4-D space).

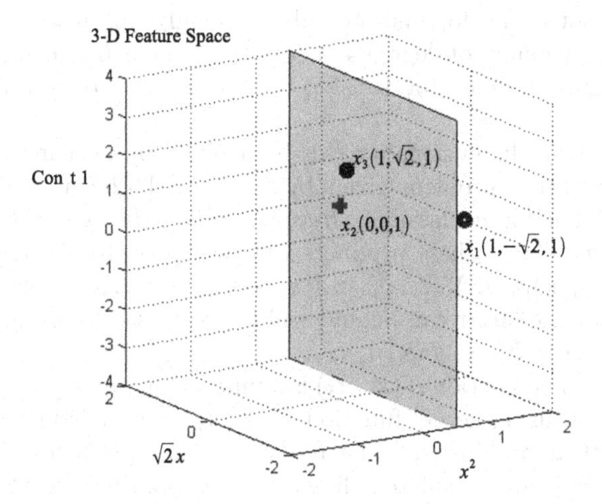

Fig. 2.13. The three points of the problem in Fig. 2.12 are linearly separable in the feature space (obtained by the mapping $\boldsymbol{\Phi}(\mathbf{x}) = [\phi_1(\mathbf{x})\ \phi_2(\mathbf{x})\ \phi_3(\mathbf{x})]^T = [x^2\ \sqrt{2}x\ 1]^T$). The separation boundary from SVMs which gives maximal margin is given by the plane $\phi_1(x) = 0.5$ as shown in the figure. Note that the decision function is in the 4-D space.

Although the use of feature mapping $\boldsymbol{\Phi}$ allows learning machines to deal with nonlinear dependency, there are two basic problems when performing such a mapping:

1. the choice of mapping $\boldsymbol{\Phi}$ that should result in a "rich" class of decision hyperplane.
2. the calculation of the scalar product $\boldsymbol{\Phi}(\mathbf{x})^T\boldsymbol{\Phi}(\mathbf{x})$ can be computationally very challenging if the number of features s (i.e., dimensionality s of a feature space) is very large.

The second problem is connected with a phenomenon called the '*curse of dimensionality*'. For example, to construct a decision surface corresponding to a polynomial of degree two in an m-D input space, a dimensionality of a feature space $s = m(m+3)/2$. In other words, a feature space is spanned by s coordinates of the form

$$
\begin{aligned}
z_1 &= x_1, \ldots, z_m = x_m \text{ (m coordinates)},\\
z_{m+1} &= (x_1)^2, \ldots, z_{2m} = (x_m)^2 \text{ (next m coordinates)},\\
z_{2m+1} &= x_1 x_2, \ldots, \; z_f = x_m x_{m-1} \text{ ($m(m-1)/2$coordinates)},
\end{aligned}
$$

and the separating hyperplane created in this space, is a second-degree polynomial in the input space [143]. Thus, constructing a polynomial of degree two only, in a 256-dimensional input space, leads to a dimensionality of a feature space $s = 33,152$. Performing a scalar product operation with vectors of such, or higher, dimensions, is not a cheap computational task. The problems become serious (and fortunately only seemingly unsolvable) if we want to construct a polynomial of degree 4 or 5 in the same 256-dimensional space leading to the construction of a decision hyperplane in a billion-dimensional feature space.

This explosion in dimensionality can be avoided by noticing that in the quadratic optimization problem given by (2.15) and (2.27a), as well as in the final expression for a classifier, *training data only appear in the form of scalar products* $\mathbf{x}_i^T\mathbf{x}_j$. These products will be replaced by scalar products $\boldsymbol{\Phi}^T(\mathbf{x})\boldsymbol{\Phi}(\mathbf{x})_i = [\phi_1(\mathbf{x}), \phi_2(\mathbf{x}), \ldots, \phi_m(\mathbf{x})][\phi_1(\mathbf{x}_i), \phi_2(\mathbf{x}_i), \ldots, \phi_m(\mathbf{x})_i]^T$ in a feature space $\mathcal{S}$, and the latter can be and will be expressed by using the *kernel function* $K(\mathbf{x}_i, \mathbf{x}_j) = \boldsymbol{\Phi}^T(\mathbf{x}_i)\boldsymbol{\Phi}(\mathbf{x}_j)$.

Note that a kernel function $K(\mathbf{x}_i, \mathbf{x}_j)$ is a function in input space. Thus, the basic advantage in using kernel function $K(\mathbf{x}_i, \mathbf{x}_j)$ is in avoiding performing a mapping $\boldsymbol{\Phi}(\mathbf{x})$ at all. Instead, the required scalar products in a feature space $\boldsymbol{\Phi}^T(\mathbf{x}_i)\boldsymbol{\Phi}(\mathbf{x}_j)$, are calculated directly by computing kernels $K(\mathbf{x}_i, \mathbf{x}_j)$ for given training data vectors in an input space. In this way, we bypass a possibly extremely high dimensionality of a feature space $\mathcal{S}$. Thus, by using the chosen kernel $K(\mathbf{x}_i, \mathbf{x}_j)$, we can construct an SVM that operates in an infinite dimensional space (such a kernel function is a Gaussian kernel function given in Table 2.2). In addition, as will be shown below, by applying kernels we do not even have to know what the actual mapping $\boldsymbol{\Phi}(\mathbf{x})$ is. A kernel is a function K such that

$$K(\mathbf{x}_i, \mathbf{x}_j) = \boldsymbol{\Phi}^T(\mathbf{x}_i)\boldsymbol{\Phi}(\mathbf{x}_j). \tag{2.33}$$

There are many possible kernels, and the most popular ones are given in Table 2.2. All of them should fulfill the so-called Mercer's conditions. The Mercer's kernels belong to a set of reproducing kernels. For further details see [101, 4, 129, 143, 81]. The simplest is a linear kernel defined as $K(\mathbf{x}_i, \mathbf{x}_j) = \mathbf{x}_i^T\mathbf{x}_j$. Below we show a few more kernels:

POLYNOMIAL KERNELS:
Let $x \in \Re^2$ i.e., $\mathbf{x} = [x_1 \;\; x_2]^T$, and if we choose $\boldsymbol{\Phi}(\mathbf{x}) = [x_1^2 \;\; \sqrt{2}x_1x_2 \;\; x_1^2]^T$ (i.e., there is an $\Re^2 \rightarrow \Re^3$ mapping), then the dot product

$$\begin{aligned}\boldsymbol{\Phi}^T(\mathbf{x}_i)\boldsymbol{\Phi}(\mathbf{x}_j) &= [x_{i1}^2 \;\; \sqrt{2}x_{i1}x_{i2} \;\; x_{i1}^2][x_{j1}^2 \;\; \sqrt{2}x_{j1}x_{j2} \;\; x_{j1}^2]^T \\ &= [x_{i1}^2x_{j1}^2 + 2x_{i1}x_{i2}x_{j1}x_{i2} + x_{i2}^2x_{j2}^2] = (\mathbf{x}_i^T\mathbf{x}_j)^2 = K(\mathbf{x}_i, \mathbf{x}_j), \text{or}\\ K(\mathbf{x}_i, \mathbf{x}_j) &= (\mathbf{x}_i^T\mathbf{x}_j)^2 = \boldsymbol{\Phi}^T(\mathbf{x}_i)\boldsymbol{\Phi}(\mathbf{x}_j)\end{aligned}$$

Note that in order to calculate the scalar product in a feature space $\boldsymbol{\Phi}^T(\mathbf{x}_i)\boldsymbol{\Phi}(\mathbf{x}_j)$, we do not need to perform the mapping $\boldsymbol{\Phi}(\mathbf{x}) = [x_1^2 \; \sqrt{2}x_1x_2 \; x_1^2]^T$ at all. Instead, we calculate this product directly in the input space by computing $(\mathbf{x}_i^T\mathbf{x}_j)^2$. This is very well known under the popular name of *the kernel trick.* Interestingly, note also that other mappings such as an

$\Re^2 \rightarrow \Re^3$ mapping given by $\boldsymbol{\Phi}(\mathbf{x}) = [x_1^2 - x_2^2 \;\; 2x_1x_2 \;\; x_1^2 + x_2^2]$, or an

$\Re^2 \rightarrow \Re^4$ mapping given by $\boldsymbol{\Phi}(\mathbf{x}) = [x_1^2 \;\; x_1x_2 \;\; x_1x_2 \;\; x_2^2]$

also accomplish the same task as $(\mathbf{x}_i^T\mathbf{x}_j)^2$.

Now, assume the following mapping

$$\boldsymbol{\Phi}(\mathbf{x}) = [1 \;\; \sqrt{2}x_1 \;\; \sqrt{2}x_2 \;\; \sqrt{2}x_1x_2 \;\; x_1^2 \;\; x_2^2],$$

i.e., there is an $\Re^2 \rightarrow \Re^5$ mapping plus bias term as the constant 6th dimension's value. Then the dot product in a feature space $\mathcal{S}$ is given as

$$\begin{aligned}\boldsymbol{\Phi}^T(\mathbf{x}_i)\boldsymbol{\Phi}(\mathbf{x}_j) &= 1 + 2x_{i1}x_{j1} + 2x_{i2}x_{j2} + 2x_{i1}x_{i2}x_{j1}x_{i2} + x_{i1}^2x_{j1}^2 + x_{i2}^2x_{j2}^2 \\ &= 1 + 2(\mathbf{x}_i^T\mathbf{x}_j) + (\mathbf{x}_i^T\mathbf{x}_j)^2 = (\mathbf{x}_i^T\mathbf{x}_j + 1)^2 = K(\mathbf{x}_i, \mathbf{x}_j), \text{ or}\\ K(\mathbf{x}_i, \mathbf{x}_j) &= (\mathbf{x}_i^T\mathbf{x}_j + 1)^2 = \boldsymbol{\Phi}^T(\mathbf{x}_i)\boldsymbol{\Phi}(\mathbf{x}_j)\end{aligned}$$

Thus, the last mapping leads to the second order *complete* polynomial.

Many candidate functions can be applied to a convolution of an inner product (i.e., for kernel functions) $K(\mathbf{x}, \mathbf{x}_i)$ in a SV machine. Each of these functions constructs a different nonlinear decision hypersurface in an input space. In the first three rows, the Table 2.2 shows the three most popular kernels in SVMs' in use today, and the inverse multiquadrics one as an interesting and powerful kernel to be proven yet. The positive definite (PD) kernels are

Table 2.2. Popular Admissible Kernels

Kernel Functions	Type of Classifier
$K(\mathbf{x},\mathbf{x}_i) = (\mathbf{x}^T\mathbf{x}_i)$	Linear, dot product, kernel, CPD[a]
$K(\mathbf{x},\mathbf{x}_i) = [(\mathbf{x}^T\mathbf{x}_i)+1]^d$	Complete polynomial of degree d, PD[b]
$K(\mathbf{x},\mathbf{x}_i) = \exp(-[\lVert\mathbf{x}-\mathbf{x}_i\rVert^2]/2\sigma^2)$	Gaussian RBF, PD[b]
$K(\mathbf{x},\mathbf{x}_i) = \tanh[(\mathbf{x}^T\mathbf{x}_i)+b]^*$	Multilayer perceptron, CPD
$K(\mathbf{x},\mathbf{x}_i) = 1/\sqrt{\lVert\mathbf{x}-\mathbf{x}_i\rVert^2+\beta}$	Inverse multiquadric function, PD

[a] Conditionally positive definite [b] Positive definite
[*] only for certain values of b

the kernels which Gramm matrix $\mathbf{G}$ (a.k.a. Grammian) calculated by using all the n training data points is positive definite (meaning all its eigenvalues are strictly positive, i.e., $\lambda_i > 0, i = 1, n$)

$$\mathbf{G} = \mathbf{K}(\mathbf{x}_i, \mathbf{x}_j) = \begin{bmatrix} k(\mathbf{x}_1,\mathbf{x}_1) & k(\mathbf{x}_1,\mathbf{x}_2) & \cdots & k(\mathbf{x}_1,\mathbf{x}_n) \\ k(\mathbf{x}_2,\mathbf{x}_1) & k(\mathbf{x}_2,\mathbf{x}_2) & \vdots & k(\mathbf{x}_2,\mathbf{x}_n) \\ \vdots & \vdots & \vdots & \vdots \\ k(\mathbf{x}_n,\mathbf{x}_1) & k(\mathbf{x}_n,\mathbf{x}_2) & \cdots & k(\mathbf{x}_n,\mathbf{x}_n) \end{bmatrix} \tag{2.34}$$

The $\mathbf{G}$ is a symmetric one. Even more, any symmetric positive definite matrix can be regarded as a kernel matrix, that is - as an inner product matrix in some space.

Finally, we arrive at the point of presenting the learning in nonlinear classifiers (in which we are ultimately interested here). The learning algorithm for a nonlinear SV machine (classifier) follows from the design of an optimal separating hyperplane in a feature space. This is the same procedure as the construction of a 'hard' (2.15) and 'soft' (2.27a) margin classifiers in an $\mathbf{x}$-space previously. In a $\boldsymbol{\Phi}(\mathbf{x})$-space, the dual Lagrangian, given previously by (2.15) and (2.27a), is now

$$L_d(\boldsymbol{\alpha}) = \sum_{i=1}^{n} \alpha_i - \frac{1}{2}\sum_{i,j=1}^{n} \alpha_i\alpha_j y_i y_j \boldsymbol{\Phi}_i^T \boldsymbol{\Phi}_j, \tag{2.35}$$

and, according to (2.33), by using chosen kernels, we should maximize the following dual Lagrangian

$$\max L_d(\alpha) = \sum_{i=1}^{n} \alpha_i - \frac{1}{2}\sum_{i,j=1}^{n} y_i y_j \alpha_i \alpha_j K(\mathbf{x}_i, \mathbf{x}_j) \tag{2.36a}$$

$$\text{s.t. } \alpha_i \geq 0, \quad i = 1, \ldots, n \quad \text{and} \tag{2.36b}$$

$$\sum_{i=1}^{n} \alpha_i y_i = 0. \tag{2.36c}$$

In a more general case, because of a noise or due to generic class' features, there will be an overlapping of training data points. Nothing but constraints for α_i change. Thus, constraints (2.36b) will be replaced by

$$0 \leq \alpha_i \leq C \quad i = 1, \ldots, n. \tag{2.36d}$$

Again, the only difference to the separable nonlinear classifier is the upper bound C on the Lagrange multipliers α_i. In this way, we limit the influence of training data points that will remain on the 'wrong' side of a separating nonlinear hypersurface. After the dual variables are calculated, the decision hypersurface $d(\mathbf{x})$ is determined by

$$d(\mathbf{x}) = \sum_{i=1}^{n} y_i \alpha_i K(\mathbf{x}, \mathbf{x}_i) + b = \sum_{i=1}^{n} v_i K(\mathbf{x}, \mathbf{x}_i) + b, \tag{2.37}$$

and the indicator function is $i_F(\mathbf{x}) = \text{sign}[d(\mathbf{x})] = \text{sign}\left[\sum_{i=1}^{n} v_i K(\mathbf{x}, \mathbf{x}_i) + b\right]$.

Note that the summation is not actually performed over all training data but rather over the support vectors, because only for them do the Lagrange multipliers differ from zero. The existence and calculation of a bias b is now not a direct procedure as it is for a linear hyperplane. Depending upon the applied kernel, the bias b can be implicitly part of the kernel function. If, for example, Gaussian RBF is chosen as a kernel, it can use a bias term as the $s + 1^{\text{st}}$ feature in $\mathcal{S}$-space with a constant output $= +1$, but not necessarily. In short, all PD kernels do not necessarily need an explicit bias term b, but b can be used. More on this can be found in [84] as well as in the [150]. Same as for the linear SVM, (2.36) can be written in a matrix notation as

$$\max L_d(\boldsymbol{\alpha}) = -0.5\boldsymbol{\alpha}^T \mathbf{H} \boldsymbol{\alpha} + \mathbf{p}^T \boldsymbol{\alpha}, \tag{2.38a}$$

$$\text{s.t.} \quad \mathbf{y}^T \boldsymbol{\alpha} = 0, \tag{2.38b}$$

$$0 \leq \alpha_i \leq C, \quad i = 1, \ldots, n, \tag{2.38c}$$

where $\boldsymbol{\alpha} = [\alpha_1, \alpha_2, \ldots, \alpha_n]^T$, $\mathbf{H}$ denotes the Hessian matrix ($H_{ij} = y_i y_j K(\mathbf{x}_i, \mathbf{x}_j)$) of this problem, and $\mathbf{p}$ is an $(n,1)$ unit vector $\mathbf{p} = \mathbf{1} = [1\ 1 \ldots 1]^T$. Note that the Hessian matrix is a dense n by n matrix. As a result, the amount of the computer memory required to solve the optimization problem is n^2. This is why the next part of the book is focused on solving the problem in an iterative way. The optimization problem (2.38) can be solved without the equality constraint (2.38b) when the Hessian matrix is positive definite (Note that if $K(\mathbf{x}_i, \mathbf{x}_j)$ is the positive definite matrix, then so is the matrix $y_i y_j K(\mathbf{x}_i, \mathbf{x}_j)$ too.). This fact is also used extensively in next chapter for deriving faster iterative learning algorithm for SVMs.

Example 2.3. The following 1-D example (just for the sake of graphical presentation) will show the creation of a linear decision function in a feature space and a corresponding nonlinear (quadratic) decision function in an input space.

Suppose we have 4 1-D data points given as $x_1 = 1,\ x_2 = 2,\ x_3 = 5,\ x_4 = 6$, with data at 1, 2, and 6 as class 1 and the data point at 5 as class 2, i.e., $y_1 = -1,\ y_2 = -1,\ y_3 = 1,\ y_4 = -1$. We use the polynomial kernel of degree 2, $K(x, y) = (xy + 1)^2$. C is set to 50, which is of lesser importance because the constraints will not be imposed in this example due to the fact that the maximal value of the dual variables alpha will be smaller than $C = 50$.

Case 1: Working with a bias term b as given in (2.37)
We first find $\alpha_i (i = 1, \ldots, 4)$ by solving dual problem (2.38) having a Hessian matrix

$$\mathbf{H} = \begin{bmatrix} 4 & 9 & -36 & 49 \\ 9 & 25 & -121 & 169 \\ -36 & -121 & 676 & -961 \\ 49 & 169 & -961 & 1369 \end{bmatrix}$$

Alphas are $\alpha_1 = 0,\ \alpha_2 = 2.499999,\ \alpha_3 = 7.333333\ \alpha_4 = 4.833333$ and the bias b will be found by using (2.17b), or by fulfilling the requirements that the values of a decision function at the support vectors should be the given y_i. The model (decision function) is given by

$$d(x) = \sum_{i=1}^{4} y_i \alpha_i K(x, x_i) + b = \sum_{i=1}^{4} v_i (x x_i + 1)^2 + b, \quad \text{or by}$$

$$d(x) = 2.4999(-1)(2x + 1)^2 + 7.3333(1)(5x + 1)^2 + 4.8333(-1)(6x + 1)^2 + b$$

$$d(x) = -0.666667x^2 + 5.333333x + b$$

Bias b is determined from the requirement that at the SV points 2, 5 and 6, the outputs must be -1, 1 and -1 respectively. Hence, $b = -9$, resulting in the decision function

$$d(x) = -0.666667x^2 + 5.333333x - 9.$$

The nonlinear (quadratic) decision function and the indicator one are shown in Fig. 2.14. Note that in calculations above 6 decimal places have been used for alpha values. The calculation is numerically very sensitive, and working with fewer decimals can give very approximate or wrong results.

The complete polynomial kernel as used in the case 1, is positive definite and there is no need to use an explicit bias term b as presented above. Thus, one can use the same second order polynomial model without the bias term b. Note that in this particular case there is no equality constraint equation that originates from an equalization of the primal Lagrangian derivative in respect to the bias term b to zero. Hence, we do not use (2.38b) while using a positive definite kernel without bias as it will be shown below in the case 2.

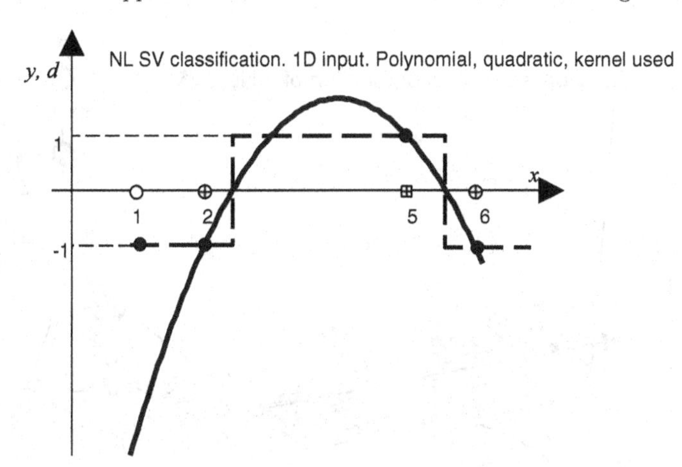

Fig. 2.14. The nonlinear decision function (solid) and the indicator function (dashed) for 1-D overlapping data. By using a complete second order polynomial the model with and without a bias term b are same.

Case 2: Working without a bias term b
Because we use the same second order polynomial kernel, the Hessian matrix $\mathbf{H}$ is same as in the case 1. The solution without the equality constraint for alphas is: $\alpha_1 = 0,\ \alpha_2 = 24.999999,\ \alpha_3 = 43.333333,\ \alpha_4 = 27.333333$. The model (decision function) is given by

$$d(x) = \sum_{i=1}^{4} y_i \alpha_i K(x,\, x_i) = \sum_{i=1}^{4} v_i (x x_i + 1)^2, \quad \text{or by}$$

$$d(x) = 24.9999(-1)(2x+1)^2 + 43.3333(1)(5x+1)^2 + 27.3333(-1)(6x+1)^2$$

$$d(x) = -0.666667x^2 + 5.333333x - 9.$$

Thus the nonlinear (quadratic) decision function and consequently the indicator function in the two particular cases are equal.

Example 2.4. XOR problems: In the next example shown by Figs. 2.15 and 2.16 we present all the important mathematical objects of a nonlinear SV classifier by using a classic XOR (*exclusive-or*) problem. The graphs show all the mathematical functions (objects) involved in a nonlinear classification. Namely, the nonlinear decision function $d(\mathbf{x})$, the NL indicator function $i_F(\mathbf{x})$, training data $(\mathbf{x}_i)$, support vectors $(\mathbf{x}_{SV})_i$ and separation boundaries.

The same objects will be created in the cases when the input vector $\mathbf{x}$ is of a dimensionality $n > 2$, but the visualization in these cases is not possible. In such cases one talks about the decision hyperfunction (hypersurface) $d(\mathbf{x})$, indicator hyperfunction (hypersurface) $i_F(\mathbf{x})$, training data $(\mathbf{x}_i)$, support vectors $(\mathbf{x}_{SV})_i$ and separation hyperboundaries (hypersurfaces).

Note the different character of a $d(\mathbf{x})$, $i_F(\mathbf{x})$ and separation boundaries in the two graphs given below. However, in both graphs all the data are

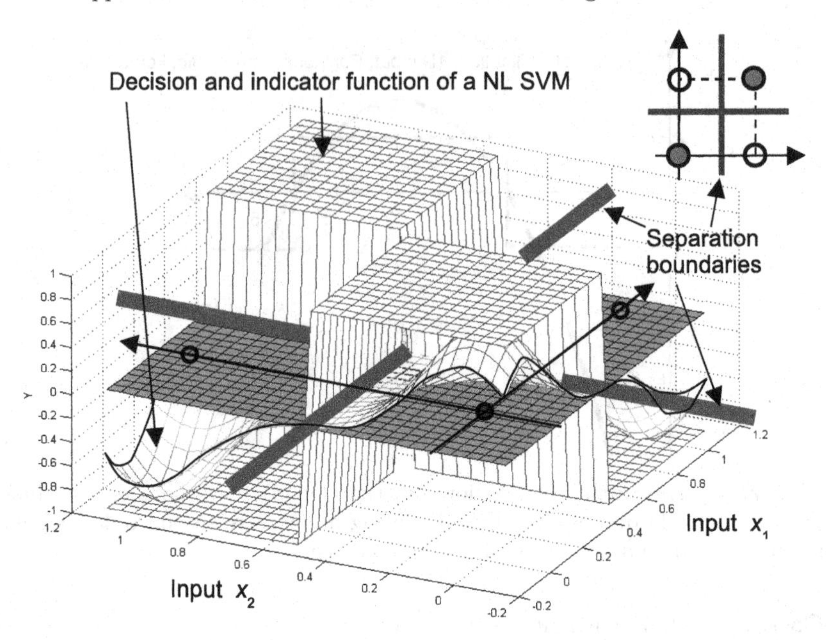

Fig. 2.15. XOR problem. Kernel functions (*2-D Gaussians*) are not shown. The nonlinear decision function, the nonlinear indicator function and the separation boundaries are shown. All four data are chosen as support vectors.

correctly classified. Fig. 2.15 shows the resulting functions for the Gaussian kernel functions, while Fig. 2.16 presents the solution for a complete second order polynomial kernel. Below, we present the analytical derivation of the (saddle like) decision function in the later (polynomial kernel) case. The analytic solution to the Fig. 2.16 for the second order polynomial kernel (i.e., for $(\mathbf{x}_i^T\mathbf{x}_j + 1)^2 = \boldsymbol{\Phi}^T(\mathbf{x}_i)\boldsymbol{\Phi}(\mathbf{x}_j)$, where

$$\boldsymbol{\Phi}(\mathbf{x}) = [1 \;\; \sqrt{2}x_1 \;\; \sqrt{2}x_2 \;\; \sqrt{2}x_1x_2 \;\; x_1^2 \; x_2^2],$$

no explicit bias and $C = \infty$) goes as follows. Inputs and desired outputs are,

$$\mathbf{x} = \begin{bmatrix} 0\,1\,1\,0 \\ 0\,1\,0\,1 \end{bmatrix}^T, \quad \mathbf{y} = \mathbf{d} = [1\;1\;-1\;-1]^T.$$

The dual Lagrangian (2.36a) has the Hessian matrix

$$\mathbf{H} = \begin{bmatrix} 1 & 1 & -1 & -1 \\ 1 & 9 & -4 & -4 \\ -1 & -4 & 4 & 1 \\ -1 & -4 & 1 & 4 \end{bmatrix}$$

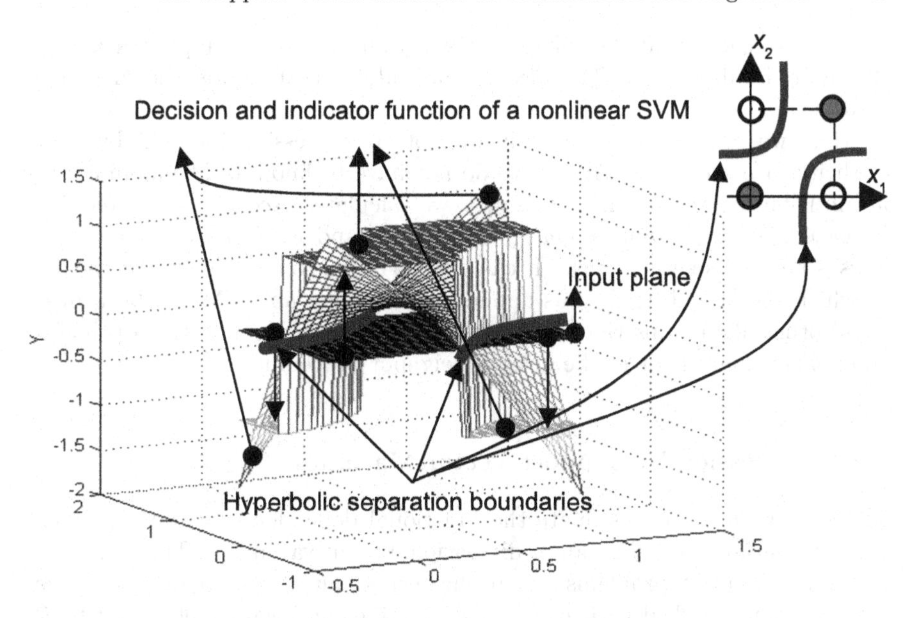

Fig. 2.16. XOR problem. Kernel function is a *2-D polynomial*. The nonlinear decision function, the nonlinear indicator function and the separation boundaries are shown. All four data are support vectors.

The optimal solution can be obtained by taking the derivative of L_d with respect to dual variables $\alpha_i (i = 1, 4)$ and by solving the resulting linear system of equations taking into account the constraints, see [84]. The solution to

$$\begin{aligned}
\alpha_1 + \alpha_2 - \alpha_3 - \alpha_4 &= 1, \\
\alpha_1 + 9\alpha_2 - 4\alpha_3 - 4\alpha_4 &= 1, \\
-\alpha_1 - 4\alpha_2 + 4\alpha_3 + \alpha_4 &= 1, \\
-\alpha_1 - 4\alpha_2 + \alpha_3 + 4\alpha_4 &= 1,
\end{aligned}$$

subject to $\alpha_i > 0, (i = 1, 4)$, is $\alpha_1 = 4.3333, \alpha_2 = 2.0000, \alpha_3 = 2.6667$ and $\alpha_4 = 2.6667$. The decision function in a 3-D space is

$$\begin{aligned}
d(\mathbf{x}) &= \sum\nolimits_{i=1}^{4} y_i \alpha_i \mathbf{\Phi}^T(\mathbf{x}_i)\mathbf{\Phi}(\mathbf{x}) \\
&= (4.3333\left[1\;0\;0\;0\;0\;0\right] + 2\left[1\;\sqrt{2}\;\sqrt{2}\;\sqrt{2}\;1\;1\right] \\
&\quad -2.6667\left[1\;\sqrt{2}\;0\;0\;1\;0\right] - 2.6667\left[1\;0\;\sqrt{2}\;0\;0\;1\right])\mathbf{\Phi}(\mathbf{x})
\end{aligned}$$

$$= [1\;\; \text{- }0.942\;\; \text{- }0.942\;\; 2.828\;\; \text{- }0.667\;\; \text{- }0.667]\,[1\;\sqrt{2}x_1\;\sqrt{2}x_2\;\sqrt{2}x_1x_2\;x_1^2\;x_2^2]^T$$

and finally

$$d(\mathbf{x}) = 1 - 1.3335x_1 - 1.3335x_2 + 4x_1x_2 - 0.6667x_1^2 - 0.6667x_2^2$$

It is easy to check that the values of $d(\mathbf{x})$ for all the training inputs in $\mathbf{x}$ equal the desired values in $\mathbf{d}$. The $d(\mathbf{x})$ is the saddle-like function shown in Fig. 2.16.

Here we have shown the derivation of an expression for $d(\mathbf{x})$ by using explicitly a mapping $\boldsymbol{\Phi}$. Again, we do not have to know what mapping $\boldsymbol{\Phi}$ is at all. By using kernels in input space, we calculate a *scalar product* required in a (*possibly high dimensional*) *feature space* and we avoid mapping $\boldsymbol{\Phi}(\mathbf{x})$. This is known as kernel 'trick'. It can also be useful to remember that the way in which the kernel 'trick' was applied in designing an SVM can be utilized in all other algorithms that depend on the scalar product (e.g., in principal component analysis or in the nearest neighbor procedure) .

2.2.4 Regression by Support Vector Machines

In the regression , we estimate the functional dependence of the dependent (output) variable $y \in \Re$ on an m-dimensional input variable $\mathbf{x}$. Thus, unlike in pattern recognition problems (where the desired outputs y_i are discrete values e.g., Boolean) we deal with *real valued* functions and we model an $\Re^m$ to $\Re^1$ mapping here. Same as in the case of classification, this will be achieved by training the SVM model on a training data set first. Interestingly and importantly, a learning stage will end in the same shape of a dual Lagrangian as in classification, only difference being in a dimensionalities of the Hessian matrix and corresponding vectors which are of a double size now e.g., $\mathbf{H}$ is a $(2n, 2n)$ matrix. Initially developed for solving classification problems, SV techniques can be successfully applied in regression, i.e., for a functional approximation problems [45, 142]. The general regression learning problem is set as follows - the learning machine is given n training data from which it attempts to learn the input-output relationship (dependency, mapping or function) $f(\mathbf{x})$. A training data set $\mathcal{X} = [\mathbf{x}(i), y(i)] \in \Re^m \times \Re, i = 1, ..., n$ consists of n pairs $(\mathbf{x}_1, y_1), (\mathbf{x}_2, y_2), \ldots, (\mathbf{x}_n, y_n)$, where the inputs $\mathbf{x}$ are m-dimensional vectors $\mathbf{x} \in \Re^m$ and system responses $y \in \Re$, are continuous values. We introduce all the relevant and necessary concepts of SVM's regression in a gentle way starting again with a *linear regression hyperplane* $f(\mathbf{x}, \mathbf{w})$ given as

$$f(\mathbf{x}, \mathbf{w}) = \mathbf{w}^T\mathbf{x} + b. \tag{2.39}$$

In the case of SVM's regression, we measure the *error of approximation* instead of the margin used in classification. The most important difference in respect to classic regression is that we use a novel loss (error) functions here. This is the Vapnik's *linear loss function* with ε-insensitivity zone defined as

$$E(\mathbf{x}, y, f) = |y - f(\mathbf{x}, \mathbf{w})|_\varepsilon = \begin{cases} 0 & \text{if } |y - f(\mathbf{x}, \mathbf{w})| \leq \varepsilon \\ |y - f(\mathbf{x}, \mathbf{w})| - \varepsilon & \text{otherwise} \end{cases}, \tag{2.40a}$$

or as,

$$E(\mathbf{x}, y, f) = \max(0, |y - f(\mathbf{x}, \mathbf{w})| - \varepsilon). \tag{2.40b}$$

Thus, the loss is equal to zero if the difference between the predicted $f(\mathbf{x}_i, \mathbf{w})$ and the measured value y_i is less than ε. In contrast, if the difference is larger than ε, this difference is used as the error. Vapnik's ε-insensitivity loss function (2.40) defines an ε tube as shown in Fig. 2.18. If the predicted value is within the tube, the loss (error or cost) is zero. For all other predicted points outside the tube, the loss equals the magnitude of the difference between the predicted value and the radius ε of the tube. The two classic error

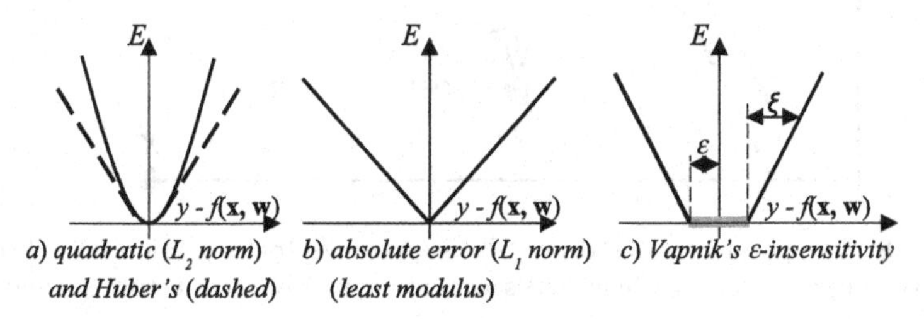

Fig. 2.17. Loss (error) functions.

functions are: a square error, i.e., L_2 norm $(y - f)^2$, as well as an absolute error, i.e., L_1 norm, least modulus $|y - f|$ introduced by Yugoslav scientist Rudjer Boskovic in 18th century [48]. The latter error function is related to Huber's error function. An application of Huber's error function results in a robust regression. It is the most reliable technique if nothing specific is known about the model of a noise. We do not present Huber's loss function here in analytic form. Instead, we show it by a dashed curve in Fig. 2.17a. In addition, Fig. 2.17 shows typical shapes of all mentioned error (loss) functions above.

Note that for $\varepsilon = 0$, Vapnik's loss function equals a least modulus function. Typical graph of a (nonlinear) regression problem as well as all relevant mathematical variables and objects required in, or resulted from, a learning unknown coefficients w_i are shown in Fig. 2.18.

We will formulate an SVM regression's algorithm for the linear case first and then, for the sake of a NL model design, we will apply mapping to a feature space, utilize the kernel 'trick' and construct a nonlinear regression hypersurface. This is actually the same order of presentation as in classification tasks. Here, for the regression, we 'measure' the empirical error term R_{emp} by Vapnik's ε-insensitivity loss function given by (2.40) and shown in Fig. 2.17c (while the minimization of the confidence term Ω will be realized through a minimization of $\mathbf{w}^T\mathbf{w}$ again). The empirical risk is given as

$$R^{\varepsilon}_{emp}(\mathbf{w}, b) = \frac{1}{n}\sum\nolimits_{i=1}^{n} \left|y_i - \mathbf{w}^T\mathbf{x}_i - b\right|_{\varepsilon} \tag{2.41}$$

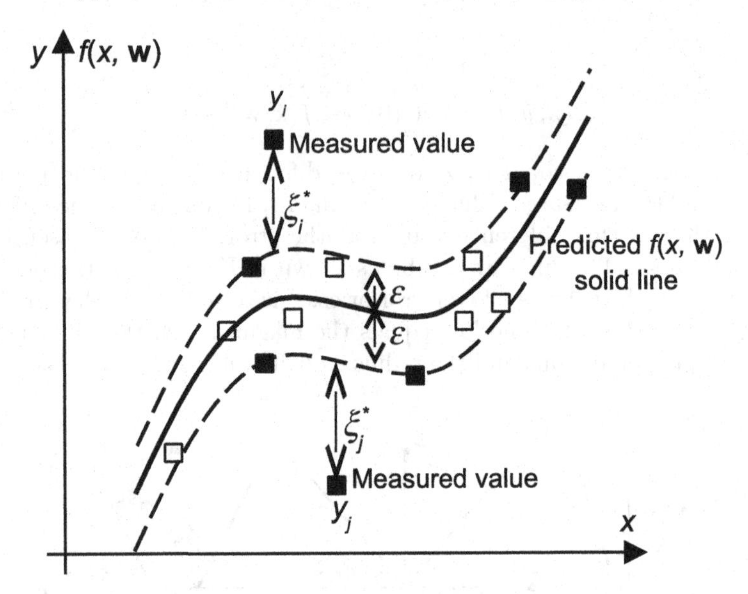

Fig. 2.18. The parameters used in (1-D) support vector regression. Filled squares data are support vectors. Hence, SVs can appear only on the tube boundary or outside the tube.

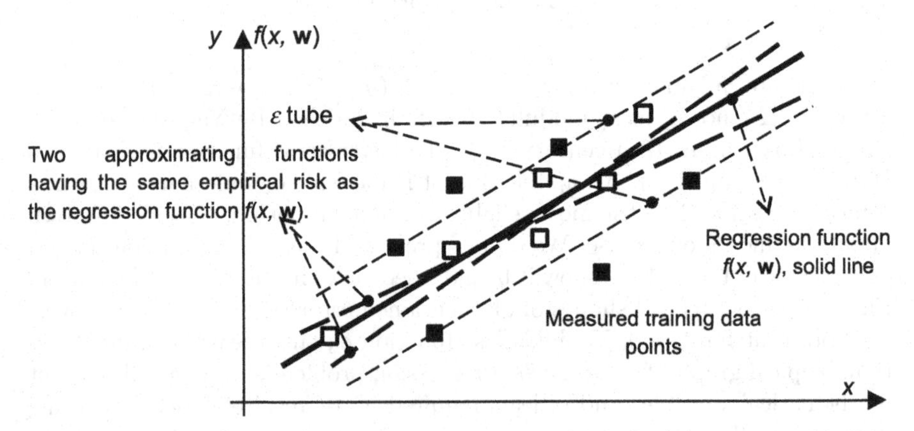

Fig. 2.19. Two linear approximations inside an ε tube (dashed lines) have the same empirical risk R^{ε}_{emp} on the training data as the regression function (solid line).

As in classification, we try to minimize both the empirical risk R^{ε}_{emp} and $\|\mathbf{w}\|^2$ simultaneously. Thus, we construct a linear regression hyperplane $f(\mathbf{x}, \mathbf{w}) = \mathbf{w}^T\mathbf{x} + b$ by minimizing

$$R = \frac{1}{2}||\mathbf{w}||^2 + C\sum_{i=1}^{n} |y_i - f(\mathbf{x}_i, \mathbf{w})|_{\varepsilon}. \tag{2.42}$$

Note that the last expression resembles the ridge regression scheme. However, we use Vapnik's ε-insensitivity loss function instead of a squared error now. From (2.40) and Fig. 2.18 it follows that for all training data outside an ε-tube,

$$|y - f(\mathbf{x}, \mathbf{w})| - \varepsilon = \xi \text{ for data 'above' an } \varepsilon\text{-tube, or} \tag{2.43a}$$

$$|y - f(\mathbf{x}, \mathbf{w})| - \varepsilon = \xi^* \text{ for data 'below' an } \varepsilon\text{-tube, or} \tag{2.43b}$$

Thus, minimizing the risk R above equals the minimization of the following risk

$$R_{\mathbf{w}, \xi, \xi^*} = \left[\frac{1}{2}||\mathbf{w}||^2 + C\left(\sum\nolimits_{i=1}^{n} \xi_i + \sum\nolimits_{i=1}^{n} \xi_i^*\right)\right] \tag{2.44a}$$

under constraints

$$y_i - \mathbf{w}^T\mathbf{x}_i - b \leq \varepsilon + \xi_i, \quad i = 1, \ldots, n \tag{2.44b}$$

$$\mathbf{w}^T\mathbf{x}_i + b - y_i \leq \varepsilon + \xi_i^*, \quad i = 1, \ldots, n \tag{2.44c}$$

$$\xi_i \geq 0, \quad \xi_i^* \geq 0 \quad i = 1, \ldots, n \tag{2.44d}$$

where ξ_i and ξ_i^* are slack variables shown in Fig. 2.18 for data points 'above' or 'below' the ε-tube respectively. Both slack variables are positive values. Lagrange multipliers α_i and α_i^* (that will be introduced during the minimization below) related to the first two sets of inequalities above, will be nonzero values for training points 'above' and 'below' an ε-tube respectively. Because no training data can be on both sides of the tube, either α_i or α_i^* will be nonzero. For data points inside the tube, both multipliers will be equal to zero. Thus $\alpha_i\alpha_i^* = 0$.

Note also that the constant C that influences a trade-off between an approximation error and the weight vector norm $\|\mathbf{w}\|$ is a design parameter that is chosen by the user. An increase in C penalizes larger errors i.e., it forces ξ_i and ξ_i^* to be small. This leads to an approximation error decrease which is achieved only by increasing the weight vector norm $\|\mathbf{w}\|$. However, an increase in $\|\mathbf{w}\|$ increases the confidence term Ω and does not guarantee a small generalization performance of a model. Another design parameter which is chosen by the user is the required precision embodied in an ε value that defines the size of an ε-tube. The choice of ε value is easier than the choice of C and it is given as either maximally allowed or some given or desired percentage of the output values y_i (say, $\varepsilon = 0.1$ of the mean value of $\mathbf{y}$).

Similar to procedures applied in the SV classifiers' design, we solve the constrained optimization problem above by forming a *primal variables* Lagrangian as follows,

$$
\begin{aligned}
L_p(\mathbf{w}, b, \xi_i, \xi_i^*, \alpha_i, \alpha_i^*, \beta_i, \beta_i^*) = & \frac{1}{2}\mathbf{w}^T\mathbf{w} + C\sum_{i=1}^{n}(\xi_i + \xi_i^*) \\
& - \sum_{i=1}^{n}(\beta_i^*\xi_i^* + \beta_i\xi_i) \\
& - \sum_{i=1}^{n}\alpha_i\left[\mathbf{w}^T\mathbf{x}_i + b - y_i + \varepsilon + \xi_i\right] \\
& - \sum_{i=1}^{n}\alpha_i^*\left[y_i - \mathbf{w}^{\mathrm{T}}\mathbf{x}_i - b + \varepsilon + \xi_i^*\right]
\end{aligned} \tag{2.45}
$$

A primal variables Lagrangian $L_p(\mathbf{w}, b, \xi_i, \xi_i^*, \alpha_i, \alpha_i^*, \beta_i, \beta_i^*)$ has to be *minimized* with respect to primal variables $\mathbf{w}, b, \xi_i$ and ξ_i^* and *maximized* with respect to nonnegative Lagrange multipliers $\alpha, \alpha_i^*, \beta_i$ and β_i^*. Hence, the function has the saddle point at the optimal solution $(\mathbf{w}_o, b_o, \xi_{io}, \xi_{io}^*)$ to the original problem. At the optimal solution the partial derivatives of L_p in respect to primal variables vanishes. Namely,

$$
\frac{\partial L_p(\mathbf{w}_o, b_o, \xi_{io}, \xi_{io}^*, \alpha_i, \alpha_i^*, \beta_i, \beta_i^*)}{\partial \mathbf{w}} = \mathbf{w}_o - \sum_{i=1}^{n}(\alpha_i - \alpha_i^*)\mathbf{x}_i = 0, \tag{2.46}
$$

$$
\frac{\partial L_p(\mathbf{w}_o, b_o, \xi_{io}, \xi_{io}^*, \alpha_i, \alpha_i^*, \beta_i, \beta_i^*)}{\partial b} = \sum_{i=1}^{n}(\alpha_i - \alpha_i^*) = 0, \tag{2.47}
$$

$$
\frac{\partial L_p(\mathbf{w}_o, b_o, \xi_{io}, \xi_{io}^*, \alpha_i, \alpha_i^*, \beta_i, \beta_i^*)}{\partial \xi_i} = C - \alpha_i - \beta_i = 0, \tag{2.48}
$$

$$
\frac{\partial L_p(\mathbf{w}_o, b_o, \xi_{io}, \xi_{io}^*, \alpha_i, \alpha_i^*, \beta_i, \beta_i^*)}{\partial \xi_i^*} = C - \alpha_i^* - \beta_i^* = 0. \tag{2.49}
$$

Substituting the KKT above into the primal L_p given in (2.45), we arrive at the problem of the *maximization of a dual variables Lagrangian* $L_d(\alpha, \alpha^*)$ below,

$$
\begin{aligned}
L_d(\alpha_i, \alpha_i^*) = & -\frac{1}{2}\sum_{i,j=1}^{n}(\alpha_i - \alpha_i^*)(\alpha_j - \alpha_j^*)\mathbf{x}_i^T\mathbf{x}_j - \varepsilon\sum_{i=1}^{n}(\alpha_i + \alpha_i^*) \\
& + \sum_{i=1}^{n}(\alpha_i - \alpha_i^*)y_i \\
= & -\frac{1}{2}\sum_{i,j=1}^{n}(\alpha_i - \alpha_i^*)(\alpha_j - \alpha_j^*)\mathbf{x}_i^T\mathbf{x}_j - \sum_{i=1}^{n}(\varepsilon - y_i)\alpha_i \\
& - \sum_{i=1}^{n}(\varepsilon + y_i)\,\alpha_i^*
\end{aligned} \tag{2.50}
$$

subject to constraints

$$
\sum_{i=1}^{n}\alpha_i^* = \sum_{i=1}^{n}\alpha_i \text{ or } \sum_{i=1}^{n}(\alpha_i - \alpha_i^*) = 0, \tag{2.51a}
$$

$$
0 \le \alpha_i \le C \quad i = 1, \ldots, n, \tag{2.51b}
$$

$$
0 \le \alpha_i^* \le C \quad i = 1, \ldots, n. \tag{2.51c}
$$

Note that the dual variables Lagrangian $L_d(\boldsymbol{\alpha}, \boldsymbol{\alpha}^*)$ is expressed in terms of Lagrange multipliers α_i and α_i^* only. However, the size of the problem, with respect to the size of an SV classifier design task, is doubled now. There are $2n$ unknown dual variables (n α_i-s and n α_i^*-s) for a linear regression and the Hessian matrix $\mathbf{H}$ of the quadratic optimization problem in the case of regression is a $(2n, 2n)$ matrix. The standard quadratic optimization problem above can be expressed in a *matrix notation* and formulated as follows:

$$\min L_d(\boldsymbol{\alpha}) = 0.5\boldsymbol{\alpha}^T \mathbf{H}\boldsymbol{\alpha} + \mathbf{p}\boldsymbol{\alpha}, \tag{2.52}$$

subject to (2.51) where $\boldsymbol{\alpha} = [\alpha_1, \alpha_2, \dots, \alpha_n, \alpha_1^*, \alpha_2^*, \dots, \alpha_n^*]^T$, $\boldsymbol{H} = [\mathbf{G} \quad -\mathbf{G}; -\mathbf{G} \quad \mathbf{G}]$, $\mathbf{G}$ is an (n,n) matrix with entries $G_{ij} = [\mathbf{x}_i^T \mathbf{x}_j]$ in a linear regression and $G_{ij} = K(\mathbf{x}_i, \mathbf{x}_j)$ for the nonlinear one, and $\mathbf{p} = [\varepsilon - y_1, \varepsilon - y_2, \dots, \varepsilon - y_n, \varepsilon + y_1, \varepsilon + y_2, \dots, \varepsilon + y_n]$ (Note that G_{ij}, as given above, is a badly conditioned matrix and we rather use $G_{ij} = [\mathbf{x}_i^T \mathbf{x}_j + 1]$ instead). Equation (2.52) is written in the form of a standard optimization routine that typically *minimizes* given objective function subject to the same constraints (2.51).

The learning stage results in n Lagrange multiplier pairs (α_i, α_i^*). After the learning, the number of SVs is equal to the number of nonzero α_i and α_i^*. However, this number does not depend on the dimensionality of input space and this is particularly important when working in very high dimensional spaces. Because at least one element of each pair (α_i, α_i^*), $i = 1, n$, is zero, the product of α_i and α_i^* is always zero,i.e. $\alpha_i \alpha_i^* = 0$. At the optimal solution the following KKT complementarity conditions must be fulfilled

$$\alpha_i \left(\mathbf{w}^{\mathrm{T}}\mathbf{x}_i + b - y_i + \varepsilon + \xi_i\right) = 0, \tag{2.53a}$$

$$\alpha_i^* \left(- \mathbf{w}^{\mathrm{T}}\mathbf{x}_i - b + y_i + \varepsilon + \xi_i^*\right) = 0, \tag{2.53b}$$

$$\beta_i \xi_i = (C - \alpha_i) \xi_i = 0, \tag{2.53c}$$

$$\beta_i^* \xi_i^* = (C - \alpha_i^*) \xi_i^* = 0. \tag{2.53d}$$

(2.53c) states that for $0 < \alpha_i < C$, $\xi_i = 0$ holds. Similarly, from (2.53d) follows that for $0 < \alpha_i^* < C$, $\xi_i^* = 0$ and, for $0 < \alpha_i, \alpha_i^* < C$, from (2.53a) and (2.53b) follows,

$$\mathbf{w}^{\mathrm{T}}\mathbf{x}_i + b - y_i + \varepsilon = 0, \tag{2.54a}$$

$$- \mathbf{w}^{\mathrm{T}}\mathbf{x}_i - b + y_i + \varepsilon = 0. \tag{2.54b}$$

Thus, for all the data points fulfilling $y - f(\mathbf{x}) = +\varepsilon$, dual variables α_i must be between 0 and C, or $0 < \alpha_i < C$, and for the ones satisfying $y - f(\mathbf{x}) = -\varepsilon$, α_i^* take on values $0 < \alpha_i^* < C$. These data points are called the *free* (or *unbounded*) support vectors. They allow computing the value of the bias term b as given below

$$b = y_i - \mathbf{w}^T \mathbf{x}_i - \varepsilon \text{ for } 0 < \alpha_i < C, \tag{2.55}$$

$$b = y_i - \mathbf{w}^T \mathbf{x}_i + \varepsilon \text{ for } 0 < \alpha_i^* < C. \tag{2.56}$$

The calculation of a bias term b is numerically very sensitive, and it is better to compute the bias b by averaging over all the *free* support vector data points.

The final observation follows from (2.53c) and (2.53d) and it tells that for all the data points outside the ε-tube, i.e., when both $\xi_i > 0$ and $\xi_i^* > 0$, both α_i and α_i^* equal C, i.e., $\alpha_i = C$ for the points above the tube and $\alpha_i^* = C$ for the points below it. These data are the so-called *bounded* support vectors. Also, for all the training data points within the tube, or when $|y - f(\mathbf{x})| < \varepsilon$, both α_i and α_i^* equal zero and they are neither the support vectors nor do they construct the decision function $f(\mathbf{x})$.

After calculation of Lagrange multipliers α_i and α_i^*, using (2.46) we can find an optimal (desired) weight vector of the *regression hyperplane* as

$$\mathbf{w}_o = \sum\nolimits_{i=1}^{n} (\alpha_i - \alpha_i^*)\mathbf{x}_i. \tag{2.57}$$

The best regression hyperplane obtained is given by

$$f(\mathbf{x}, \mathbf{w}, b) = \mathbf{w}_o^T\mathbf{x} + b = \sum\nolimits_{i=1}^{n} (\alpha_i - \alpha_i^*)\mathbf{x}_i^T\mathbf{x} + b. \tag{2.58}$$

More interesting, more common and the most challenging problem is to aim at solving the *nonlinear regression tasks.* A generalization to nonlinear regression is performed in the same way the nonlinear classifier is developed from the linear one, i.e., by carrying the mapping to the feature space, or by using kernel functions instead of performing the complete mapping which is usually of extremely high (possibly of an infinite) dimension. Thus, the nonlinear regression function in an input space will be devised by considering a linear regression hyperplane in the *feature space.*

We use the same basic idea in designing SV machines for creating a *nonlinear regression function.* First, a mapping of input vectors $\mathbf{x} \in \Re^m$ into vectors $\boldsymbol{\Phi}(x)$ of a higher dimensional *feature space* $\mathcal{S}$ (where $\boldsymbol{\Phi}$ represents mapping: $\Re^m \rightarrow \Re^s$) takes place and then, we solve a linear regression problem in this feature space. A mapping $\boldsymbol{\Phi}(\mathbf{x})$ is again the chosen in advance, or fixed, function. Note that an input space ($\mathbf{x}$-space) is spanned by components x_i of an input vector $\mathbf{x}$ and a feature space $\mathcal{S}$ ($\boldsymbol{\Phi}$-space) is spanned by components $\phi_i(\mathbf{x})$ of a vector $\boldsymbol{\Phi}(\mathbf{x})$. By performing such a mapping, we hope that in a $\boldsymbol{\Phi}$-space, our learning algorithm will be able to perform a linear regression hyperplane by applying the linear regression SVM formulation presented above. We also expect this approach to again lead to solving a quadratic optimization problem with inequality constraints in the feature space. The (linear in a feature space $\mathcal{S}$) solution for the regression hyperplane $f = \mathbf{w}^T\boldsymbol{\Phi}(\mathbf{x}) + b$, will create a nonlinear regressing hypersurface in the original input space. The most popular kernel functions are polynomials and RBF with Gaussian kernels. Both kernels are given in Table 2.2.

In the case of the nonlinear regression, the learning problem is again formulated as the maximization of a dual Lagrangian (2.52) with the Hessian matrix $\mathbf{H}$ structured in the same way as in a linear case, i.e. $\mathbf{H} = [\mathbf{G} \quad -\mathbf{G}; -\mathbf{G} \quad \mathbf{G}]$ but with the changed Grammian matrix $\mathbf{G}$ that is now given as

$$\mathbf{G} = \begin{bmatrix} G_{11} & \cdots & G_{1n} \\ \vdots & G_{ii} & \vdots \\ G_{n1} & \cdots & G_{nn} \end{bmatrix}$$

where the entries $G_{ij} = \boldsymbol{\Phi}^T(\mathbf{x}_i)\boldsymbol{\Phi}(\mathbf{x}_j) = K(\mathbf{x}_i, \mathbf{x}_j), i, j = 1,\ n.$

After calculating Lagrange multiplier vectors $\boldsymbol{\alpha}$ and $\boldsymbol{\alpha}^*$, we can find an optimal weighting vector of the *kernels expansion* as

$$\mathbf{v}_o = \boldsymbol{\alpha} - \boldsymbol{\alpha}^* \tag{2.59}$$

Note however the difference in respect to the linear regression where the expansion of a regression function is expressed by using the optimal weight vector $\mathbf{w}_o$. Here, in a NL SVMs' regression, the optimal weight vector $\mathbf{w}_o$ could often be of infinite dimension (which is the case if the Gaussian kernel is used). Consequently, we neither calculate $\mathbf{w}_o$ nor we have to express it in a closed form. Instead, we create the best nonlinear regression function by using the weighting vector $\mathbf{v}_o$ and the kernel (Grammian) matrix $\mathbf{G}$ as follows,

$$f(\mathbf{x}, \mathbf{w}) = \mathbf{G}\mathbf{v}_0 + b \tag{2.60}$$

In fact, the last result follows from the very setting of the learning (optimizing) stage in a feature space where, in all the equations above from (2.44b) to (2.58), we replace $\mathbf{x}_i$ by the corresponding feature vector $\boldsymbol{\Phi}(\mathbf{x}_i)$. This leads to the following changes:

- instead $G_{ij} = \mathbf{x}_i^T\mathbf{x}_j$ we get $G_{ij} = \boldsymbol{\Phi}^T(\mathbf{x}_i)\boldsymbol{\Phi}(\mathbf{x}_j)$ and, by using the kernel function $K(\mathbf{x}_i, \mathbf{x}_j) = \boldsymbol{\Phi}^T(\mathbf{x}_i)\boldsymbol{\Phi}(\mathbf{x}_j)$, it follows that $G_{ij} = K(\mathbf{x}_i, \mathbf{x}_j)$.
- similarly, (2.57) and (2.58) change as follows:

$$\mathbf{w}_o = \sum_{i=1}^{n} (\alpha_i - \alpha_i^*)\boldsymbol{\Phi}(\mathbf{x}_i) \quad \text{and,} \tag{2.61}$$

$$\begin{aligned} f(\mathbf{x}, \mathbf{w}, b) = \mathbf{w}_o^T\boldsymbol{\Phi}(\mathbf{x}) + b &= \sum_{i=1}^{n}(\alpha_i - \alpha_i^*)\boldsymbol{\Phi}^T(\mathbf{x}_i)\boldsymbol{\Phi}(\mathbf{x}) + b \\ &= \sum_{i=1}^{n}(\alpha_i - \alpha_i^*)K(\mathbf{x}_i, \mathbf{x}) + b. \end{aligned} \tag{2.62}$$

If the bias term b is explicitly used as in (2.60) then, for a NL SVMs' regression, it can be calculated from the upper SVs as,

$$\begin{aligned} b &= y_i - \sum_{j=1}^{N free\ upper\ SVs} (\alpha_j - \alpha_j^*)\boldsymbol{\Phi}^T(\mathbf{x}_j)\boldsymbol{\Phi}(\mathbf{x}_i) - \varepsilon \\ &= y_i - \sum_{j=1}^{N free\ upper\ SVs} (\alpha_j - \alpha_j^*)K(\mathbf{x}_i, \mathbf{x}_j) - \varepsilon, \text{ for } 0 < \alpha_i < C \end{aligned} \tag{2.63}$$

or from the lower ones as,

$$b = y_i - \sum_{j=1}^{N free\, lower\, SVs} (\alpha_j - \alpha_j^*) \mathbf{\Phi}^T(\mathbf{x}_j)\mathbf{\Phi}(\mathbf{x}_i) + \varepsilon$$
$$= y_i - \sum_{j=1}^{N free\, lower\, SVs} (\alpha_j - \alpha_j^*) K(\mathbf{x}_i, \mathbf{x}_j) + \varepsilon, \text{ for } 0 < \alpha_i^* < C. \tag{2.64}$$

Note that $\alpha_j^* = 0$ in (2.63) and so is $\alpha_j = 0$ in (2.64). Again, it is much better to calculate the bias term b by an averaging *over all* the *free* support vector data points.

There are a few learning parameters in constructing SV machines for regression. The three most relevant are the insensitivity zone ε, the penalty parameter C (that determines the trade-off between the training error and VC dimension of the model), and the shape parameters of the kernel function (variances of a Gaussian kernel, order of the polynomial, or the shape parameters of the inverse multiquadrics kernel function). All three parameters' sets should be selected by the user. To this end, the most popular method for their selection is a cross-validation. Unlike in a classification, for not too noisy data (primarily without huge outliers), the penalty parameter C could be set to infinity and the modeling can be controlled by changing the insensitivity zone ε and shape parameters only.

The *example* below shows how an increase in an insensitivity zone ε has smoothing effects on modeling highly noise polluted data. Increase in ε means a reduction in requirements on the accuracy of approximation. It decreases the number of SVs leading to higher data compression too. This can be readily followed in the lines and Fig. 2.20 below.

Example 2.5. The task here is to construct an SV machine for modeling measured data pairs. The underlying function (known to us but, not to the SVM) is a sinus function multiplied by the square one (i.e., $f(x) = x^2 \sin(x)$) and it is corrupted by 25% of normally distributed noise with a zero mean. Analyze the influence of an insensitivity zone ε on modeling quality and on a compression of data, meaning on the number of SVs. Fig. 2.20 shows that for a very noisy data a decrease of an insensitivity zone ε (i.e., shrinking of the tube shown by dashed line) approximates the noisy data points more closely. The related more and more wiggly shape of the regression function can be achieved only by including more and more support vectors. However, being good on the noisy training data points easily leads to an overfitting. The cross-validation should help in finding correct ε value, resulting in a regression function that filters the noise out but not the true dependency and which, consequently, approximate the underlying function as close as possible. The approximation function shown in Fig. 2.20 is created by 9 and 18 weighted Gaussian basis functions for $\varepsilon = 1$ and $\varepsilon = 0.75$ respectively. These supporting functions are not shown in the figure. However, the way how the learning algorithm selects SVs is an interesting property of support vector machines and in Fig. 2.21 we also present the supporting Gaussian functions.

Note that the selected Gaussians lie in the 'dynamic area' of the function in Fig. 2.21. Here, these areas are close to both the left hand and the right

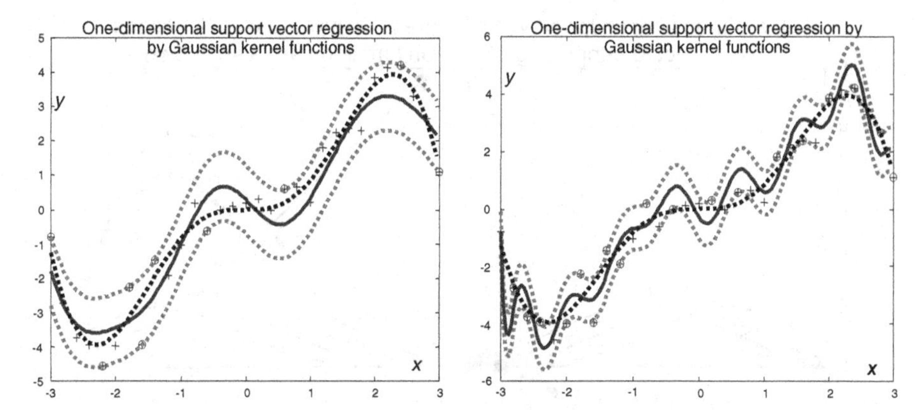

Fig. 2.20. The influence of an insensitivity zone ε on the model performance. A nonlinear SVM creates a regression function f with Gaussian kernels and models a highly polluted (25% noise) function $x^2 \sin(x)$ (dotted). 31 training data points (plus signs) are used. *Left*: $\varepsilon = 1$; 9 SVs are chosen (encircled plus signs). *Right*: $\varepsilon = 0.75$; the 18 chosen SVs produced a better approximation to noisy data and, consequently, there is the tendency of overfitting.

hand boundary. In the middle, the original function is pretty flat and there is no need to cover this part by supporting Gaussians. The learning algorithm realizes this fact and simply, it does not select any training data point in this area as a support vector. Note also that the Gaussians are not weighted in Fig. 2.21 , and they all have the peak value of 1. The standard deviation of Gaussians is chosen in order to see Gaussian supporting functions better. Here, in Fig. 2.21, $\sigma = 0.6$. Such a choice is due the fact that for the larger σ values the basis functions are rather flat and the supporting functions are covering the whole domain as the broad umbrellas. For very big variances one can't distinguish them visually. Hence, one can't see the true, bell shaped, basis functions for the large variances.

2.3 Implementation Issues

In both the classification and the regression the learning problem boils down to solving the QP problem subject to the so-called 'box-constraints' and to the equality constraint in the case that a model with a bias term b is used. The SV training works almost perfectly for not too large data basis. However, when the number of data points is large (say $n > 2,000$) the QP problem becomes extremely difficult to solve with standard QP solvers and methods. For example, a classification training set of 50,000 examples amounts to a Hessian matrix $\mathbf{H}$ with $2.5 * 10^9$ (2.5 billion) elements. Using an 8-byte floating-point representation we need 20,000 Megabytes = 20 Gigabytes of memory [109]. This cannot be easily fit into memory of present standard computers, and

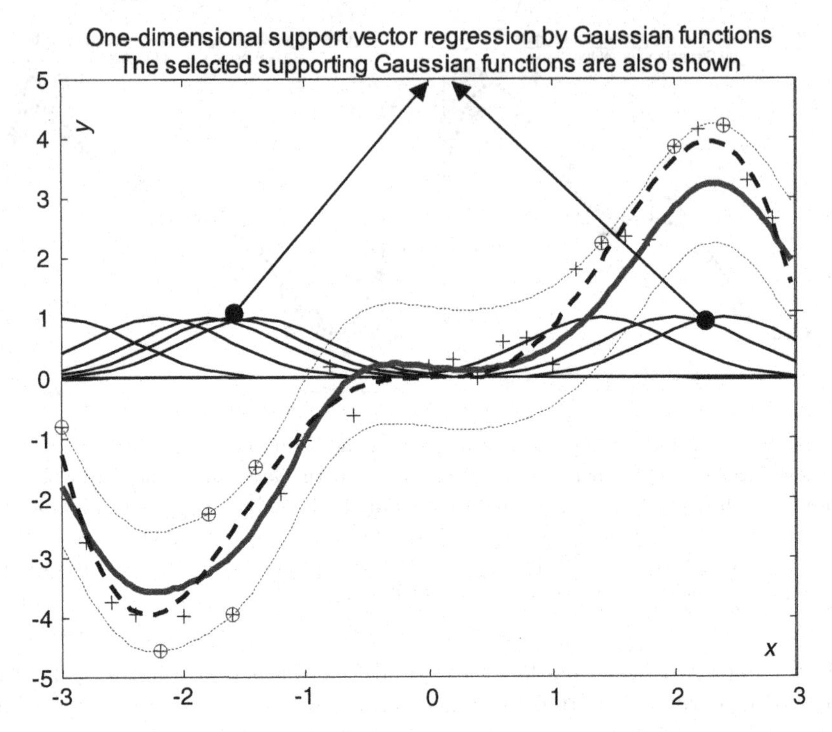

Fig. 2.21. Regression function f created as the sum of 8 weighted Gaussian kernels. A standard deviation of Gaussian bells $\sigma = 0.6$. Original function (dashed line) is $x^2 \sin(x)$ and it is corrupted by 25% noise. 31 training data points are shown as plus signs. Data points selected as the SVs are encircled. The 8 selected supporting Gaussian functions are centered at these data points.

this is the single basic disadvantage of the SVM method. There are three approaches that resolve the QP for large data sets. Vapnik in [144] proposed the chunking method that is the decomposition approach. Another decomposition approach is suggested in [109]. The sequential minimal optimization [115] algorithm is of different character and it seems to be an 'error back propagation' for an SVM learning. A systematic exposition of these various techniques is not given here, as all three would require a lot of space. However, the interested reader can find a description and discussion about the algorithms mentioned above in next chapter and [84, 150]. The Vogt and Kecman's chapter [150] discusses the application of an active set algorithm in solving small to medium sized QP problems. For such data sets and when the high precision is required the active set approach in solving QP problems seems to be superior to other approaches (notably to the interior point methods and to the sequential minimal optimization (SMO) algorithm). Next chapter introduces the efficient iterative single data algorithm (ISDA) for solving huge data sets (say more than 100,000 or 500,000 or over 1 million training data pairs). It seems that ISDA is the fastest algorithm at the moment for such large data sets still ensuring the convergence to the global minimum (see the comparisons with SMO in Sect.3.4). This means that the ISDA provides the exact, and not the approximate, solution to original dual problem.

Let us conclude the presentation of SVMs part by summarizing the basic constructive steps that lead to the SV machine.

A training and design of a support vector machine is an iterative algorithm and it involves the following steps:

1. define your problem as the classification or as the regression one,
2. preprocess your input data: select the most relevant features, scale the data between [-1, 1], or to the ones having zero mean and variances equal to one, check for possible outliers (strange data points),
3. select the kernel function that determines the hypothesis space of the decision and regression function in the classification and regression problems respectively,
4. select the 'shape', i.e., 'smoothing' parameter of the kernel function (for example, polynomial degree for polynomials and variances of the Gaussian RBF kernels respectively),
5. choose the penalty factor C and, in the regression, select the desired accuracy by defining the insensitivity zone ε too,
6. solve the QP problem in n and $2n$ variables in the case of classification and regression problems respectively,
7. validate the model obtained on some previously, during the training, unseen test data, and if not pleased iterate between steps 4 (or, eventually 3) and 7.

The optimizing part 6 is computationally extremely demanding. First, the Hessian matrix $\mathbf{H}$ scales with the size of a data set - it is an (n,n) and an $(2n, 2n)$ matrix in classification and regression respectively. Second, unlike in classic original QP problems $\mathbf{H}$ is very dense matrix and it is usually badly conditioned requiring a regularization before any numeric operation. Regularization means an addition of a small number to the diagonal elements of $\mathbf{H}$. Luckily, there are many reliable and fast QP solvers. A simple search on an Internet will reveal many of them. Particularly, in addition to the classic ones such as MINOS or LOQO for example, there are many more free QP solvers designed specially for the SVMs. The most popular ones are - the LIBSVM, SVMlight, SVM Torch, mySVM and SVM Fu. All of them can be downloaded from their corresponding sites.A user friendly software implementation of the ISDA that can handle huge data set can be download from the web site of this book `www.learning-from-data.com` . Good educational software in MATLAB named LEARNSC, with a very good graphic presentations of all relevant objects in a SVM modeling, can be downloaded from the second author's book site `www.support-vector.ws` too.

Finally we mention that there are many alternative formulations and approaches to the QP based SVMs described above. Notably, they are the linear programming SVMs [94, 53, 128, 59, 62, 83, 81, 82], μ-SVMs [123] and least squares support vector machines [134]. Their description is far beyond this chapter and the curious readers are referred to references given above.

3

Iterative Single Data Algorithm for Kernel Machines from Huge Data Sets: Theory and Performance

3.1 Introduction

One of the mainstream research fields in learning from empirical data by support vector machines (SVMs), and solving both the classification and the regression problems is an implementation of the iterative (incremental) learning schemes when the training data set is huge. The challenge of applying SVMs on huge data sets comes from the fact that the amount of computer memory required for solving the quadratic programming (QP) problem presented in the previous chapter increases drastically with the size of the training data set n. Depending on the memory requirement, all the solvers of SVMs can be classified into one of the three basic types as shown in Fig. 3.1 [150]. Direct methods (such as interior point methods) can efficiently obtain solution in machine precision, but they require at least $\mathcal{O}(n^2)$ of memory to store the Hessian matrix of the QP problem. As a result, they are often used to solve small-sized problems which require high precision. At the other end of the spectrum are the working-set (decomposition) algorithms whose memory requirements are only $\mathcal{O}(n + q^2)$ where q is the size of the working-set (for the ISDAs developed in this book, q is equal to 1). The reason for the low memory footprint is due to the fact that the solution is obtained iteratively instead of directly as in most of the QP solvers. They are the only possible algorithms for solving large-scale learning problems, but they are not suitable for obtaining high precision solutions because of the iterative nature of the algorithm. The relative size of the learning problem depends on the computer being used. As a result, a learning problem will be regarded as a "large" or "huge" problem in this book if the Hessian matrix of its unbounded SVs ($\mathbf{H}_{S_f S_f}$ where S_f denotes the set of free SVs) cannot be stored in the computer memory. Between the two ends of the spectrum are the active-set algorithms [150] and their memory requirements are $\mathcal{O}(N_{FSV}^2)$, i.e. they depend on the number of unbounded support vectors of the problem. The main focus of this book is to develop efficient algorithms that can solve large-scale QP problems for SVMs in practice. Although many applications in engineering also require the

T.-M. Huang et al.: *Kernel Based Algorithms for Mining Huge Data Sets*, Studies in Computational Intelligence (SCI) **17**, 61–95 (2006)
www.springerlink.com

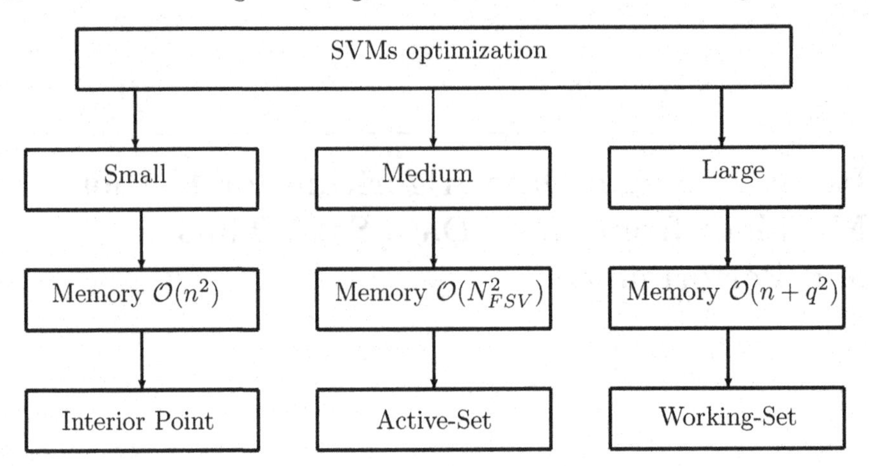

Fig. 3.1. QP optimization methods for different training data size.

solving of large-scale QP problems (and there are many solvers available), the QP problems induced by SVMs are different from these applications. In the case of SVMs, the Hessian matrix of (2.38a) is extremely dense, whereas in most of the engineering applications, the optimization problems have relatively sparse Hessian matrices. This is why many of the existing QP solvers are not suitable for SVMs and new approaches need to be invented and developed. Among several candidates that avoid the use of standard QP solvers, the two learning approaches which recently have drawn the attention are the Iterative Single Data Algorithm (ISDA), and the Sequential Minimal Optimization (SMO) [69, 78, 115, 148].

The most important characteristic of ISDAs are that they work on one data point at a time (per-pattern based learning) to improve the objective function. The Kernel AdaTron (KA) is the earliest ISDA for SVMs proposed in [54], which uses kernel functions to map data into SVMs' high dimensional feature space and performs AdaTron learning [12] in the feature space. The Platt's SMO algorithm [115] is a natural continuation of the various decomposition algorithms (chunking in [144], decomposition algorithms in [108, 78]) which operate on a working-set of two data points at a time. Because the solution for a working-set of two can be found analytically, an SMO algorithm does not invoke standard QP solvers. Due to its analytical foundation the SMO approach is particularly popular and at the moment the widely used, analyzed and still continuously improving algorithm. At the same time, the KA algorithm, although providing similar results in solving classification problems (in terms of both the accuracy and the training computation time required), did not attract that many devotees. There are two basic reasons for that. First, until recently [147], the KA seemed to be restricted to the classification problems only and second, it "lacked" the fleur of the strong theory (despite its beautiful "simplicity" and strong convergence proofs). The KA is based on a gradient ascent technique and this fact might also have distracted some researchers being aware of problems with gradient ascent approaches faced with a possibly ill-conditioned kernel matrix. Another very important ISDA is the no-bias version of the SMO algorithm developed recently in [148]. The development of this ISDA is based on the idea that the equality constraint of the SVMs optimization problem can be removed when a positive definite kernel is used. At the same time, it also follows similar idea as in the original SMO for working-set selection. This chapter is organized as follows:

- Section 3.2: The equality of two seemingly different ISDAs, which are a KA method and a without-bias version of SMO learning algorithm in designing the SVMs having positive definite kernels are derived and shown. The equality is valid for both the nonlinear classification and the nonlinear regression tasks, and it sheds a new light on these seemingly different learning approaches. Other learning techniques related to the two above-mentioned approaches are also introduced, such as the classic Gauss-Seidel coordinate ascent procedure and its derivative known as the successive over-relaxation algorithm as a viable and usually faster training algorithms for performing nonlinear classification and regression tasks.
- Section 3.3: In the later part of this chapter, a method of including the explicit bias term b efficiently into the ISDA will be presented. The explicit bias term b is included into the ISDA, because many simulations show that it leads to less support vectors. As a result, this version of the ISDAs can reduce the training time.
- Section 3.4: A comparison in performance between different ISDAs derived in this book and the popular SVM software LIBSVM is presented.
- Section 3.5: Efficient implementation of the ISDA software is discussed.
- Section 3.6: This section concludes the presentations here and discusses possible avenues for the future research in the area of developing faster iterative solver for SVMs.

3.2 Iterative Single Data Algorithm for Positive Definite Kernels without Bias Term b

In terms of representational capability, when applying Gaussian kernels, SVMs are similar to radial basis function networks. At the end of the learning, they produce a decision function of the following form

$$d(\mathbf{x}) = (\sum_{i=1}^{n} v_i K(\mathbf{x}, \mathbf{x}) + b). \qquad (3.1)$$

However, it is well known that positive definite kernels (such as the most popular and the most widely used RBF Gaussian kernels as well as the complete polynomial ones in Table 2.2) do not require bias term b [49, 85]. This means that the SVM learning problems should maximize (2.36a) with box constraints (2.36d) in classification and maximize (2.50) with box constraints (2.51b) and (2.51c) in regression (Note that in nonlinear regression the $\mathbf{x}_i\mathbf{x}_j$ terms in (2.50) should be replaced with $K(\mathbf{x}_i, \mathbf{x}_j)$). As a result, the optimization problem for SVMs classification without the explicit bias term b is as follows:

$$\max L_d(\alpha) = \sum_{i=1}^{n} \alpha_i - \frac{1}{2} \sum_{i,j=1}^{n} y_i y_j \alpha_i \alpha_j K(\mathbf{x}_i, \mathbf{x}_j), \tag{3.2a}$$

$$\text{s.t. } C \geq \alpha_i \geq 0, \quad i = 1, \dots, n. \tag{3.2b}$$

Hence, the dual optimization problem for nonlinear SVMs regression without the explicit bias term b is as follows:

$$\max \ L_d(\alpha_i, \alpha_i^*) = -\frac{1}{2} \sum_{i,j=1}^{n} (\alpha_i - \alpha_i^*)(\alpha_j - \alpha_j^*) K(\mathbf{x}_i, \mathbf{x}_j) - \varepsilon \sum_{i=1}^{n} (\alpha_i + \alpha_i^*)$$

$$+ \sum_{i=1}^{n} y_i (\alpha_i - \alpha_i^*) \tag{3.3a}$$

$$\text{s.t.} \quad 0 \leq \alpha_i \leq C \quad i = 1, \dots, n, \tag{3.3b}$$

$$0 \leq \alpha_i^* \leq C \quad i = 1, \dots, n. \tag{3.3c}$$

In this section, the KA and the SMO algorithms will be presented for such a fixed (i.e., no-) bias design problem and compared for the classification and regression cases. The equality of the two learning schemes and the resulting models will be established. Originally, in [115], the SMO classification algorithm was developed for solving (2.36a) including the equality constraint (2.36c) related to the bias b. In these early publications (on the classification tasks only) the case when bias b is a fixed variable was also mentioned but the detailed analysis of a fixed bias update was not accomplished. The algorithms here extend and develop a new method to regression problems too.

3.2.1 Kernel AdaTron in Classification

The classic AdaTron algorithm as given in [12] is developed for a linear classifier. As mentioned previously, the KA is a variant of the classic AdaTron algorithm in the feature space of SVMs. The KA algorithm solves the maximization of the dual Lagrangian (3.2a) by implementing the gradient ascent algorithm. The update $\Delta\alpha_i$ of the dual variables α_i is given as:

$$\Delta\alpha_i = \eta_i \frac{\partial L_d}{\partial \alpha_i} = \eta_i \left(1 - y_i \sum\nolimits_{j=1}^{n} \alpha_j y_j K(\mathbf{x}_i, \mathbf{x}_j)\right) = \eta_i \left(1 - y_i d_i\right), \tag{3.4}$$

The update of the dual variables α_i is given as

$$\alpha_i \leftarrow \min\{\max\{\alpha_i + \Delta\alpha_i, 0\}, C\} \quad i = 1, \dots, n. \tag{3.5}$$

In other words, the dual variables α_i are clipped to zero if $(\alpha_i + \Delta\alpha_i) < 0$. In the case of the soft nonlinear classifier ($C < \infty$) α_i are clipped between zero and C, $(0 \leq \alpha_i \leq C)$. The algorithm converges from any initial setting for the Lagrange multipliers α_i.

3.2.2 SMO without Bias Term b in Classification

Recently [148] derived the update rule for multipliers α_i that includes a detailed analysis of the Karush-Kuhn-Tucker (KKT) conditions for checking the optimality of the solution. (As referred above, a fixed bias update was mentioned only in Platt's papers). The no-bias SMO algorithm can be broken down into three different steps as follows:

1. The first step is to find the data points or the α_i variables to be optimized. This is done by checking the KKT complementarity conditions of the α_i variables. An α_i that violates the KKT condition will be referred to as a KKT violator. If there are no KKT violators in the entire data set, the optimal solution for (3.2) is found and the algorithm will stop. The α_i need to be updated if:

$$\alpha_i < C \quad \wedge \quad y_i E_i < -\tau, \quad \text{or} \tag{3.6}$$

$$\alpha_i > 0 \quad \wedge \quad y_i E_i > \tau \tag{3.7}$$

   where $E_i = d_i - y_i$ denotes the difference between the value of the decision function d_i (i.e., it is a SVM output) at the point $\mathbf{x}_i$ and the desired target (label) y_i and τ is the precision of the KKT conditions which should be fulfilled.
2. In the second step, the α_i variables that do not fulfill the KKT conditions will be updated. The following update rule for α_i was proposed in [148]:

$$\Delta\alpha_i = -\frac{y_i E_i}{K(\mathbf{x}_i, \mathbf{x}_i)} = -\frac{y_i d_i - 1}{K(\mathbf{x}_i, \mathbf{x}_i)} = \frac{1 - y_i d_i}{K(\mathbf{x}_i, \mathbf{x}_i)} \tag{3.8}$$

   After an update, the same clipping operation as in (3.5) is performed

$$\alpha_i \leftarrow \min\{\max\{\alpha_i + \Delta\alpha_i, 0\}, C\} \quad i = 1, \ldots, n. \tag{3.9}$$

3. After the updating of an α_i variable, the $y_j E_j$ terms in the KKT conditions of all the α_j variables will be updated by the following rules:

$$y_j E_j = y_j E_j^{old} + (\alpha_i - \alpha_i^{old}) K(\mathbf{x}_i, \mathbf{x}_j) y_j \quad j = 1, \ldots, n \tag{3.10}$$

   The algorithm will return to Step 1 in order to find a new KKT violator for updating.

Note the equality of the updating term between KA (3.4) and (3.8) of SMO without the bias term when the learning rate in (3.4) is chosen to be $\eta =$

$1/K(\mathbf{x}_i, \mathbf{x}_i)$. Because SMO without-bias-term algorithm also uses the same clipping operation in (3.9), both algorithms are strictly equal. This equality is not that obvious in the case of a 'classic' SMO algorithm with bias term due to the heuristics involved in the selection of active points which should ensure the largest increase of the dual Lagrangian L_d during the iterative optimization steps.

3.2.3 Kernel AdaTron in Regression

The first extension of the Kernel AdaTron algorithm for regression is presented in [147] as the following gradient ascent update rules for α_i and α_i^*,

$$\begin{aligned}\Delta\alpha_i &= \eta_i \frac{\partial L_d}{\partial \alpha_i} = \eta_i \left(y_i - \varepsilon - \sum\nolimits_{j=1}^{n} (\alpha_j - \alpha_j^*) K(\mathbf{x}_j, \mathbf{x}_i) \right) = \eta_i \left(y_i - \varepsilon - f_i \right) \\ &= -\eta_i \left(E_i + \varepsilon \right) \qquad (3.11a) \\ \Delta\alpha_i^* &= \eta_i \frac{\partial L_d}{\partial \alpha_i^*} = \eta_i \left(-y_i - \varepsilon + \sum\nolimits_{j=1}^{n} (\alpha_j - \alpha_j^*) K(\mathbf{x}_j, \mathbf{x}_i) \right) = \eta_i \left(-y_i - \varepsilon + f_i \right) \\ &= \eta_i \left(E_i - \varepsilon \right), \qquad (3.11b)\end{aligned}$$

where E_i is an error value given as a difference between the output of the SVM f_i and desired value y_i. The calculation of the gradient above does not take into account the geometric reality that no training data can be on both sides of the tube. In other words, it does not use the fact that either α_i or α_i^* or both will be nonzero, i.e. that $\alpha_i \alpha_i^* = 0$ must be fulfilled in each iteration step. Below the gradients of the dual Lagrangian L_d accounting for geometry will be derived following [85]. This new formulation of the KA algorithm strictly equals the SMO method given below in Sect. 3.2.4 and it is given as

$$\begin{aligned}\frac{\partial L_d}{\partial \alpha_i} &= -K(\mathbf{x}_i, \mathbf{x}_i)\alpha_i - \sum\nolimits_{j=1, j \neq i}^{n} (\alpha_j - \alpha_j^*) K(\mathbf{x}_j, \mathbf{x}_i) + y_i - \varepsilon + K(\mathbf{x}_i, \mathbf{x}_i)\alpha_i^* \\ &\quad - K(\mathbf{x}_i, \mathbf{x}_i)\alpha_i^* \\ &= -K(\mathbf{x}_i, \mathbf{x}_i)\alpha_i^* - (\alpha_i - \alpha_i^*) K(\mathbf{x}_i, \mathbf{x}_i) - \sum\nolimits_{j=1, j \neq i}^{n} (\alpha_j - \alpha_j^*) K(\mathbf{x}_j, \mathbf{x}_i) \\ &\quad + y_i - \varepsilon \\ &= -K(\mathbf{x}_i, \mathbf{x}_i)\alpha_i^* + y_i - \varepsilon - f_i = -\left(K(\mathbf{x}_i, \mathbf{x}_i)\alpha_i^* + E_i + \varepsilon \right).\end{aligned} \qquad (3.12)$$

For the α^* multipliers, the value of the gradient is

$$\frac{\partial L_d}{\partial \alpha_i^*} = -K(\mathbf{x}_i, \mathbf{x}_i)\alpha_i + E_i - \varepsilon \qquad (3.13)$$

The update value for α_i is now

$$\Delta\alpha_i = \eta_i \frac{\partial L_d}{\partial \alpha_i} = -\eta_i \left(K(\mathbf{x}_i, \mathbf{x}_i)\alpha_i^* + E_i + \varepsilon \right), \tag{3.14a}$$

$$\alpha_i \leftarrow \alpha_i + \Delta\alpha_i = \alpha_i + \eta_i \frac{\partial L_d}{\partial \alpha_i} = \alpha_i - \eta_i \left(K(\mathbf{x}_i, \mathbf{x}_i)\alpha_i^* + E_i + \varepsilon \right) \tag{3.14b}$$

For the learning rate $\eta = 1/K(\mathbf{x}_i, \mathbf{x}_i)$ the gradient ascent learning KA is defined as,

$$\alpha_i \leftarrow \alpha_i - \alpha_i^* - \frac{E_i + \varepsilon}{K(\mathbf{x}_i, \mathbf{x}_i)} \tag{3.15a}$$

Similarly, the update rule for α_i^* is

$$\alpha_i^* \leftarrow \alpha_i^* - \alpha_i + \frac{E_i - \varepsilon}{K(\mathbf{x}_i, \mathbf{x}_i)} \tag{3.15b}$$

Same as in the classification, α_i and α_i^* are clipped between zero and C,

$$\alpha_i \leftarrow \min(\max(0, \alpha_i + \Delta\alpha_i), C) \quad i = 1, \ldots, n \tag{3.16a}$$

$$\alpha_i^* \leftarrow \min(\max(0, \alpha_i^* \Delta\alpha_i^*), C) \quad i = 1, \ldots, n \tag{3.16b}$$

3.2.4 SMO without Bias Term b in Regression

The first algorithm for the SMO without-bias-term in regression, together with a detailed analysis of the KKT conditions for checking the optimality of the solution is derived in [148]. The following learning rules for the Lagrange multipliers α_i and α_i^* updates were proposed

$$\alpha_i \leftarrow \alpha_i - \alpha_i^* - \frac{E_i + \varepsilon}{K(\mathbf{x}_i, \mathbf{x}_i)} \tag{3.17a}$$

$$\alpha_i^* \leftarrow \alpha_i^* - \alpha_i + \frac{E_i - \varepsilon}{K(\mathbf{x}_i, \mathbf{x}_i)} \tag{3.17b}$$

The equality of equations (3.15a, b) and (3.17a, b) is obvious when the learning rate, as presented above in (3.15a, b), is chosen to be $\eta = 1/K(\mathbf{x}_i, \mathbf{x}_i)$. Thus, in both the classification and the regression, the optimal learning rate is not necessarily equal for all training data pairs. For a Gaussian kernel, $\eta = 1$ is same for all data points, and for a complete nth order polynomial each data point has a different learning rate $\eta = 1/K(\mathbf{x}_i \mathbf{x}_i)$. Similar to classification, a joint update of α_i and α_i^* is performed only if the KKT conditions are violated by at least τ i.e. if

$$\alpha_i < C \wedge \varepsilon + E_i < -\tau, \text{or} \tag{3.18a}$$

$$\alpha_i > 0 \wedge \varepsilon + E_i > \tau, \text{or} \tag{3.18b}$$

$$\alpha_i^* < C \wedge \varepsilon - E_i < -\tau, \text{or} \tag{3.18c}$$

$$\alpha_i^* > 0 \wedge \varepsilon - E_i > \tau \tag{3.18d}$$

After the changes, the same clipping operations as defined in (3.16) are performed

$$\alpha_i \leftarrow \min(\max(0, \alpha_i + \Delta\alpha_i), C) \quad i = 1, \ldots, n \tag{3.19a}$$
$$\alpha_i^* \leftarrow \min(\max(0, \alpha_i^* + \Delta\alpha_i^*), C) \quad i = 1, \ldots, n \tag{3.19b}$$

After an update of $\alpha_i^{(*)}$ (i.e. α_i or α_i^*), the algorithm updates the KKT conditions as follows [148]:

$$(E_j + \varepsilon) = (E_j + \varepsilon)^{old} + (\alpha_i - \alpha_i^* - \alpha_i^{old} + \alpha_i^{*old})K(\mathbf{x}_i, \mathbf{x}_j) \quad j = 1, \ldots, n \tag{3.20a}$$
$$(\varepsilon - E_j) = (\varepsilon - E_j)^{old} - (\alpha_i - \alpha_i^* - \alpha_i^{old} + \alpha_i^{*old})K(\mathbf{x}_i, \mathbf{x}_j) \quad j = 1, \ldots, n. \tag{3.20b}$$

The KA learning as formulated in this chapter and the SMO algorithm without-bias-term for solving regression tasks are strictly equal in terms of both the number of iterations required and the final values of the Lagrange multipliers. The equality is strict despite the fact that the implementation is slightly different [85, 84]. In every iteration step, namely, the KA algorithm updates both weights α_i and α_i^* without checking whether the KKT conditions are fulfilled or not, while the SMO performs an update according to conditions (3.18).

3.2.5 The Coordinate Ascent Based Learning for Nonlinear Classification and Regression Tasks

When positive definite kernels are used, the learning problem for both tasks is the same. The matrix notation of the optimization problem is derived by removing the equality constraint (2.38b) and only keep the box constraints as follows:

$$\max L_d(\boldsymbol{\alpha}) = -0.5\boldsymbol{\alpha}^T \mathbf{H} \boldsymbol{\alpha} + \mathbf{p}^T \boldsymbol{\alpha}, \tag{3.21a}$$
$$\text{s.t.} \quad 0 \leq \alpha_i \leq C, \quad i = 1, \ldots, k, \tag{3.21b}$$

where, in the classification $k = n$ and the Hessian matrix $\mathbf{H}$ is an (n, n) symmetric positive definite matrix, while in regression $k = 2n$ and $\mathbf{H}$ is a $(2n, 2n)$ symmetric semipositive definite one.

Note that the constraints (3.21b) define a convex subspace over which the convex dual Lagrangian should be maximized. It is very well known that the vector $\boldsymbol{\alpha}$ may be looked at as the solution of a system of linear equations:

$$\mathbf{H}\boldsymbol{\alpha} = \mathbf{p} \tag{3.22}$$

subject to the same constraints as given by (3.21b). Thus, it may seem natural to solve (3.22), subject to (3.21b), by applying some of the well known and established techniques for solving a general linear system of equations. The size of the training data set and the constraints (3.21b) eliminate direct techniques. Hence, one has to resort to iterative approaches in solving the problems above. There are three possible iterative avenues that can be followed. They are; the use of the Non-Negative Least Squares (NNLS) technique [90], an application of the Non-Negative Conjugate Gradient (NNCG) method [66] and the implementation of Gauss-Seidel (GS) i.e., the related Successive Over-Relaxation technique (SOR). The first two methods originally solve for the non-negative constraints only. Thus, they are not suitable in solving 'soft' tasks, when penalty parameter $C < \infty$ is used, i.e., when there is an upper bound on maximal value of α_i. Nevertheless, in the case of nonlinear regression, one can apply NNLS and NNCG by taking $C = \infty$ and compensating (i.e. smoothing or 'softening' the solution) by increasing the sensitivity zone ε. However, the two methods (namely NNLS and NNCG) are not suitable for solving soft margin ($C < \infty$) classification problems in their present form, because there is no other parameter that can be used in 'softening' the margin. As part of this book, the NNCG is extended to take both upper and lower bounds into account. This development can be found in Sect. 5.6.3.

Now, the extension of Gauss-Seidel (GS) and Successive Over-Relaxation (SOR) to both nonlinear classification and to nonlinear regression tasks will be shown. The Gauss-Seidel method solves (3.22) by using the i-th equation to update the i-th unknown doing it iteratively, i.e. starting in the k-th step with the first equation to compute the α_1^{k+1}, then the second equation is used to calculate the α_2^{k+1} by using new α_1^{k+1} and $\alpha_i^k (i > 2)$ and so on. The iterative learning takes the following form,

$$\begin{aligned}
\alpha_i^{k+1} &= \left(p_i - \sum_{j=1}^{i-1} H_{ij}\alpha_j^{k+1} - \sum_{j=i+1}^{n} H_{ij}\alpha_j^k \right) / H_{ii} \\
&= \alpha_i^k - \frac{1}{H_{ii}} \underbrace{\left(\sum_{j=1}^{i-1} H_{ij}\alpha_j^{k+1} + \sum_{j=i}^{n} H_{ij}\alpha_j^k - p_i \right)}_{r_i} \qquad (3.23) \\
&= \alpha_i^k + \frac{1}{H_{ii}} \left. \frac{\partial L_d}{\partial \alpha_i} \right|_{k+1}
\end{aligned}$$

where we use the fact that the term within a second bracket (called the residual r_i in mathematics' references) is the i-th element of the gradient of a dual Lagrangian L_d given in (3.21a) at the $k+1$st iteration step. The (3.23) above shows that GS method is a coordinate gradient ascent procedure as well as the KA and the SMO are. The KA and SMO for positive definite kernels equal the GS! Note that the optimal learning rate used in both the KA algorithm and in the SMO without-bias-term approach is exactly equal to the coefficient $1/H_{ii}$ in a GS method. Based on this equality, the convergence theorem for the KA, SMO and GS (i.e., SOR) in solving (3.21a) subject to constraints (3.21b) can be stated and proved as follows [85]:

Theorem 3.1. *For SVMs with positive definite kernels, the iterative learning algorithms KA i.e., SMO i.e., GS i.e., SOR, in solving nonlinear classification and regression tasks* (3.21a) *subject to constraints* (3.21b), *converge starting from any initial choice* $\boldsymbol{\alpha}_0$.

Proof. The proof is based on the very well known theorem of convergence of the GS method for symmetric positive definite matrices in solving (3.22) without constraints [107]. First note that for positive definite kernels, the matrix $\mathbf{H}$ created by terms $y_i y_j K(\mathbf{x}_i, \mathbf{x}_j)$ in the second sum in (3.2a), and involved in solving classification problem, is also positive definite. In regression tasks $\mathbf{H}$ is a symmetric positive semidefinite (meaning still convex) matrix, which after a mild regularization given as $(\mathbf{H} \leftarrow \mathbf{H} + \lambda\mathbf{I}, \lambda \approx 1e-12)$ becomes positive definite one. *(Note that the proof in the case of regression does not need regularization at all, but there is no space here to go into these details).* Hence, the learning without constraints (3.21b) converges, starting from any initial point $\boldsymbol{\alpha}_0$, and each point in an n-dimensional search space for multipliers α_i is a viable starting point ensuring a convergence of the algorithm to the maximum of a dual Lagrangian L_d. This, naturally, includes all the (starting) points within, or on a boundary of, any convex subspace of a search space ensuring the convergence of the algorithm to the maximum of a dual Lagrangian L_d over the given subspace. The constraints imposed by (3.21b) preventing variables α_i to be negative or bigger than C, and implemented by the clipping operators above, define such a convex sub-space. Thus, each "clipped" multiplier value α_i defines a new starting point of the algorithm guaranteeing the convergence to the maximum of L_d over the subspace defined by (3.21b). For a convex constraining subspace such a constrained maximum is unique Q.E.D.

It should be mentioned in passing that both KA and SMO (i.e. GS and SOR) for positive definite kernels have been successfully applied for many problems (see references given here, as well as many other, benchmarking the mentioned methods on various data sets). The standard extension of the GS method is the method of successive over-relaxation that can reduce the number of iterations required by proper choice of relaxation parameter ω significantly. The SOR method uses the following update rule

$$\alpha_i^{k+1} = \alpha_i^k - \omega \frac{1}{H_{ii}} \left(\sum_{j=1}^{i-1} H_{ij} \alpha_j^{k+1} + \sum_{j=i}^{n} H_{ij} \alpha_j^k - p_i \right) = \alpha_i^k + \omega \frac{1}{H_{ii}} \left. \frac{\partial L_d}{\partial \alpha_i} \right|_{k+1} \tag{3.24}$$

and similarly to the KA, SMO, and Gauss-Seidel, its convergence is guaranteed. With a proper choice of the relaxation parameter ω, the performance of the algorithms can be speed up significantly. Furthermore, the improvement in performance is more noticeable in the regression than in the classification. The typical range of ω used in this book is between 1.5 and 1.9. Note that the algorithm will not converge if ω is equal or greater than 2.

Example 3.2. To help visualize the connection between solving QP problem (3.21) and system of linear equation (3.22) using ISDA, consider a simple 2D QP problem as follows:

$$\begin{aligned} \max \quad Q(\mathbf{x}) &= -\frac{1}{2}\mathbf{x}^T\mathbf{H}\mathbf{x} + \mathbf{p}^T\mathbf{x} \\ &= -\frac{1}{2}[x_1, x_2]\begin{bmatrix} 1 & -0.5 \\ -0.5 & 1 \end{bmatrix}\begin{bmatrix} x_1 \\ x_2 \end{bmatrix} + [2, 2]\begin{bmatrix} x_1 \\ x_2 \end{bmatrix} \end{aligned} \tag{3.25}$$

For an unconstrained problem, the solution is found by solving the following system of linear equations:

$$\begin{bmatrix} 1 & -0.5 \\ -0.5 & 1 \end{bmatrix}\begin{bmatrix} x_1 \\ x_2 \end{bmatrix} = \begin{bmatrix} 2 \\ 2 \end{bmatrix} \tag{3.26}$$

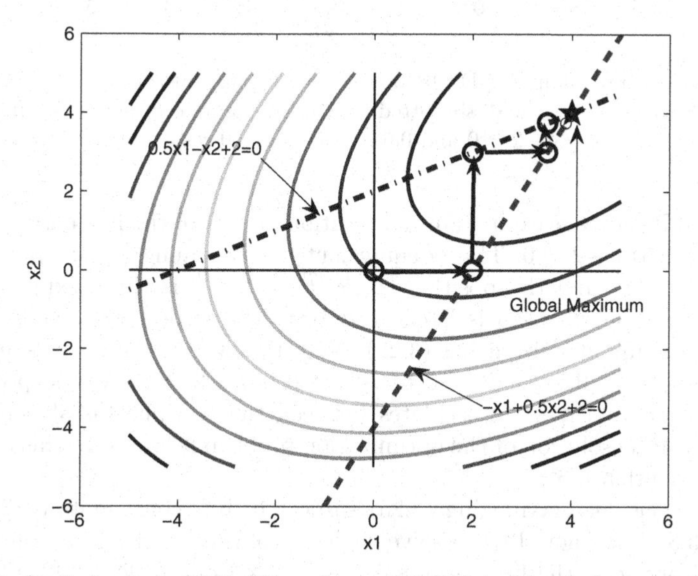

Fig. 3.2. Optimization of 2-D QP problem with ISDA. The dashed and the dash-dotted lines in the figure represent system of linear equations $-x_1 + 0.5x_2 + 2 = 0$ and $0.5x_1 - x_2 + 2 = 0$ respectively.

In other words, the solution of the problem (3.25) is also equal to the intersection of straight line $-x_1 + 0.5x_2 + 2 = 0$ and $0.5x_1 - x_2 + 2 = 0$ as shown in Fig. 3.2. Fig. 3.2 shows how ISDA solves such a problem starting from the initial position of $\mathbf{x} = [0, 0]^T$. The update rule (3.23) is first applied

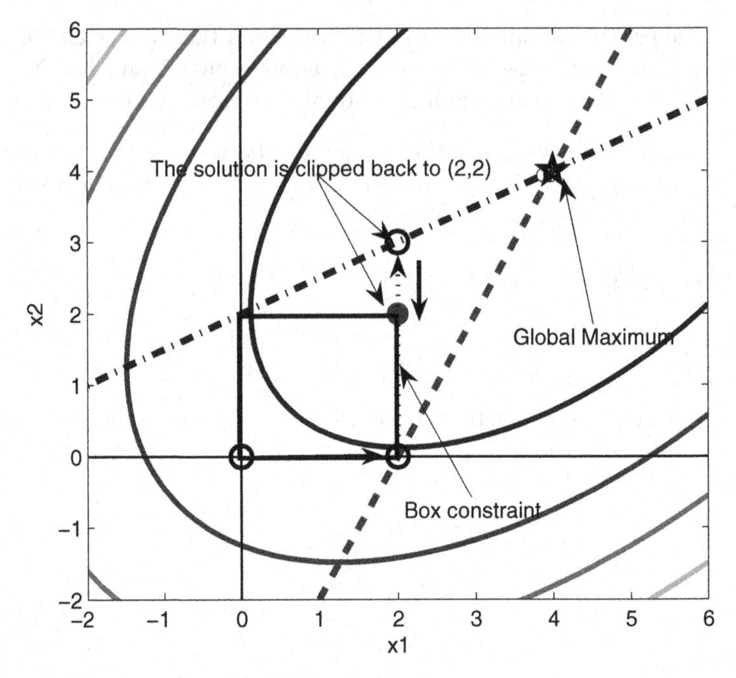

Fig. 3.3. Optimization of 2-D QP problem with box constraint $0 \leq \mathbf{x} \leq 2$ using ISDA. The dashed and the dash-dotted lines in the figure represent system of linear equations $-x_1 + 0.5x_2 + 2 = 0$ and $0.5x_1 - x_2 + 2 = 0$ respectively.

to the x_1. It moves $\mathbf{x}$ from its initial position of zero to the line corresponding to $-x_1 + 0.5x_2 + 2 = 0$. This is equivalent to performing the steepest ascent on direction x_1 and the position $[x_1, x_2]^T = [2, 0]^T$ is the maximum along the x_1 direction when x_2 is fixed at zero. In the second step, x_1 is kept at 2 and x_2 is updated by using (3.23). Note that the updating rule moves $\mathbf{x}$ from line $-x_1 + 0.5x_2 + 2 = 0$ to $-x_1 + 0.5x_2 + 2 = 0$. As the process of optimization continues, $\mathbf{x}$ oscillates between the two lines as shown in the figure until the solution of the optimization problem is found. In this case, the optimal solution is $\mathbf{x} = [4, 4]^T$.

Now, consider maximizing (3.25) subject to box constraints $0 \leq \mathbf{x} \leq 2$. Figure 3.3 shows how ISDAs solve such a problem. Similarly to the unconstrained problem, the first step which moves $\mathbf{x}$ from $[0, 0]^T$ to $[2, 0]^T$ is unchanged. However, after an updating of x_2 in the second step, the vector $\mathbf{x}$ becomes $[2, 3]^T$ and it is no longer inside the box constraints due to $x_2 > 2$. As a result, x_2 is clipped back to 2 (i.e. $\mathbf{x}$ is equal to $[2, 2]^T$ after clipping) in order to fulfill the box constraints. At $\mathbf{x} = [2, 2]^T$ the algorithm can no longer make any improvement without going outside the box constraints. As a result $\mathbf{x} = [2, 2]^T$ is the solution of the constrained optimization problem.

3.2.6 Discussion on ISDA Without a Bias Term b

Both the KA and the SMO algorithms were recently developed and introduced as alternatives to solve quadratic programming problem while training support vector machines on huge data sets. It was shown that when using a positive definite kernels the two algorithms are identical in their analytic form and numerical implementation. In addition, for positive definite kernels both algorithms are strictly identical with a classic iterative Gauss-Seidel (optimal coordinate ascent) learning and its extension successive over-relaxation. Until now, these facts were blurred mainly due to different pace in posing the learning problems and due to the heavy heuristics involved in the SMO implementation that shadowed an insight into the possible identity of the methods [85, 84]. It is shown that in the so-called no-bias SVMs, both the KA and the SMO procedure are the coordinate ascent based methods and they are dubbed as ISDA. Hence, they are the inheritors of all good and bad "genes" of a gradient approach and both algorithms have the same performance.

In the next section, the ISDAs with explicit bias term b will be presented. The explicit bias term b is incorporated into the ISDAs by adding a constant i/k to the kernel matrix. This technique and its connection with the classical penalty method in optimization are presented in the book as the the novel contributions to the SVMs field. The motivations for incorporating bias term into the ISDAs are to improve the versatility and the performance of the algorithms [69]. The ISDA without bias term developed in this section can only deal with positive definite kernel, which may be a limitation in applications where positive a semi-definite kernel such as a linear kernel is more desirable. As will be discussed shortly, ISDA with explicit bias term b also seems to be faster in terms of training time.

3.3 Iterative Single Data Algorithm with an Explicit Bias Term b

Before presenting iterative algorithms with bias term b, some recent presentations of the bias b utilization are discussed. As mentioned previously, for positive definite kernels there is no need for bias b. However, one can use it and this means implementing a different kernel. In [116] it was also shown that when using positive definite kernels, one can choose between two types of solutions for both classification and regression. The first one uses the model without bias term (i.e.,$d(\mathbf{x}) = \sum_{j=1}^{n} v_j K(\mathbf{x}, \mathbf{x}_j)$), while the second SVM uses an explicit bias term b. For the second one $d(\mathbf{x}) = \sum_{j=1}^{n} v_j K(\mathbf{x}, \mathbf{x}_j) + b$ and it was shown that $d(\mathbf{x})$ is a function resulting from a minimization of the functional shown below

$$I[d] = \sum_{j=1}^{n} V(y_j, d(\mathbf{x}_j)) + \lambda \|d\|_{K^*}^2 \tag{3.27}$$

where $K^* = K - a$ (for an appropriate constant a) and K is an original kernel function (more details can be found in [116]). This means that by adding a constant term to a positive definite kernel function K, one obtains the solution to the functional $I[d]$ where K^* is a conditionally positive definite kernel. Interestingly, a similar type of model was also presented in [95]. However, their formulation is done for the linear classification problems only. They reformulated the optimization problem by adding the $b^2/2$ term to the cost function, so $\frac{2}{\|\mathbf{w}b\|}$ is maximized instead of the margin $\frac{2}{\|\mathbf{w}\|}$. This is equivalent to an addition of 1 to each element of the original kernel matrix $\mathbf{K}$. As a result, they changed the original classification dual problem to the optimization of the following one

$$L_d(\alpha) = \sum_{i=1}^{n} \alpha_i - \frac{1}{2} \sum_{i,j=1}^{n} y_i y_j \alpha_i \alpha_j (K(\mathbf{x}_i, \mathbf{x}_j) + 1). \tag{3.28}$$

3.3.1 Iterative Single Data Algorithm for SVMs Classification with a Bias Term b

In the previous section for the SVMs models when positive definite kernels are used without a bias term b, the learning algorithms (originating from minimization of a primal Lagrangian in respect to the weights w_i) for classification and regression (in a dual domain) were solved with box constraints only. However, there remains an open question - how to apply the proposed ISDA scheme for the SVMs that do use explicit bias term b. The motivation for developing the ISDAs for the SVMs with an explicit bias term b originates from the fact that the use of an explicit bias b seems to lead to the SVMs with less support vectors. This fact can often be very useful for both the data (information) compression and the speed of learning. An iterative learning algorithm for the classification SVMs (2.31) with an explicit bias b, subjected to the equality constraint (2.36c) will be presented in this section.

There are three major avenues (procedures, algorithms) possible in solving (2.36a) with constraints (2.36d) and (2.36c).

The first one is the standard SVMs algorithm which imposes the equality constraint (2.36c) during the optimization and in this way ensures that the solution never leaves a feasible region. As mentioned in Sect. 2.2.1 the last term in (2.14a) will vanish if the derivate $\partial L_p/\partial b$ (2.12b) is substituted into (2.14a) for nonlinear classification as below,

$$L_d(\alpha) = \sum_{i=1}^{n} \alpha_i - \frac{1}{2} \sum_{i,j=1}^{n} y_i y_j \alpha_i \alpha_j K(\mathbf{x}_i, \mathbf{x}_j) - \underbrace{\sum_{i=1}^{n} \alpha_i y_i b}_{=0}. \tag{3.29}$$

Now, the solution will have to fulfill the equality constraint (2.15c). After the dual problem is solved, the bias term b is calculated by using free (or unbounded) Lagrange multipliers as in (2.28) for linear SVMs. For the nonlinear

SVMs, a similar expression for calculation of bias b is used:

$$b_o = \frac{1}{N_{FSV}} \sum_{s=1}^{N_{FSV}} (y_s - \sum_{j=1}^{n} \alpha_j y_j K(\mathbf{x}_s, \mathbf{x}_j)), \quad s = 1, \ldots, N_{FSV}. \tag{3.30}$$

The original SMO algorithm tries to fulfill the equality constraint (2.15c) by optimizing two α_i variables at a time because the minimal number of variables in order to keep the equality constraint fulfilled (2.15c) is two.

There are two other possible ways how the ISDA works for the SVMs containing an explicit bias term b. In the first method, the cost function (2.10a) is augmented with the term $0.5kb^2$ (where $k \geq 0$) and this step changes the primal Lagrangian L_p from

$$L_p(\mathbf{w}, b, \alpha) = \frac{1}{2}\mathbf{w}^T\mathbf{w} - \sum\nolimits_{i=1}^{n} \alpha_i \left\{ y_i \left[\mathbf{w}^T \Phi(\mathbf{x}_i) + b \right] - 1 \right\}, \tag{3.31}$$

to the following one

$$L_p(\mathbf{w}, b, \alpha) = \frac{1}{2}\mathbf{w}^T\mathbf{w} - \sum\nolimits_{i=1}^{n} \alpha_i \left\{ y_i \left[\mathbf{w}^T \Phi(\mathbf{x}_i) + b \right] - 1 \right\} + k\frac{b^2}{2}. \tag{3.32}$$

The derivative $\partial L_p / \partial b$ also changes from (2.12b) to the following one

$$\frac{\partial L_p}{\partial b} = 0 \quad b = \frac{1}{k} \sum_{i=1}^{n} \alpha_i y_i. \tag{3.33}$$

After forming (3.31) as well as using (3.33) and $\partial L_p / \partial \mathbf{w}_o$ (2.26a) in the Φ space (changing $\mathbf{x}_i$ in (2.26a) to $\Phi(\mathbf{x}_i)$), one obtains the dual problem without an explicit bias b,

$$\begin{aligned} L_d(\alpha) &= \sum_{i=1}^{n} \alpha_i - \frac{1}{2} \sum_{i,j=1}^{n} y_i y_j \alpha_i \alpha_j K(\mathbf{x}_i, \mathbf{x}_j) \\ &\quad - \frac{1}{k} \sum_{i,j=1}^{n} \alpha_i y_i \alpha_j y_j + \frac{1}{2k} \sum_{i,j=1}^{n} \alpha_i y_i \alpha_j y_j \\ &= \sum_{i=1}^{n} \alpha_i - \frac{1}{2} \sum_{i,j=1}^{n} y_i y_j \alpha_i \alpha_j (K(\mathbf{x}_i, \mathbf{x}_j) + \frac{1}{k}) \end{aligned} \tag{3.34}$$

Actually, the optimization of a dual Lagrangian is reformulated for the SVMs with a bias b by applying "tiny" changes $1/k$ only to the original matrix $\mathbf{K}$ as illustrated in (3.34). Hence, for the nonlinear classification problems ISDA stands for an iterative solving of the following linear system

$$\mathbf{H}_k \boldsymbol{\alpha} = \mathbf{1}_n \tag{3.35a}$$

$$\text{s.t.} \quad 0 \leq \alpha_i \leq C, \quad i = 1, \ldots, n \tag{3.35b}$$

where $H_k(\mathbf{x}_i, \mathbf{x}_j) = y_i y_j (K(\mathbf{x}_i, \mathbf{x}_j) + 1/k)$, $\mathbf{1}_n$ is an n-dimensional vector containing ones and C is a penalty factor equal to infinity for a hard margin classifier. Note that during the updates of α_i, the bias term b must not be used because it is implicitly incorporated within the $\mathbf{H}_k$ matrix. Only after the solution vector $\boldsymbol{\alpha}$ in (3.35a) is found, the bias b should be calculated either by using unbounded Lagrange multipliers α_i as given in (3.30), or by implementing the equality constraint from $\partial L_p / \partial b = 0$ and given in (3.33) as

$$b = \frac{1}{k} \sum_{i=1}^{N_{sv}} \alpha_i y_i. \tag{3.36}$$

Note, however, that all the Lagrange multipliers, meaning both bounded (clipped to C) and unbounded (smaller than C) must be used in (3.36). Thus, using the SVMs with an explicit bias term b means that, in the ISDA proposed above, the original kernel is changed, i.e., another kernel function is used. This means that the α_i values will be different for each k chosen, and so will be the value for b. The final SVM as given in (2.31) is produced by original kernels. Namely, $d(\mathbf{x})$ is obtained by adding the sum of the weighted original kernel values and corresponding bias b. The approach of adding a small change to the kernel function can also be associated with a classic penalty function method in optimization as follows below.

To illustrate the idea of the penalty function, let us consider the problem of maximizing a function $f(x)$ subject to an equality constraint $g(x) = 0$. To solve this problem using classical penalty function method, the following penalty function is formulated:

$$\max \;\; P(x, \rho) = f(x) - \frac{1}{2} \rho \left\| g(x) \right\|_2^2, \tag{3.37}$$

where ρ is the penalty parameter and $\|g(x)\|_2^2$ is the square of the L_2 norm of the function $g(x)$. As the penalty parameter ρ increases towards infinity, the size of the $g(x)$ is pushed towards zero, hence the equality constraint $g(x) = 0$ is fulfilled. Now, consider the standard SVMs' dual problem, which is maximizing (2.36a) subject to box constraints (2.36d) and the equality constraint (2.36c). By applying the classical penalty method (3.37) to the equality constraint (2.36c), the following quadratic penalty function can be formed as

$$\begin{aligned}
P(x, \rho) &= L_d(\alpha) - \frac{1}{2} \rho \left\| \sum_{i=1}^{n} \alpha_i y_i \right\|_2^2 \\
&= \sum_{i=1}^{n} \alpha_i - \frac{1}{2} \sum_{i,j=1}^{n} y_i y_j \alpha_i \alpha_j K(\mathbf{x}_i, \mathbf{x}_j) - \frac{1}{2} \rho \sum_{i,j=1}^{n} y_i y_j \alpha_i \alpha_j \\
&= \sum_{i=1}^{n} \alpha_i - \frac{1}{2} \sum_{i,j=1}^{n} y_i y_j \alpha_i \alpha_j (K(\mathbf{x}_i, \mathbf{x}_j) + \rho)
\end{aligned} \tag{3.38}$$

The expression above is exactly equal to (3.34) when ρ equals $1/k$. Thus, the parameter $1/k$ in (3.34) for the first method of adding bias into the ISDAs can be regarded as a penalty parameter of enforcing equality constraint (2.36c) in the original SVMs dual problem. Also, for a large value of $1/k$, the solution will have a small L_2 norm of $\sum_{i=1}^{n} \alpha_i y_i$ (2.36c). In other words, as k approaches zero a bias b converges to the solution of the standard QP method that enforces the equality constraints. However, we do not use the ISDA with small parameter k values here, because the condition number of the matrix $\mathbf{H}_k$ increases as $1/k$ rises. Furthermore, the strict fulfillment of (2.36c) may not be needed in obtaining a good SVM.

Example 3.3. The same 1-D problem in Example 2.2 will be used again to show how the solution of ISDA with $1/k$ added to the kernel converges to the solution from a standard QP solver which enforces the equality constraint (2.36c). Figure 3.4 shows the same three data points from Example 2.2 mapped

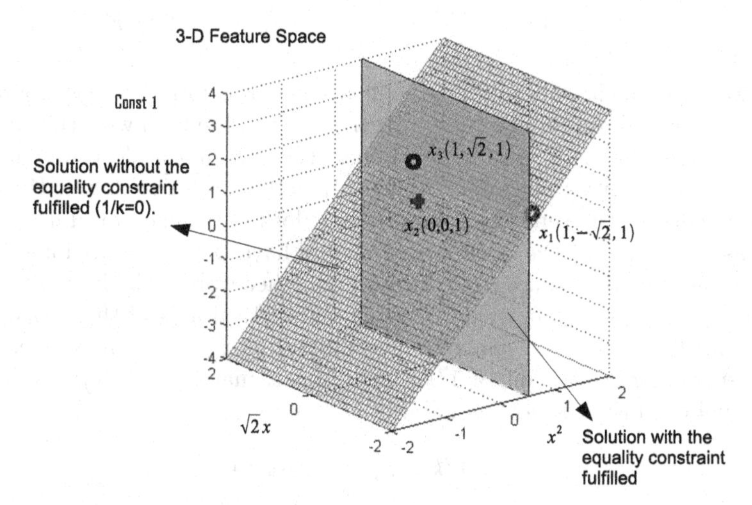

Fig. 3.4. Two different decision boundaries, one from the decision function with the equality constraint (2.36c) fulfilled and another from the decision function without the equality constraint fulfilled ($1/k = 0$).

into the same 3-D feature space of the second order complete polynomial ($\boldsymbol{\Phi}(\mathbf{x}) = [\phi_1(\mathbf{x})\ \phi_2(\mathbf{x})\ \phi_3(\mathbf{x})]^T = [x^2\ \sqrt{2}x\ 1]^T$). There are two separation boundaries shown in the figure. The separation plane that is perpendicular to the $\sqrt{2}x$-x^2 plane is the solution from a standard QP solver with the equality constraint (2.36c) satisfied. Another separation plane is from the ISDA algorithm without the bias term b (i.e. $1/k = 0$). Both planes separate the data points perfectly. The decision function with the equality constraint fulfilled is given as:

$$d(x) = w_3\phi_3(x) + w_2\phi_2(x) + w_1\phi_1(x) + b = -2\phi_1(x) + \underbrace{1}_{b=1}. \quad (3.39)$$

As a result, the separation boundary of the decision function (3.39) is given as:

$$\underbrace{w_3}_{w_3=0}\phi_3(x) + \underbrace{w_2}_{w_2=0}\phi_2(x) + \underbrace{w_1}_{w_1=-2}\phi_1(x) + \underbrace{b}_{b=1} = 0 \quad (3.40)$$

The decision function without the equality constraint fulfilled is equal to:

$$d(x) = w_3\phi_3(x) + w_2\phi_2(x) + w_1\phi_1(x) + b = -2\phi_1(x) + \underbrace{1}_{w_3=1}. \quad (3.41)$$

As a result, the separation boundary is given as:

$$\begin{aligned} \underbrace{w_3}_{w_3=1}\phi_3(x) + \underbrace{w_2}_{w_2=0}\phi_2(x) + \underbrace{w_1}_{w_1=-2}\phi_1(x) + \underbrace{b}_{b=0} &= 0 \\ \phi_3(x) = \frac{-w_2\phi_2(x) - w_1\phi_1(x) - b}{w_3} &= 2\phi_1(x) \end{aligned} \quad (3.42)$$

The two solutions have exactly the same decision function, but the separation boundaries are different. The margin of the decision function with the equality constraint fulfilled is larger (2/2=1) than the one of the decision function without the equality constraint fulfilled ($M = 2/\sqrt{2^2 + 1^2} = 0.89$). However, the difference in the separation boundaries does not make any difference in this case, because all the data points always have $\phi_3(x) = 1$ as long as the same kernel function $(x_i x_j + 1)^2$ is used. In other words, all the data points are not going to move up or down along the vertical axis of the 3-D feature space and the separation can be made on the $\sqrt{2}x - x^2$ plane alone. Now, consider adding a constant of $1/k$ to the kernel matrix of $(x_i x_j + 1)^2$, then the kernel expansion is now as follow:

$$(x_i x_j + 1)^2 + 1/k = x_i^2 x_j^2 + 2x_i x_j + 1 + 1/k. \quad (3.43)$$

This means that the feature mapping is now equal to $\boldsymbol{\Phi}(\mathbf{x}) = [x^2\ \sqrt{2}x\ \sqrt{1 + 1/k}]^T$, i.e. the three points in Fig. 3.4 will be lifted up by an amount of $\sqrt{1/k + 1} - 1$. Instead of having constant of 1 at the feature $\phi_3(x)$ (or the vertical axis in Fig. 3.4), these points will have $\sqrt{1 + 1/k}$. This corresponds to the fact that adding a positive constant into the kernel matrix is the equivalent of using another kernel shown in [116], i.e., the data points are mapped into a different feature space. As $1/k$ gets larger and larger, the decision boundary made by the ISDA will be closer to the vertical one that is constructed from the decision function with the equality constraint fulfilled. This effect can be observed graphically in Fig. 3.5. In Fig. 3.6, the relationship between the size of the margin and the constant $1/k$ added to the kernel function is shown. It is clear that the solution of ISDA with large $1/k$ will converge to the solution

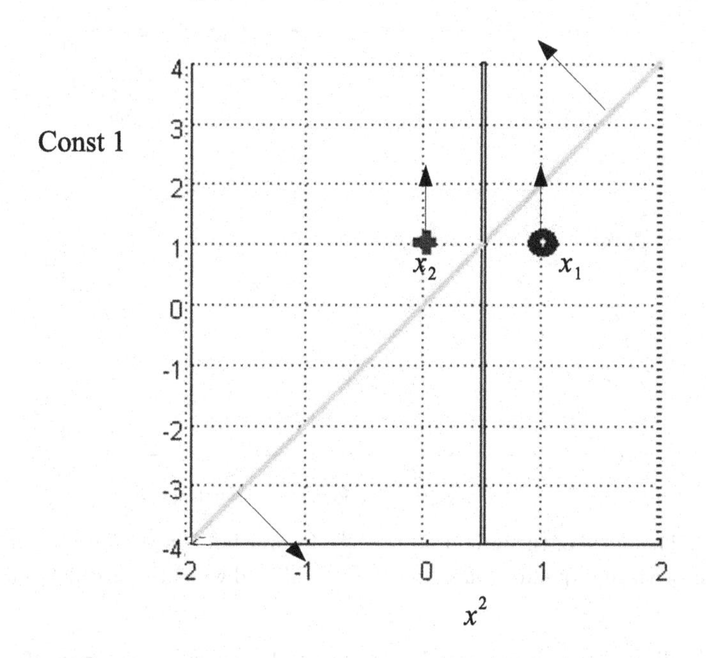

Fig. 3.5. The x^2-const1 plane of the 3-D feature space in Fig. 3.4. The vertical thick line in the figure is the decision boundary from the decision function (calculated by a standard QP solver) with equality constraint fulfilled. The thin line is the decision boundary from the decision function without the equality constraint fulfilled (calculated by ISDA without bias). The arrows in the figure indicate how the points and the decision boundary will change as $1/k$ increases. The solution from the ISDA converges to the solution of the standard QP solver as $1/k$ approaches to infinite.

which has equality constraint fulfilled. Similarly, the explicit bias term b from the ISDA with a larger $1/k$ will also converge to the solution with the equality constraint fulfilled.

In the next section, it will be shown that in classifying the *MNIST* data with Gaussian kernels, the value $k = 10$ proved to be a very good one, justifying all the reasons for its introduction (fast learning, small number of support vectors and good generalization).

The second method in implementing the ISDA for SVMs with the bias term b is to work with original cost function (2.36) and keep imposing the equality constraints during the iterations as suggested in [147]. The learning starts with $b = 0$ and after each epoch the bias b is updated by applying a secant method as follows

$$b^k = b^{k-1} - \omega^{k-1} \frac{b^{k-1} - b^{k-2}}{\omega^{k-1} - \omega^{k-2}} \tag{3.44}$$

where $\omega = \sum_{i=1}^{n} \alpha_i y_i$ represents the value of equality constraint after each epoch. In the case of the regression SVMs, (3.44) is used by implementing the

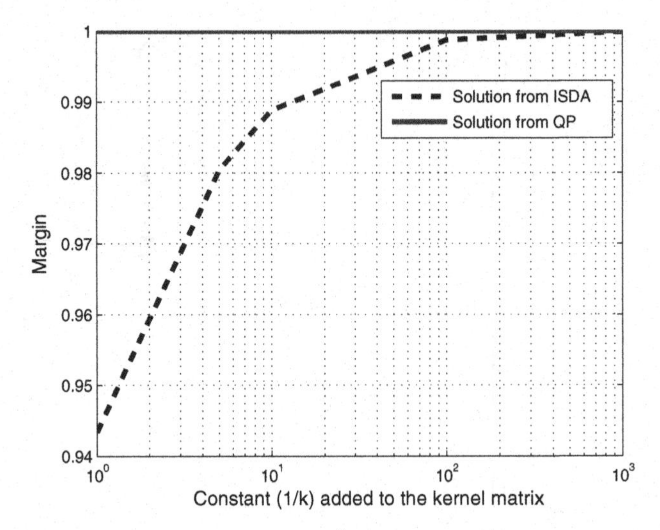

Fig. 3.6. The norm of the weight vector $\frac{2}{\|\mathbf{w}\|}$ from ISDA converges to the one from a QP solver with the equality constraint (2.36c) fulfilled when the size of $1/k$ increases.

corresponding regression's equality constraints, namely $\omega = \sum_{i=1}^{n}(\alpha_i - \alpha_i^*)$. This is different from [147] where an iterative update after each data pair is proposed. In the SVMs regression experiments that are done in this book such an updating led to an unstable learning. Also, in an addition to change expression for ω, both the $\mathbf{K}$ matrix, which is now $(2n, 2n)$ matrix, and the right hand side of (3.35a) which becomes $(2n, 1)$ vector, should be changed too and formed as given in [85].

3.4 Performance of the Iterative Single Data Algorithm and Comparisons

To measure the relative performance of different ISDAs, all the algorithms are run on a *MNIST* dataset with 576-dimensional inputs [44] with RBF Gaussian kernels, and compared to the performance of ISDAs software developed in this book with LIBSVM V2.4 [27] which is one of the fastest and the most popular SVM solvers at the moment based on the SMO type of an algorithm. The *MNIST* dataset consists of 60,000 training and 10,000 test data pairs. To make sure that the comparison is based purely on the nature of the algorithm rather than on the differences in implementation, the encoding of the algorithms are the same as LIBSVM's ones in terms of caching strategy (LRU-Least Recent Used), and in respect of data structure, heuristics for shrinking and stopping criterions (more details on implementing ISDA will be discussed in the next section). The only significant difference is that instead of two heuristic rules for selecting and updating two data points at each iteration step aiming at the maximal improvement of the dual objective function, our ISDA selects the worse KKT violator only and updates its α_i at each step. Also, in order to speed up the LIBSVM's training process, the original LIBSVM routine was modified to perform faster by reducing the numbers of complete KKT checking without any deterioration of accuracy. All the routines were written and compiled in Visual C++ 6.0, and all simulations were run on a 2.4 GHz P4 processor PC with 1.5 Gigabyte of memory under the operating system Windows XP Professional. The shape parameter σ^2 of an RBF Gaussian kernel and the penalty factor C are set to be 0.3 and 10 [44]. The stopping criterion τ and the size of the cache used are 0.01 and 250 Megabytes. The simulation results of different ISDAs against both LIBSVM are presented in Tables 3.1 and 3.2, and in Fig. 3.7. The first and the second column of the tables show the performance of the original and modified LIBSVM respectively. The last three columns show the results for the single data point learning algorithms with various values of constant $1/k$ added to the kernel matrix in (3.35a). For $k = \infty$, ISDA is equivalent to the SVMs without bias term, and for $k = 1$, it is the same as the classification formulation proposed in [95]. Table 3.1 illustrates the running time for each algorithm. The ISDA with $k = 10$ was the quickest and required the shortest average time (T_{10}) to complete the training. The average time needed for the original LIBSVM is almost $2T_{10}$ and the average time for a modified version of LIBSVM is 10.3% bigger than T_{10}. This is contributed mostly to the simplicity of the ISDA. One may think that the improvement achieved is minor, but it is important to consider the fact that approximately more than 50% of the CPU time is spent on the final checking of the KKT conditions in all simulations.

Table 3.1. Simulation times for different algorithms

	LIBSVM Original	LIBSVM Modified	Iterative Single Data Algorithm (ISDA) $k = 1$	$k = 10$	$k = \infty$
Class	Time(sec)	Time(sec)	Time(sec)	Time(sec)	Time(sec)
0	1606	885	800	794	1004
1	740	465	490	491	855
2	2377	1311	1398	1181	1296
3	2321	1307	1318	1160	1513
4	1997	1125	1206	1028	1235
5	2311	1289	1295	1143	1328
6	1474	818	808	754	1045
7	2027	1156	2137	1026	1250
8	2591	1499	1631	1321	1764
9	2255	1266	1410	1185	1651
Time, hr	5.5	3.1	3.5	2.8	3.6
Time Increase	+95.3%	+10.3%	+23.9%	0	+28.3%

During the checking, the algorithm must calculate the output of the model at each datum in order to evaluate the KKT violations. This process is unavoidable if one wants to ensure the solution's global convergence, i.e. that all the data do satisfy the KKT conditions with precision τ indeed. Therefore, the reduction of time spent on iterations is approximately double the figures shown. Note that the ISDA slows down for $k < 10$ here. This is a consequence of the fact that with a decrease in k there is an increase of the condition number of a matrix $\mathbf{K}_k$, which leads to more iterations in solving (3.35a). At the same time, implementing the no-bias SVMs, i.e., working with $k = \infty$, also slows the learning down due to an increase in the number of support vectors needed when working without bias b. Table 3.2 presents the numbers of support vectors selected. For the ISDA, the numbers reduce significantly when the explicit bias term b is included. One can compare the numbers of SVs for the case without the bias b ($k = \infty$) and the ones when an explicit bias b is used (cases with $k = 1$ and $k = 10$). Because identifying less support vectors definitely speeds up the overall training, the SVMs implementations with an explicit bias b are faster than the version without bias. In terms of a generalization, or a performance on a test data set, all algorithms had very similar results and this demonstrates that the ISDAs produce models that are as good as the standard QP, i.e., SMO based, algorithms (see Fig. 3.7). The percentages of the errors on the test data are shown in Fig. 3.7. Notice the extremely low error percentages on the test data sets for all numerals. In the next section, the implementation details for developing ISDA software in order to produce the results in this section will be discussed.

Table 3.2. Number of support vectors for each algorithm.

	LIBSVM Original	LIBSVM Modified	Iterative $k = 1$	Single Data $k = 10$	Algorithm $k = \infty$
Class	# SV (BSV)[a]	# SV (BSV)	# SV (BSV)	# SV (BSV)	# SV (BSV)
0	2172 (0)	2172 (0)	2162 (0)	2132 (0)	2682 (0)
1	1440 (4)	1440 (4)	1429 (4)	1453 (4)	2373 (4)
2	3055 (0)	3055 (0)	3047 (0)	3017 (0)	3327 (0)
3	2902 (0)	2902 (0)	2888 (0)	2897 (0)	3723 (0)
4	2641 (0)	2641 (0)	2623 (0)	2601 (0)	3096 (0)
5	2900 (0)	2900 (0)	2884 (0)	2856 (0)	3275 (0)
6	2055 (0)	2055 (0)	2042 (0)	2037 (0)	2761 (0)
7	2651 (4)	2651 (4)	3315 (4)	2609 (4)	3139 (4)
8	3222 (0)	3222 (0)	3267 (0)	3226 (0)	4224 (0)
9	2702 (2)	2702 (2)	2733 (2)	2756 (2)	3914 (2)
Av.# of SV	2574	2574	2639	2558	3151

[a] Bounded Support Vectors

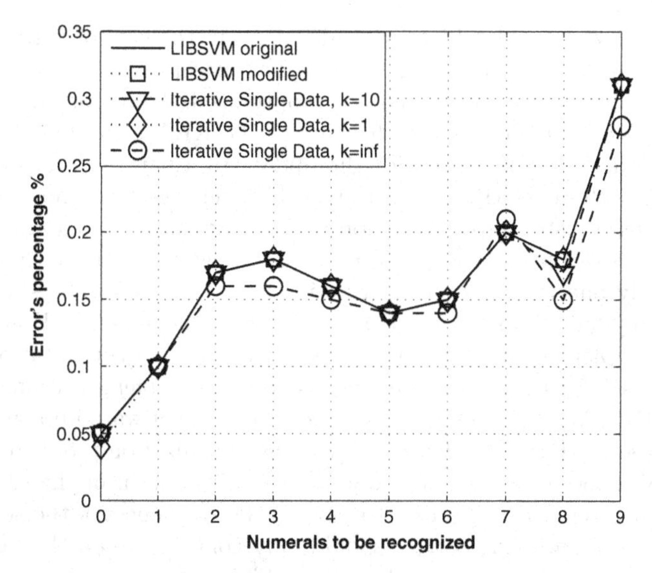

Fig. 3.7. The percentage of an error on the test data.

3.5 Implementation Issues

Despite the fact that ISDA is simpler than SMO (as shown in the previous sections), implementing it efficiently in order to compare with the state of the art SMO-based SVMs solver is a challenging problem. The important issues for implementing ISDAs efficiently will be discussed in this section. To help better understanding of these important details, the MATLAB codes of ISDAs developed in this book for classification and regression are shown in Appendix B and C. Note that implementing ISDA in a medium level computer languages such as C++ can have much better performance than in MATLAB because of the iterative nature of the algorithms. The MATLAB implementations presented in Appendix B and C are used only for illustrating the details of the concepts presented here. These important concepts will be discussed in the following subsections. Pseudo-code of ISDAs for classification and regression for large-scale problems will also be given shortly. The ISDA software developed in this book is available at *www.learning-from-data.com.*

3.5.1 Working-set Selection and Shrinking of ISDA for Classification

As discussed previously, the ISDA is classified as the working-set algorithm for solving SVMs. The working-set is referred to as the set of data points whose α_i will be changed [78], i.e. the algorithm is only going to work or optimize on selected α_i keeping the non working-set α_i fixed. In the original SMO algorithm, the working-set is equal to two in order to satisfy the equality constraint (2.36c) for classification and (2.51a) in the case of regression. The ISDA presented in the previous sections has a working-set of one, because each time only a single data point's α_i is optimized. An important step in the original SMO algorithms is the checking of the KKT conditions in order to decide whether there is a need for the update to a pair of data points. This procedure is often referred to as the working-set selection and it is crucial for obtaining a faster implementation. In the original SMO algorithms, there are two heuristics for selecting two α_i variables, whereas in the case of ISDA there is only one rule or even no rule in the case of KA. The simplest KA algorithm updates every data point in a sequential manner. However, KA algorithm implemented in this way often results in long computational times when comparing with the state of the art software implementations of SVMs such as LIBSVM [27]. This is because most of the software implementations available use the fact that the solution of the optimization problem (2.38) for a good SVM should be sparse (i.e. only a small fraction of the α_i variables are different from zero). In order to explore the sparseness of the solution for a faster convergence, the solvers should only concentrate on the points that are either support vectors or at least have high potential to become support vectors. This led to the idea of shrinking [78] the optimization problem (3.21) for speeding up the SVMs algorithm.

During the early stage of the training, it is often possible to guess which α_i variables are going to be on the upper bound C (bounded SVs) or on the lower bound 0 (non-SVs) at the final solution. Because good SVMs' models normally have a large proportion of non-SVs, and the bounded SVs cannot be optimized any further, the idea of shrinking from [78] is to reduce the size of the optimization problem (3.21) gradually by removing both bounded SVs and the possible non-SVs during the optimization process. However, the correct sets of non-SVs and bounded ones cannot be found without solving the optimization problem (3.21) heuristics are used to shrink the optimization problem. In general, the algorithms try to guess which variables are more likely to stay as either non-SVs or bounded ones in the final solution of (3.21) by considering the history of these variables in the last h iterations [78]. As a result of shrinking, the algorithm needs to solve the following reduced optimization problem (in an inner loop, see below):

$$\max L_d(\boldsymbol{\alpha}_A) = -0.5\boldsymbol{\alpha}_A^T \mathbf{H}_{AA}\boldsymbol{\alpha}_A + \mathbf{f}_A^T\boldsymbol{\alpha}_A, \tag{3.45a}$$

$$\text{s.t.} \quad 0 \leq (\alpha_A)_t \leq C, \quad t = 1, \ldots, q, \tag{3.45b}$$

where the set A (with size q) is composed of the indices corresponding to the remaining α_i that are not removed by the heuristic for shrinking. This means that the solver needs to perform the optimization on the α_A which may be only 10% or 20% of the entire data set. Consequently, the size of the optimization problem will be only $0.2n$ by $0.2n$ instead of n by n for classification tasks.

Currently, there are two approaches for working-set selection and shrinking in ISDA. The first approach is similar to the original SMO [115] which consists of an inner and an outer loop, and will be referred to as the two-loops approach. The outer loop first iterates over the entire data set in order to determine whether each data point violates the KKT conditions (3.6). Once a data point violates the KKT conditions, it will be optimized using (3.4) and clipped according to (3.5). The outer loop makes one pass through the entire data set, and then the algorithm will switch to the inner loop. The inner loop only iterates through the data points whose α_i are neither 0 nor C (i.e. it iterates through the unbounded or free SVs). Thus, the inner loop solves the QP problem exactly by making repeated passes over the reduced data set A until all the unbounded SVs satisfied the KKT conditions within τ. The outer loop then iterates over the entire data set again. The algorithm keeps alternating between the outer (single pass through the entire data set) and the inner (multiple passes through on the unbounded SVs) loop until all the data points obey KKT conditions within τ. The switching between the outer and the inner loop is a form of shrinking, because the scope of the algorithm for finding the working-set reduces from n data points to N_{FSV}, i.e. the algorithm solves the reduced optimization problem (3.45) in the inner loop.

The two-loops approach is generally faster than the one having only the outer loop, because the use of the inner loop allows the algorithm to concentrate on the SVs which are generally a small fraction of the entire data set. However, a major drawback of this approach is that one pass through the entire data set will normally produce excessive number of unbounded SVs. In other words, sweeping through the outer loops produces the number of unbounded SVs which can be orders of magnitude larger than the number that results at the end of the following inner loop. This problem also occurs in SMO and it is often referred to as the "Intermediate SVs Bulge" in [43]. In Fig. 3.8, the numbers of unbounded SVs after the outer and inner loops are shown during the course of solving a subset of *MNIST* with 8000 data. The solid line shows the numbers of unbounded SVs after one pass through the entire data set (i.e. the number of unbounded SVs before entering the inner loop), whereas the dashed line shows the numbers of unbounded SVs after the inner loop. This problem is more serious at the beginning of the iteration and it can be shown by the ratio of 1.35 (1571/1162) between the number of unbounded SVs before and after first entering the inner loop. [43] proposed a method known as "digesting" to resolve this problem for the SMO algorithm. The main idea of digesting is to apply a direct QP solver to the unbounded SVs in the outer loop when the numbers of them exceed a certain size (e.g. 1000 SVs). This procedure had the effect of reducing (digesting) the excessive number of unbounded SVs and has been shown to improve the performance of the SMO algorithm. However, a more complicated QP solver is required for this approach and the actual speed improvement depends also on the efficiency of the QP solver. Note that neither the two-loop approach nor the related digesting step are used in the final implementation of the ISDAs software in this book. The method used here is described below.

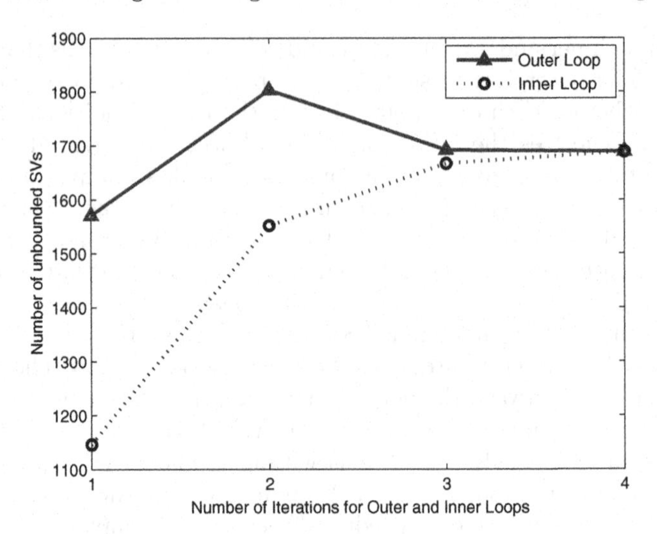

Fig. 3.8. The number of unbounded SVs after outer and inner loops for two loops approach. The simulation is performed on the class 8 of the first 8000 data points of the *MNIST* data set. The σ^2 and C are equal to 0.3 and 10 respectively.

The second working-set selection that is similar to the ones used in [78], LIBSVM [27] and BSVM [68] is employed in this book for implementing ISDA classification and it is shown in Algorithm 3.1. The idea of the decomposition in [78] is to find a steepest feasible direction $\mathbf{d}$ of descent which has only z non-zero elements. Note that in the LIBSVM $z = 2$, in BSVM z is previously selected number and in ISDA $z = 1$.

This led to the minimization of the following optimization problem [78]:

$$\min \quad V(\mathbf{d}) = g(\boldsymbol{\alpha}^t)^T \mathbf{d} \tag{3.46a}$$

$$\text{s.t.} \quad \mathbf{y}^T \mathbf{d} = 0 \tag{3.46b}$$

$$d_i \geq 0 \quad \text{for} \quad i : \alpha_i = 0 \tag{3.46c}$$

$$d_i \leq 0 \quad \text{for} \quad i : \alpha_i = C \tag{3.46d}$$

$$-\mathbf{1} \leq \mathbf{d} \leq \mathbf{1} \tag{3.46e}$$

$$|\{d_i : d_i \neq 0\}| = z \tag{3.46f}$$

Because ISDA optimizes only one variable at a time and the equality constraint (2.36c) is not required, the steepest direction can be found by comparing the absolute value of the derivative $\partial L_d / \partial \alpha_i$ for all the variables that violate the KKT conditions. In other words, the algorithm optimizes the worst KKT violator in each step in order to have the greatest improvement in the objective function (3.21a). As a result, the working-set selection for classification can be formulated as follows:

$$z = j \quad \text{if} \, |y_j E_j| > |y_k E_k| \;\; (k \neq j), \quad j, k \in J_{vio} \tag{3.47}$$

Algorithm 3.1 ISDA Classification

Consider a binary classification problem with input matrix $\mathbf{X}^0 = [\mathbf{x}_1, \mathbf{x}_2, \dots \mathbf{x}_n]^T$ where n is the number of samples, the class labels $\mathbf{y} = [y_1, y_2, \dots, y_n]^T$ and $y \in [+1, -1]$. Let the set I denotes indices corresponding to all the training data, and the set $A (A \subset I)$ contains the indices α_i in the reduced optimization problem (3.45).

1. Set $A = I = [1 : n]$ and A^C(the complement of A in I) is empty.
2. Find z which is the index of the worst KKT violator in the set A using the KKT conditions (3.6) (repeated here for your convenience):
$$\alpha_i < C \quad \wedge \quad y_i E_i < -\tau, \quad \text{or} \quad \alpha_i > 0 \quad \wedge \quad y_i E_i > \tau,$$
where $E_i = d_i - y_i$, y_i stands for the desired value and d_i is the SVM output value, and utilizing the working-set selection (3.47).
3. If $|y_z E_z|$ of the worst KKT violator α_z was smaller than τ, then the reduced optimization problem (3.45) is solved. Go to step 8. Otherwise continue to Step 4.
4. Update α_z using (3.8) as follow,
$$\Delta\alpha_z = \frac{1 - y_z d_z}{K(\mathbf{x}_z, \mathbf{x}_z)},$$
and clip α_z using (3.9) $\alpha_z \leftarrow \min\{\max\{\alpha_z + \Delta\alpha_z, 0\}, C\}$.
5. Update the KKT conditions $y_i E_i$ for all the i in the set A as follows:
$$y_i E_i = y_i E_i^{old} + (\alpha_z - \alpha_z^{old}) K(\mathbf{x}_z, \mathbf{x}_i) y_i \quad i \in A.$$
6. Repeat Step 2 to 5 for h times. (Default value for h is 1000).
7. Shrink the set A by removing all indices of the non-KKT violators whose α_i are also equal to zero in the set A. In other words, the new set A will consist of all the KKT violators currently in A and all the SVs. It is given as follows:
$$A^{new} = J_{vio} \cup \mathbf{s}, \quad J_{vio} \subset A, \tag{3.48}$$
where J_{vio} contains the indices corresponding to the α_i variables that violate the KKT conditions (3.6) in the set A and $\mathbf{s}$ contains the indices of SVs. Check the KKT conditions $y_z E_z$ of the worst KKT violators α_z. Find A^C which is the complement of A in I. If $|y_z E_z|$ is larger than τ, go to step 6.
8. Recompute $y_i E_i = y_i (d_i - y_i)$ for the data points whose indices i are in set A^C, where A^C is the complement of A in I. In other words, recompute $y_i E_i$ for all the α_i whose indices are not in A.
9. Find the index z of the worst KKT violator in set A^C, if $|y_z E_z|$ was smaller than τ then $\boldsymbol{\alpha}_A$ is the solution of the original optimization problem. The algorithm should stop. If set A^C is empty, then the $\boldsymbol{\alpha}_A$ is also the solution of the original optimization (3.21). Otherwise set $A = I$ and go to step 7.

---

where the set J_{vio} contains the indices corresponding to the α_i variables that violate the KKT conditions in the set A. This working-set selection works quite well with ISDA. It often reduces the computational time significantly and requires fewer iterations in comparison with the two loops approach without the digesting part. This is because the probability of the worst KKT violator being a support vector is much higher than sequentially picking up an α_i variable for updating. As a result, this approach does not have the problem of intermediate SV bulge. However, the most expensive part of this approach is finding the worst KKT violator, because $y_i E_i$ of the entire data set needs to be updated for all the α_i variables using (3.10) in order to find the worst KKT violator. Consequently, the entire row of the kernel matrix needs to be evaluated for α_i. This is more expensive than optimizing only one α_i per iteration in the two-loops approach where only E_i needs to be computed from all the SVs, but updating the worst KKT violator makes a larger improvement on the objective function.

To speed up the algorithm, the idea of shrinking the data set is also used here. Instead of trying to find the worst KKT violator of the entire data set, the scope of searching for the worst KKT violator is reduced gradually. This is done by removing the non-KKT violators whose α_i are also equal to zero after every h iterations as shown in Step 7 of the Algorithm 3.1. Initially, the algorithm works on the complete data set. After shrinking, the algorithm tries to solve the reduced optimization problem (3.45) by searching the worst KKT violator in the set A only. Consequently, this improves the performance of the algorithm, because only the KKT conditions of the variables α_A that are in the set A with size q need to be updated after each iteration.

Set A is shrunken dynamically after every h iterations as shown in Step 7 of the Algorithm 3.1. This is due to the fact that at the beginning almost all the data points will be KKT violators, i.e. set A is close to the complete data set. As a result, the size of A should be reduced every h iterations for speeding up and A may be changed before the solution of (3.45) is reached. Note that this is different from the two-loops approach mentioned after the equation (3.45) above, where the set A is fixed and (3.45) is solved by the inner loop. The algorithm continues to shrink A until all the α_i in the set A obey KKT conditions within τ (as shown in Step 7 of the Algorithm 3.1). In other words, the algorithm will stop shrinking when the α_i in the set A reach the solution of the optimization problem (3.45) and there are no KKT violators in the set A. The solution of the optimization problem (3.45) may or may not solve the original problem (3.21). Consequently, the algorithm checks the KKT conditions of the α_i that are not in the shrunken set A (i.e. in set A^C) in order to make sure the rest of the α_{A^C} satisfied KKT conditions. Because Step 5 of the Algorithm 3.1 only updates the $y_i E_i$ of KKT conditions that are in set A, all the $y_i E_i$ in set A^C need to be recomputed from the beginning as follows:

$$y_i E_i = y_i\Big(\sum_{j \in \mathbf{s}} y_j \alpha_j K(\mathbf{x}_i, \mathbf{x}_j)\Big) - 1 \quad i \in A^C, \tag{3.49}$$

where $\mathbf{s}$ corresponds to the indices of SVs. This is a very time-consuming checking, especially for large-scale problems. For example, checking KKT conditions for data points whose indices are in A^C needs about 881 seconds for class 8 of the *MNIST* data set, more than 50% of the total computation time. In the best-case scenario, the algorithm only needs to perform one such KKT checking when the solution of (3.45) is equal to the one of the globally optimal solutions. However, most of the simulations show that it is more likely that the solution (3.45) is not equal to the globally optimal one at first, i.e. there are still KKT violators outside A. In this case, the algorithm will form a new A by including all the KKT violators as shown in Step 9 and 7. Thus, a new (3.45) is optimized by the solver. This process of increasing the size of A is referred to as "unshrinking" [27]. The shrinking and unshrinking of the set A is repeated until there are no KKT violators in the complete data set. This means that Algorithm 3.1 normally cannot finish when reaching Step 9 for the first time, and Step 8 needs to be executed twice. This was a major bottleneck for the performance of the algorithm. In the earlier development stage of the ISDA software, it was found that ISDA always needs at least two such checkings, whereas LIBSVM only needs approximately one and a half checkings for the *MNIST* data set. This is because LIBSVM performs such checking (i.e. Step 8 of Algorithm 3.1) first when the reduced problem (3.45) is solved in precision of 10τ rather than τ to compensate for its aggressive shrinking strategy (More details can be found in [27]). This makes ISDA rather disadvantageous in terms of performance. To overcome this problem an additional caching strategy which only increases the memory requirement by $2n$ are implemented. They are discussed in detail in the following section. Note that the default number of steps h for performing shrinking is 1000 (same as in LIBSVM in [27]) in the ISDAs program developed. It proved to be a good setting justifying all the reasons for its introduction.

3.5.2 Computation of the Kernel Matrix and Caching of ISDA for Classification

In the SMO algorithms, the KKT conditions y_iE_i and part of the kernel matrix are usually cached in the memory to speed up the learning. In most of the large-scale SVMs problems, the most of the time is spent on the computation of the kernel matrix entries. As a result, most of the SVM solvers allow users to specify the amount of memory for caching parts of the kernel matrix after they are computed, to speed up the computation. Because the size of the cache can only store parts of the kernel matrix $\mathbf{K}$, some strategies are required for maintaining the cache. The most common caching strategy for SMO-based routines is the Least Recent Used (LRS) strategy [27] and it is also used for the software implementation of ISDAs. The idea of this strategy is to store the elements of the kernel matrix that are computed or used recently and to remove the elements that are the oldest or the least recently used from the cache. In each iteration, in order to find the worst KKT violator, one row of a kernel matrix needs to be computed in order to update y_iE_i for checking KKT conditions. The row of the kernel matrix will be computed and stored if it is currently not in the cache. When the cache is filled, the oldest row of the kernel matrix in the cache will be removed in order to allow the most recently computed row of kernel matrix to be stored. This simple strategy generally works well, and the computational time reduces as a larger cache is used.

As mentioned in the previous section, ISDA normally requires more complete KKT checkings (Step 8 of the Algorithm 3.1) than the SMO algorithms. This slows down the algorithm significantly as each checking can take more than 50% of the total training time. However, extensive simulations have shown that after the first execution of Step 8 in Algorithm 3.1, ISDA tends to update only a few α_i variables, i.e. the ones that are newly added to A from Step 9 to 7. This opens up the possiblity of speeding up the following executions of Step 8 by caching two additional variables. At the current implementation of ISDA, the KKT conditions y_iE_i are cached at two different points. First, the recent KKT conditions y_iE_i are always cached in the memory as in the SMO algorithms. In addition to that, all the y_iE_i are also stored in different variables denoted as $(y_iE_i)^{pre}$ after the execution of Step 8. $(y_iE_i)^{pre}$ will be kept fixed until the next execution of Step 8. Furthermore, the true changes $\Delta\alpha_i^{tr}$ (after clipping) of all the α_i are also cached and they are reset to zero after the execution of Step 8. $\Delta\alpha_z^{tr}$ will be updated after each updating of α_z as follows:

$$\Delta\alpha_z^{tr} \leftarrow \Delta\alpha_z^{tr} + (\alpha_z - \alpha_z^{old}) \tag{3.50}$$

Instead of using (3.49) for computing the y_iE_i terms from the beginning using all the SVs, the cached $\Delta\alpha_i^{tr}$ are then used to determine which α_i are changed after the previous execution of Step 8. These α_i with $\Delta\alpha_i^{tr} \neq 0$ are used for computing y_iE_i for $i \in A^C$ at current execution of Step 8 from $(y_iE_i)^{pre}$ as follows:

$$y_iE_i = (y_iE_i)^{pre} + \sum_{j \in J_{ch}} \Delta\alpha_j^{tr} y_i K(\mathbf{x}_i, \mathbf{x}_j) \quad \text{for all} \quad i \in A^C, \tag{3.51}$$

where J_{ch} denotes the indices corresponding to $\Delta\alpha_j^{tr}$ that are different from zeros (i.e $J_{ch} = \{j|\Delta\alpha_j^{tr} \neq 0\}$). This approach can reduce the execution time of Step 8 significantly if the number of element in J_{ch} is relatively small in comparison to the total number of SVs. This is often the case after Step 8 is executed once. In other words, when all the $\Delta\alpha_j^{tr}$ of SVs are different from zero, there is no improvement on using (3.51). As a result, (3.49) should be used to compute y_iE_i from scratch, because (3.49) will give a better precision. In general, (3.49) will be used for the first execution of Step 8 and (3.51) will be used for the following ones. Without using (3.51), the simulation time for class 8 of *MNIST* can be as long as 2300 seconds because the use of (3.49) for both first and second executions of Step 8 is much slower than the results presented in Table 3.1: 1321 seconds with $k = 10$. This uses (3.51) for the second execution of Step 8 which takes approximately 40 seconds to complete. This is also why the LIBSVM is modified to use this strategy in order to make the comparison in Table 3.1 fair. The strategy only requires $2n$ additional memory for storage and the amount of speeding up is very significant.

While computing the Gaussian RBF kernel matrix, the ISDA software uses the Intel BLAS routines to accelerate the computation for dense data set. A data set is considered to be dense if most of the features in its input vectors are non-zero. For the calculation of the distance $\|\mathbf{x}_i - \mathbf{x}_j\|^2$, the following identity from [115] is used very efficiently:

$$\|\mathbf{x}_i - \mathbf{x}_j\|^2 = \mathbf{x}_i \cdot \mathbf{x}_i - 2\mathbf{x}_i \cdot \mathbf{x}_j + \mathbf{x}_j \cdot \mathbf{x}_j. \tag{3.52}$$

Note that during each iteration, a single row of the kernel matrix should be calculated and this determines the way how the level 2 BLAS routine is implemented. The terms $\mathbf{x}_i\mathbf{x}_i$ and $\mathbf{x}_j\mathbf{x}_j$ are precomputed and stored in the memory. The cross terms $\mathbf{x}_i\mathbf{x}_j$ are calculated in a batch using the level 2 BLAS routine *cblas_dgemv* as follow. For a given $\mathbf{x}_i$, *cblas_dgemv* is used to compute the product of $\mathbf{X}^T\mathbf{x}_i$ where $\mathbf{X} = [\mathbf{x}_1, \mathbf{x}_2, \ldots, \mathbf{x}_j]$ where $j = 1, \ldots, q$ for the reduced optimization problem (3.45). The amount of speeding up using the BLAS routine is quite significant. For a class 8 of *MNIST* data set, the ISDA with BLAS requires only 798 seconds to complete, which is much faster than the result of 1384 seconds shown in Table 3.1 (Recall that for fair comparison with LIBSVM the BLAS was not used for ISDA in creating results given in Table 3.1). If the data set is sparse, the sparse input vector $\mathbf{x}_i$ is stored in the following format, $\mathbf{x}_i = (idx,\ valx,\ length = \ numx)$, where idx contains the indices of all the elements in $\mathbf{x}_i$ that are nonzero, $numx$ is the number of nonzero elements and $valx$ is a vector of length $numx$ that holds the values of all the nonzero elements in $\mathbf{x}_i$. To compute the dot product between two sparse vectors, the same approach as in [115] is used here. The following pseudo-code from [115] shows how to compute the dot product for two sparse vectors $(id1,\ val1,\ length = num1)$ and $(id2,\ val2,\ length = \ num2)$.

```
p1 = 0, p2 = 0, dot = 0
while (p1 < num1 && p2 < num2)
{a1 = id1[p1], a2 = id2[p2]
if (a1 == a2)
{
dot += val1[p1]*val2[p2]
p1++, p2++
}
else if (a1 > a2)
p2++
else
p1++;}
```

3.5.3 Implementation Details of ISDA for Regression

The implementation of ISDA regression in this book is shown in Algorithm 3.2. Similarly to ISDA classification, the working-set selection for ISDA regression tries to find the worst KKT violator among all the α_i and α_i^*. As a result, it can be formulated using KKT conditions (3.18) as follows:

$$u = k \quad \text{if } |E_u + \varepsilon| > |E_k + \varepsilon| \, (k \neq u), \quad u, k \in U_v \tag{3.53a}$$

$$l = k \quad \text{if } |\varepsilon - E_l| > |\varepsilon - E_k| \, (k \neq l), \quad l, k \in L_v \tag{3.53b}$$

$$z = l \quad \text{if } |\varepsilon - E_l| > |E_u + \varepsilon| \quad \text{else} \quad z = u. \tag{3.53c}$$

where set U_v corresponds to indices of α_i that violate the KKT conditions (3.18a, b), set L_v contains the indices of the α_i^* variables that violate the KKT conditions (3.18c ,d) and z is the index of the worst KKT violators.

The working-set selection above tries to compare the worst KKT violators from α_i and α_i^* (i.e. both above and below the ε-tube) and picks up the worst one. As in classification, the optimization problem is reduced after every h iterations by selecting only the KKT violators that are currently in set A and the all SVs to form a new A. The only difference is that the reduced optimization problem (3.45) is now $2q$ instead of q as in the classification. This is due to the fact that both α_i and α_i^* are included in $\boldsymbol{\alpha}_A$. The checking of KKT conditions can still cost a considerable amount of computational time and the same strategy for speeding up the complete KKT checking in classification is used in regression. The ISDA software will recompute $E_i + \varepsilon$ and $\varepsilon - E_i$ for all the indices i in A^C from the beginning when first-time executing Step 9 of Algorithm (3.2) as follow:

$$E_i + \varepsilon = \sum_{j \in \mathbf{s}} (\alpha_j - \alpha_j^*) K(\mathbf{x}_j, \mathbf{x}_i) - y_i + \varepsilon \quad \text{for} \quad i \in A^C \tag{3.54}$$

$$\varepsilon - E_i = y_i + \varepsilon - \sum_{j \in \mathbf{s}} (\alpha_j - \alpha_j^*) K(\mathbf{x}_j, \mathbf{x}_i) \quad \text{for} \quad i \in A^C \tag{3.55}$$

where $\mathbf{s}$ corresponds to all the indices of SVs. Again, $E_i + \varepsilon$ and $\varepsilon - E_i$ are cached at two different points as in the classification. One cache stores both $E_i + \varepsilon$ and $\varepsilon - E_i$ variables at current iteration and another holds the values of the two variables after Step 9 is executed. They will be denoted as $(E_i + \varepsilon)^{pre}$ and $(\varepsilon - E_i)^{pre}$. Similarly, the true total changes of $(\alpha_i - \alpha_i^*)$ are also cached and they are denoted as $\Delta(\alpha_i - \alpha_i^*)^{tr}$ and reset to zero after execution of Step 9. They will be updated after each update of $\alpha_i^{(*)}$ (after clipping) as follows:

$$\Delta(\alpha_i - \alpha_i^*)^{tr} \leftarrow \Delta(\alpha_i - \alpha_i^*)^{tr} + (\alpha_i - \alpha_i^{old} - \alpha_i^* + \alpha_i^{*old}) \tag{3.56}$$

These additional variables can be used again to accelerate the execution of Step 9 by updating $E_i + \varepsilon$ and $\varepsilon - E_i$ that are not in A^C as follows:

Algorithm 3.2 ISDA Regression

Consider a regression problem with $n \times m$ input matrix $\mathbf{X}^0 = [\mathbf{x}_1, \mathbf{x}_2, \ldots \mathbf{x}_n]^T$ where m is the number of features, n is the number of samples, the outputs $\mathbf{y} = [y_1, y_2, \ldots, y_n]^T$ and $y \in \Re$. Let set I denote indices corresponding to all the training data, and set $A (A \subset I)$ contains the indices corresponding to $\alpha_i^{(*)}$ that are in the reduced optimization problem (3.45). A^C is the complement of A in I.

1. Let $A = I = [1 : n]$.
2. Find z which is the index of the worst KKT violator in set A using the KKT conditions (3.18) and the working-set selection (3.53).
3. If the data point $\mathbf{x}_z$ is above the ε tube (i.e. $f_z < y_z$) and $|E_z + \varepsilon|$ was smaller than τ, then the reduced optimization problem (3.45) is solved. Go to step 9.
4. If the data point $\mathbf{x}_z$ is below the ε tube (i.e. $f_z > y_z$) and $|E_z - \varepsilon|$ was smaller than τ, then the reduced optimization problem (3.45) is solved. Go to step 9.
5. If the data point $\mathbf{x}_z$ is above the ε tube then update α_z using $\alpha_z \leftarrow \alpha_z - \alpha_z^* - \frac{E_z + \varepsilon}{K(\mathbf{x}_z, \mathbf{x}_z)}$ (3.17a), else update α_z^* with $\alpha_i^* \leftarrow \alpha_z^* - \alpha_z + \frac{E_z - \varepsilon}{K(\mathbf{x}_z, \mathbf{x}_z)}$ (3.17b). Then clip $\alpha_z^{(*)}$ using $\alpha_z^{(*)} \leftarrow \min(\max(0, \alpha_z^{(*)} + \Delta\alpha_z^{(*)}), C)$ (3.19).
6. Update the KKT conditions $(E_i + \varepsilon)$ using (3.20a) (i.e. $(E_i + \varepsilon) = (E_i + \varepsilon)^{old} + (\alpha_z - \alpha_z^* - \alpha_z^{old} + \alpha_z^{*old}) K(\mathbf{x}_z, \mathbf{x}_i)$) and $(\varepsilon - E_i)$ using (3.20b) (i.e. $(\varepsilon - E_i) = (\varepsilon - E_i)^{old} - (\alpha_z - \alpha_z^* - \alpha_z^{old} + \alpha_z^{*old}) K(\mathbf{x}_z, \mathbf{x}_i)$) for all the indices i in set A.
7. Repeat Step 2 to 6 for h times.
8. Shrink set A by removing all indices of the non-KKT violators whose α_i and α_i^* are both equal to zero in the set A. In other words, the new set A is formed by including all the KKT violators that are currently in the set A and all the SVs. This is given as follows,

$$A^{new} = U_v \cup L_v \cup \mathbf{s}, \quad U_v, L_v \subset A \tag{3.57c}$$

   where, set U_v corresponds to indices of α_i that violate the KKT conditions (3.18a, b) in A, set L_v contains the indices of the α_i^* variables that violate the KKT conditions (3.18c ,d) in A and $\mathbf{s}$ contains the indices of SVs. Find the worst KKT violator in the new A. Check the KKT conditions $|E_z + \varepsilon|$ for $f_z < y_z$ or $|E_z - \varepsilon|$ for $f_z > y_z$ of the worst KKT violators $\alpha_z^{(*)}$ in A. If $|E_z + \varepsilon|$ (for $f_z < y_z$) or $|E_z - \varepsilon|$ (for $f_z > y_z$) is larger than τ, go to Step 7. Otherwise, find A^C which is the complement of A in I.
9. Recompute both $E_i + \varepsilon$ and $E_i - \varepsilon$ for the data points whose indices i are in set A^C, where A^C is the complement of A in I. In other words, recompute $E_i + \varepsilon$ and $E_i - \varepsilon$ for all the $\alpha_i^{(*)}$ whose indices are not in A.
10. Find the index z of the worst KKT violator in set A^C. If any one of the following three conditions is fulfilled, then the solution of the original optimization problem is $\boldsymbol{\alpha}_{oA}$ (the solution of the reduced optimization problem (3.45)) and the algorithm should stop: 1. if $f_z < y_z$ and $|E_z + \varepsilon|$ was smaller than τ, 2. if $f_z > y_z$ and $|\varepsilon - E_z|$ was smaller than τ, or 3. if set A^C is empty. Otherwise set $A = I$ and go to Step 8.

$$E_i + \varepsilon \leftarrow (E_i + \varepsilon)^{pre} + \sum_{j \in J_{ch}} (\Delta(\alpha_j - \alpha_j^*)^{tr}) K(\mathbf{x}_i, \mathbf{x}_j) \quad \text{for} \quad i \in A^C, \tag{3.57a}$$

$$\varepsilon - E_i \leftarrow (\varepsilon - E_i)^{pre} - \sum_{j \in J_{ch}} (\Delta(\alpha_j - \alpha_j^*)^{*tr}) K(\mathbf{x}_i, \mathbf{x}_j) \quad \text{for} \quad i \in A^C, \tag{3.57b}$$

where J_{ch} corresponds to indices of $\Delta(\alpha_j - \alpha_j^*)^{tr}$ and are different from zero. The same strategy is used as in the classification for deciding whether (3.57) or (3.54) should be used for updating of $E_i + \varepsilon$ and $\varepsilon - E_i$. If all of the $\Delta(\alpha_i - \alpha_i^*)^{tr}$ of SVs are different from zero, then there is no improvement by using (3.57). As a result, (3.54) should be used because it can give more precise calculation on $E_i + \varepsilon$ and $\varepsilon - E_i$.

As mentioned previously, iterative solvers such as ISDA and SMO are more efficient to achieve precision around $\tau \approx 1e-3$. They are ideal for the SVMs classification because the precision of $1e-3$ is sufficiently smaller than 1 which is the magnitude of the decision function on unbounded SVs. In contrast, the precision τ in regression has to be sufficiently smaller than the ε which can be very small. As shown in [149], the active set approach has better performance than iterative approach such as LIBSVM [27] and ISDAs in regression problems where a large C and a high precision (i.e. small τ) are required. As a result, ISDA for regression developed in this book should be used only when the direct or active-set [149] approach can not solve the problem. In other words, the kernel matrix of unbounded SVs is too large to be stored in memory.

3.6 Conclusions

In the first part of this chapter, a new learning algorithms, dubbed ISDA,are introduced as an alternatives in solving the quadratic programming problems while training support vector machines on huge data sets. ISDA is an iterative algorithm which optimizes one variable in each step. This is because of the fact that for positive definite kernel, the bias term b in the SVMs formulation is not required.

The equality of two seemingly different ISDA methods, a KA method and version of SMO learning algorithm without bias, is derived for designing the SVMs having positive definite kernels. In addition, for positive definite kernels both algorithms are strictly identical with a classic iterative Gauss-Seidel (optimal coordinate ascent) learning and its extension successive over-relaxation.

In the later part of this chapter, the use, calculation and impact of incorporating an explicit bias term b in SVMs trained with the ISDA has been introduced. Furthermore, important issues for implementing ISDA on large-scale problems have been also presented. The simulation results show that models generated by ISDA (either with or without the bias term b) are as good as the standard SMO based algorithms in terms of a generalization performance. Moreover, ISDA with an appropriate k value are faster than the standard SMO algorithms on large-scale classification problems ($k = 10$ worked particularly well in all our simulations using Gaussian RBF kernels). This is due to both the simplicity of ISDA and the decrease in the number of SVs chosen after an inclusion of an explicit bias b in the model. The simplicity of ISDA is the consequence of the fact that the equality constraints (2.36c) do not need to be fulfilled during the training stage. In this way, the second-choice heuristics is avoided during the iterations. Thus, the ISDA is an extremely good tool for solving large scale SVMs problems containing huge training data sets because it is faster than, and it delivers "same" generalization results as, the other standard (SMO) based algorithms. The fact that an introduction of an explicit bias b means solving the problem with different kernel suggests that it may be hard to tell in advance for what kind of previously unknown multivariable decision (regression) function the models with bias b may perform better, or may be more suitable, than the ones without it. As it is often the case, the real experimental results, their comparisons and the new theoretical developments should probably be able to tell one day. As for the future development of ISDA, a possible directions may be to improve the performance of the algorithm when high precision is required for regression problems.

4

Feature Reduction with Support Vector Machines and Application in DNA Microarray Analysis

4.1 Introduction

Recently, huge advances in DNA microarrays have allowed scientists to test thousands of genes in normal or tumor tissues on a single array and check whether those genes are active, hyperactive or silent. Therefore, there is an increasing interest in changing the criterion of tumor classification from morphologic to molecular [11, 99]. In this perspective, the problem can be regarded as a classification problem in machine learning, in which the class of a tumor tissue with feature vector $\mathbf{x} \in \Re^m$ is determined by a classifier. Each dimension, or a feature, in $\mathbf{x}$ holds the expression value of a particular gene which is obtained from DNA microarray experiment. The classifier is constructed by inputting n feature vectors of known tumor tissues into machine learning algorithms. To construct an accurate and reliable classifier with every gene included is not a straightforward task because in practice the number of tissue samples available for training is smaller (a few dozens) than the number of features (a few thousands). In such a case, the classification space is nearly empty and it is difficult to construct a classifier that generalizes well. An analogy to this problem is the task of figuring out in the dark the shape of a footpath with a limited number of streetlights that are far apart. Each streetlight is like the sample or data point in the learning problem and the distance between the streetlights can be viewed as the dimensionality of the problem. With high dimensional inputs and a limited number of data points (i.e. data are sparse), meaning the distance between the streetlights is longer (several kilometers), it is impossible to guess the shape of the footpath accurately. Therefore, there is a need to select a handful of the most decisive genes in order to shrink the classification space and to improve the performance. Or, the distance between the streetlights needs to be shrunk first before guessing the shape of the footpath.

SVMs-based feature selection algorithm has been referred to as Recursive Feature Elimination with Support Vector Machines (RFE-SVMs) and it has been introduced as well as applied to a gene selection for a cancer classification

T.-M. Huang et al.: *Kernel Based Algorithms for Mining Huge Data Sets*, Studies in Computational Intelligence (SCI) **17**, 97–123 (2006)
www.springerlink.com

[61]. In the first part of this chapter, the results of improving the performance of RFE-SVMs by working on two, often neglected, aspects of the algorithm implementation which may affect the overall performance of the RFE-SVMs are presented. These two aspects are - the selection of a proper value for the penalty parameter C and the preprocessing of the microarray data. The C parameter plays an important role for SVMs in preventing an overfitting but its effects on the performance of RFE-SVMs are still unexplored. In terms of the microarray data preprocessing, this book will only focus on the part after the gene expressions have been calculated for each array. For example, for an Affymetric array, only the preprocessing of the gene expressions which have been calculated by passing the raw data from a scanner to MAS5 and filtration procedures will be investigated.

To assess the relative performance of the RFE-SVMs to other learning algorithms, comparisons with the nearest shrunken centroid classification algorithm are carried out. The nearest shrunken centroid is one of the well-known techniques developed in [137] from the field of statistics. The reason for selecting this method is not only its popularity and robustness but also because of the fact that the underlying philosophy behind this algorithm is completely different from SVMs. Namely, the nearest shrunken centroid method is based on the assumption that all the samples are normally distributed around its centroid, whereas SVMs make no such assumption about the underlying distribution of the model. Therefore, it is interesting to see which philosophy works better in the context of DNA microarray classification. Particularly, interesting fact is that most of the simulations will be performed with a small number of available training data. This is because each sample in the data sets of microarray studies can take several months or years to collect. As a result, the feature reduction techniques and the classifiers used in this type of application should achieve relatively good performance when faced with small number of samples.

As suggested in [119], it should be mentioned that the use of a more complicated kernel may not yield better performance in a classification of microarray data. This is because the original classification space is already high dimensional and even a linear kernel will easily overfit the data in such a space. As a result, it is appropriate to limit ourselves only on the linear form of RFE-SVMs in this work.

Another major part of this book is to determine whether gene selection algorithms such as RFE-SVMs can help the biologist in finding the right set of genes causing a certain disease. Generally, this is done by submitting a list of genes to the experts in the field for a qualitative assessment. In this work, a quantitative assessment is achieved by comparing the genes selected by different algorithms. This is done by using the Rankgene software [133] which implements eight different classification, i.e. selection approaches.

This chapter is organized as follows:

- Section 4.2: The basics of microarray technology is presented.

- Section 4.3 reviews RFE-SVMs and some prior work in this area.
- Section 4.4 presents the results on the influence of the C parameter on a correct selection of relevant features.
- Section 4.5 investigates the results of the improved RFE-SVMs on two medical data sets (colon cancer and lymphoma data set) and it compares two standard preprocessing steps in a microarray analysis for RFE-SVMs.
- Section 4.6: The performance of RFE-SVMs and the nearest shrunken centroid on two medical data sets (colon cancer and lymphoma data set) are compared.
- Section 4.7 discusses and compares in a detail the genes selected by RFE-SVMs and by eight other different selection approaches implemented within the Rankgene software.
- Section 4.8 is a concluding one in which several possible avenues for further research in the area of DNA microarray in cancer diagnosis are suggested.

4.2 Basics of Microarray Technology

Since this chapter is mainly related to feature reduction using SVMs in DNA microarray analysis, it is essential to understand the basic steps involved in a microarray experiment and why this technology has become a major tool for biologists to investigate the function of genes and their relations to a particular disease.

In an organism, proteins are responsible for carrying out many different functions in the life-cycle of the organism. They are the essential part of many biological processes. Each protein consists of chain of amino acids in a specific order and it has unique functions. The order of amino acids is determined by the DNA sequences in the gene which codes for a specific proteins. To produce a specific protein in a cell, the gene is transcribed from DNA into a messenger RNA (mRNA) first, then the mRNA is converted to a protein via translation.

To understand any biological process from a molecular biology perspective, it is essential to know the proteins involved. Currently, unfortunately, it is very difficult to measure the protein level directly because there are simply too many of them in a cell. Therefore, the levels of mRNA are used as a surrogate measure of how much a specific protein is presented in a sample, i.e. it gives an indication of the levels of gene expression. The idea of measuring the level of mRNA as a surrogate measure of the level of gene expression dates back to 1970s [21, 99], but the methods developed at the time allowed only a few genes to be studied at a time. Microarrays are a recent technology which allows mRNA levels to be measured in thousands of genes in a single experiment.

The microarray is typically a small glass slide or silicon wafer, upon which genes or gene fragment are deposited or synthesized in a high-density manner. To measure thousands of gene expressions in a sample, the first stage in making of a microarray for such an experiment is to determine the genetic materials to be deposited or synthesized on the array. This is the so-called probe selection stage, because the genetic materials deposited on the array are going to serve as probes to detect the level of expressions for various genes in the sample. For a given gene, the probe is generally made up from only part of the DNA sequence of the gene that is unique, i.e. each gene is represented by a single probe. Once the probes are selected, each type of probe will be deposited or synthesized on a predetermined position or "spot" on the array. Each spot will have thousands of probes of the same type, so the level of intensity pick up at each spot can be traced back to the corresponding probe. It is important to note that a probe is normally single stranded (denatured) DNA, so the genetic material from the sample can bind with the probe.

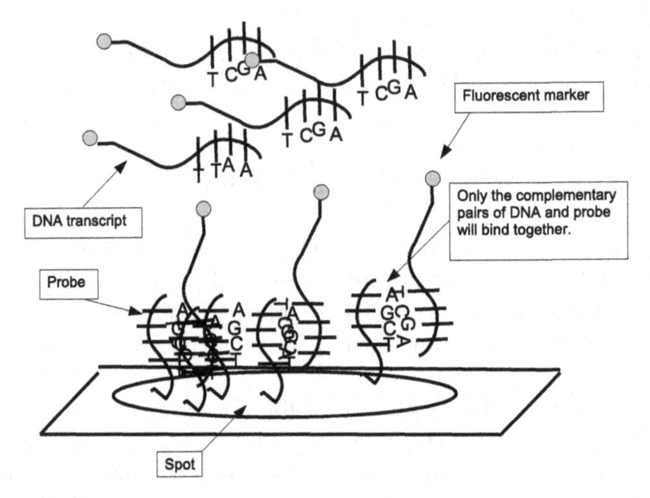

Fig. 4.1. The figure shows the core part of the microarray technology where a DNA transcript with a fluorescence marker will only bind with the probe that is complementary to the transcript during the hybridization.

Once the microarray is made, the next stage is to prepare the sample. The mRNA in the sample is first extracted and purified. Then mRNA is reverse transcribed into single stranded DNA and a fluorescent marker is attached to each transcript. The core of the microarray technologies is reliant on the fact that the single stranded DNA transcript (with fluorescent marker from the sample) will only bind with the probe (part of the DNA sequences of the gene) that is complementary with the transcript during the process of hybridization (as shown in Fig. 4.1). This means that the binding will only occur if the DNA transcript from the sample is coming from the same gene as the probe. By measuring the amount of fluorescence in each spot (with thousands of the same probes) using a scanner, the level of expression of each gene can be measured (as shown in Fig. 4.2). Many preprocessing steps need to be done before the raw data from the scanner can be converted into gene expression (i.e. into a number representing a gene's expression in a sample).

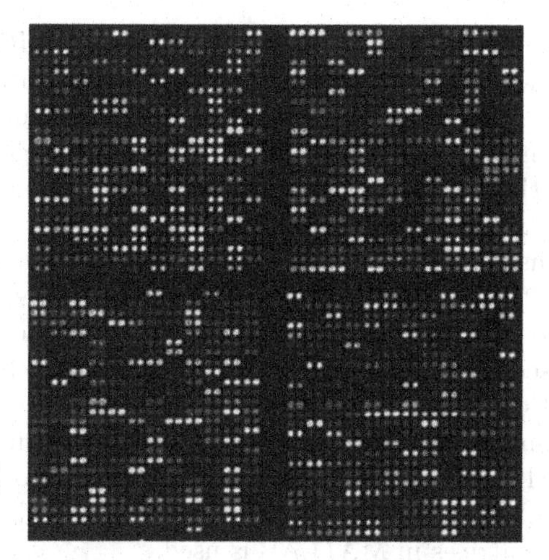

Fig. 4.2. The output of an DNA microarray experiment. The intensity level at each spot indicates how much the corresponding gene's expression is in the sample. More fluorescence on a given spot indicates that the corresponding gene is more expressed.

As mentioned previously, these preprocessing steps will be neglected here. The starting point of this book is after the gene expression profile of each sample is given and the task is to find the most relevant genes that give the best performance in classification of different diseases. The next section will present some prior work in this area of feature reduction for DNA microarray and the basic concept of RFE-SVMs.

4.3 Some Prior Work

4.3.1 Recursive Feature Elimination with Support Vector Machines

As mentioned in Chap. 2, maximization of a margin has been proven to perform very well in many real world applications and makes SVMs one of the most popular machine learning algorithms at the moment. Since the margin is the criterion for developing one of the best-known classifiers, it is natural to consider using it as a measure of relevancy of genes or features. This idea of using margin for gene selection was first proposed in [61]. It was achieved by coupling recursive features elimination with linear SVMs (RFE-SVMs) in order to find a subset of genes that maximizes the performance of the classifiers. In a linear SVM, the decision function is given as $f(x) = \mathbf{w}^T\mathbf{x} + b$ or $f(x) = \sum_{k=1}^{n} w_k x_k + b$. For a given feature x_k, the size of the absolute value of its weight w_k shows how significantly does x_k contributes to the margin of the linear SVMs and to the output of a linear classifier. Hence, w_k is used as a feature ranking coefficient in RFE-SVMs. In the original RFE-SVMs, the algorithm first starts constructing a linear SVMs classifier from the microarray data with n number of genes. Then the gene with the smallest w_k^2 is removed and another classifier is trained on the remaining $n - 1$ genes. This process is repeated until there is only one gene left. A gene ranking is produced at the end from the order of each gene being removed. The most relevant gene will be the one that is left at the end. However, for computational reasons, the algorithm is often implemented in such a way that several features are reduced at the same time. In such a case, the method produces a feature subset ranking, as opposed to a feature ranking. Therefore, each feature in a subset may not be very relevant individually, and it is the feature subset that is to some extent optimal [61]. The linear RFE-SVMs algorithm is presented in Algorithm 4.1 and the presentation here follows closely to [61]. Note that in order to simplify the presentation of the Algorithm 4.1, the standard syntax for manipulating matrices in MATLAB is used.

Algorithm 4.1 Recursive Feature Elimination with Support Vector Machines

Consider a binary classification problem with $n \times m$ input matrix $\mathbf{X}^0 = [\mathbf{x}_1, \mathbf{x}_2, \ldots \mathbf{x}_n]^T$ where m is the number of feature, n is the number of samples, the class labels $\mathbf{y} = [y_1, y_2, \ldots, y_n]^T$ and $y \in [+1, -1]$. Let $\mathbf{s}$ denote the set of surviving features' indices (i.e. features that are still left to be eliminated) and $\mathbf{r}$ denote feature ranking list.

1. Set $\mathbf{s} = [1, 2, \ldots, m]$ and $\mathbf{r}$ is an empty set.
2. Construct the training examples $\mathbf{X}$ with features in $\mathbf{s}$, as $\mathbf{X} = \mathbf{X}^0(:, \mathbf{s})$ in MATLAB.
3. Train the SVMs classifier $\alpha = \text{SVM-train}(\mathbf{X}, \mathbf{y})$.
4. Compute for all the features $i \in \mathbf{s}$ their corresponding weight $w_i = \sum_{k=1}^{n} \alpha_k y_k X_{ki}^0$.
5. Compute the ranking criteria $c_i = (w_i)^2$, for all $i \in \mathbf{s}$.
6. Find the set of features $\mathbf{f} = [f_1, f_2, \ldots, f_p]$ from $\mathbf{s}$ which has the p-smallest ranking criterions.
7. Update the feature ranked list $\mathbf{r} = [\mathbf{s}(\mathbf{f}), \mathbf{r}]$.
8. Eliminate the set of features $\mathbf{f}$ which has the p-smallest ranking criterions $\mathbf{s} = \text{setdiff}(\mathbf{s}, \mathbf{f})$ (Find the indices that are in $\mathbf{s}$ but not in $\mathbf{f}$).
9. Repeat step 2 to 8 until $\mathbf{s}$ is empty. Finally output the feature ranked list $\mathbf{r}$.

4.3.2 Selection Bias and How to Avoid It

As shown in [61], the leave-one-out error rate of RFE-SVMs can reach as low as zero percent with only 16 genes on the well-known colon cancer data set from [8]. However, as it was later pointed out in [11], the simulation results in [61] did not take selection bias into account. The leave-one-out error presented in [61] was measured using the classifier constructed from the subset of genes that were selected by RFE-SVMs using the complete data set. It gives a too optimistic an assessment of the true prediction error, because the error is calculated internally. To take the selection bias into account, one needs to apply the gene selection and the learning algorithm on a training set to develop a classifier, and only then to perform an external cross-validation on a test set that had not been seen during the selection stage on a training data set. As shown in [11, 99], the selection bias can be quite significant and the test error based on 50% training and 50% test can be as high as 17.5% for the colon cancer data set. Another important observation from [11] is that there are no significant improvements when the number of genes used for constructing the classifier is reduced: the prediction errors are relatively constant until approximately 64 or so genes. These observations indicate that the performance and the usefulness of RFE-SVMs may be in question. However, the influence of the parameter C was neglected in [11] which restricts the results obtained. As a major part of this work, the problem is further investigated by changing (reducing) the parameter C in RFE-SVMs, in order to explore and to show the full potentials of RFE-SVMs.

4.4 Influence of the Penalty Parameter C in RFE-SVMs

As discussed previously, the formulation presented in (2.10) is often referred to as the "hard" margin SVMs, because the solution will not allow any point to be inside, or on the wrong side of the margin and it will not work when classes are overlapped and noisy. This shortcoming led to the introduction of the slack variables ξ and the C parameter to (2.10a) for relaxing the margin by making it 'soft' to obtain the formulation in (2.24). In the soft margin SVMs, C parameter is used to enforce the constraints (2.24b). If C is infinitely large, or larger than the biggest α_i calculated, the margin is basically 'hard'. If C is smaller than the biggest original α_i, the margin is 'soft'. As seen from (2.27b) all the $\alpha_j > C$ will be constrained to $\alpha_j = C$ and corresponding data points will be inside, or on the wrong side of, the margin. In most of the work related to RFE-SVMs e.g.,[61, 119], the C parameter is set to a number that is sufficiently larger than the maximal α_i, i.e. a hard margin SVM is implemented within such an RFE-SVMs model. Consequently, it has been reported that the performance of RFE-SVMs is insensitive to the parameter C. However, Fig. 4.3 [72] shows how C may influence the selection of more relevant features in a toy example where the two classes (stars * and pluses +) can be perfectly separated in a feature 2 direction only. In other words, the feature 1 is irrelevant for a perfect classification here.

As shown in Fig. 4.3, although a hard margin SVMs classifier can make perfect separation, the ranking of the features based on w_i can be inaccurate.

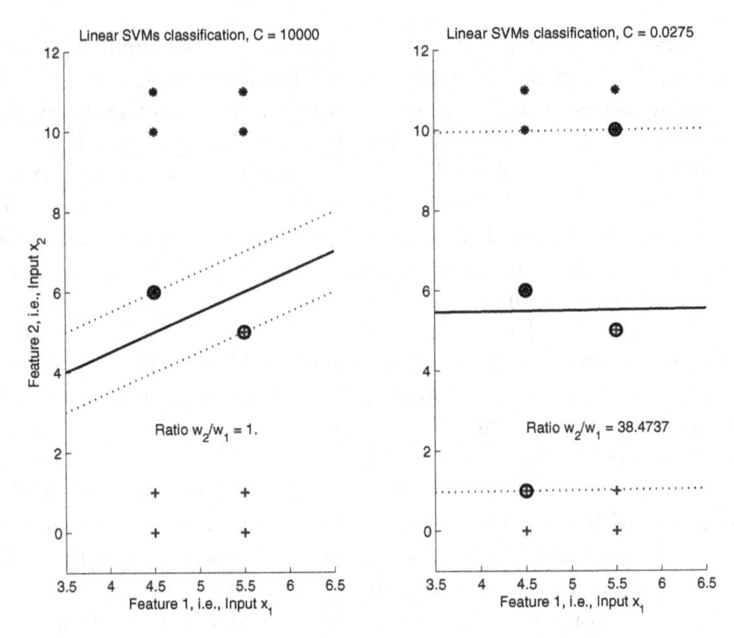

Fig. 4.3. A toy example shows how C may be influential in a feature selection. With C equal to 10000, both features seem to be equally important according to the feature ranking coefficients (namely, $w_1 = w_2$). With $C = 0.025$, a request for both a maximal and a "hard" margin is relaxed and the feature 2 becomes more relevant than feature 1, because w_2 is larger than w_1 ($w_2/w_1 = 38.4$). While the former choice $C = 10000$ enforces the largest margin and all the data to be outside of it, the later one ($C = 0.025$) enforces the feature "relevance" and gives a better separation boundary because the two classes can be perfectly separated in a feature 2 direction only. Note that support vectors are marked by circle. Notice also the two SVs inside the margin in the later case.

The C parameter also affects the performance of the SVMs if the classes overlap each other. In the following section, the gene selection based on an application of the RFE-SVMs having various C parameters in the cases of two medicine data sets is presented.

4.5 Gene Selection for the Colon Cancer and the Lymphoma Data Sets

4.5.1 Results for Various C Parameters

In this section, the selection of relevant genes for the two known data sets in the gene microarray literature is presented. The colon data set was analyzed initially in [8] and the lymphoma data was first analyzed in [7]. The colon data set is composed of 62 samples (22 normal and 40 cancerous) with 2000 genes' expressions in each sample. The training and the test sets are obtained by splitting the dataset into two equal groups of 31 elements, while ensuring that each group has 11 normal and 20 cancerous tissues. The RFE-SVMs is only applied to the training set in order to select relevant genes and to develop classifiers. After the training the classifiers are used on the test data set in order to estimate the error rate of the algorithms. 50 trials were carried out with random split for estimating the test error rate. (This is the same approach as in [11]). A simple preprocessing step is performed on the colon data set to make sure each sample is treated equally and to reduce the array effects. Standardization is achieved by normalizing each sample to the one with zero mean and with a standard deviation of one [72]. To speed up the gene selection process, 25% of the genes are removed at each step until less than 100 genes remain still to be ranked. Then the genes are removed one at a time. Because there were 50 trails, the gene ranking is based on averaging over fifity runs. In other words, gene rankings produce from each run are combining together to become one ranking. The simulation results for the colon data set are shown in Fig. 4.4 [72].

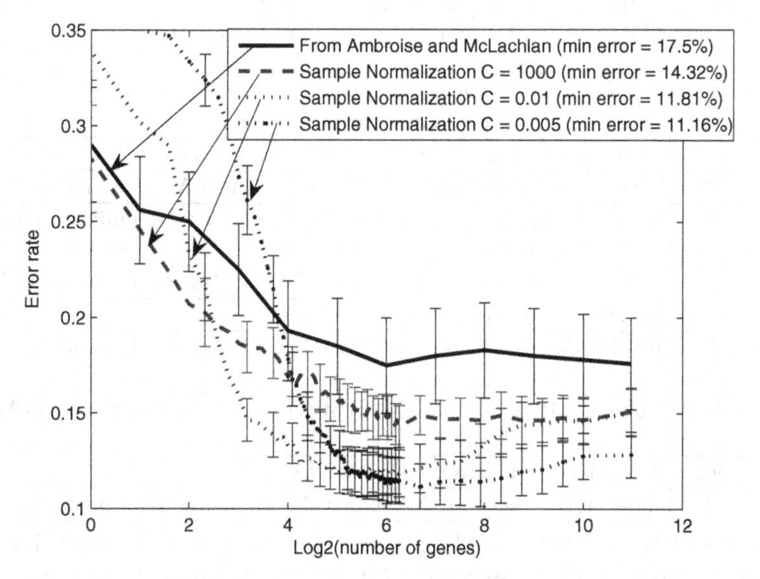

Fig. 4.4. Simulation result on the colon cancer data set with various C parameters. The error bar represents the 95% confidence interval.

The Ambroise and McLachlan's curve in Fig. 4.4 is directly taken from [11], and it is unclear what C value is used in their paper. However, note that the performance curve for $C = 1000$ is very similar to the one from [11]. In both cases, the error rates are virtually unchanged as the number of genes is reduced down to 16 genes. By comparing the error rates for various

C parameters, it is clear that changing the parameter C has a significant influence on the performance of RFE-SVMs in this data set. The error rate is reduced from the previous 17.5% as reported in [11] to 11.16% when C is equal to 0.005. Also, for $C = 0.01$, the gene selection procedure improves the performance of the classifier. This trend can be observed by looking at the error rate reduction from initially around 15% at 2000 genes to 11.8% with 2^6 genes. A similar trend can be observed when $C = 0.005$, but the error rate reduction is not as significant as in the previous case. This is because the error rate of the linear SVMs with $C = 0.005$ is already low at the very begining, when all the genes are used. The simulations shown demonstrate that tuning the C parameter can reduce the amount of over-fitting on the training data even in such a high dimensional space with small number of samples. Note that the minimal error rate here is 11.16% and this coincides with the suggestion in [11] that there are some wrongly labelled data in the data set. This makes colon cancer data more difficult to classify than the lymphoma data set presented next.

Table 4.1. Top 10 genes for colon cancer data obtained by RFE-SVM with C=0.005. Genes are ranked in order of decreasing importance.

Ranking	GAN[a]	Description
1	J02854	Myosin Regulatory light chain 2
2	X86693	H.sapiens mRNA for hevin like protein
3	H06524	GELSOLIN PRECURSOR, PLASMA (HUMAN);
4	M36634	Human vasoactive intestinal peptide (VIP) mRNA, complete cds.
5	M76378	Human cysteine-rich protein (CRP) gene, exons 5 and 6.
6	M63391	Human desmin gene, complete cds.
7	R87126	MYOSIN HEAVY CHAIN, NONMUSCLE (Gallus gallus)
8	T92451	TROPOMYOSIN, FIBROBLAST AND EPITHELIAL MUSCLE-TYPE (HUMAN)
9	T47377	S-100P PROTEIN (HUMAN).
10	Z50753	H.sapiens mRNA for GCAP-II/uroguanylin precursor.

[a] Gene Accession Number: The Gene accession number is the unique identifier assigned to the entire sequence record (i.e. entire DNA sequence of a gene) when the record is submitted to GenBank.

In terms of the gene selected, comparison can be made between Table 4.1 and Table 4.4 to show the differences caused by using different penalty parameter C, namely $C = 0.005$ and $C = 0.01$. Except for the first gene in the table, the rank of all the other genes is different. However, same seven genes are selected into the top 10 genes by both C values.

The second benchmarking data set is the lymphoma data set first analyzed in [7]. This version of the lymphoma data set is also the same as the one used in [34]. It is composed of 62 samples (42 diffuse large B-Cell (DLCL), 8 follicular lymphoma (FL) and 12 chronic lymphocytic lymphoma (CL)) with 4000 genes expressions in each sample. This is a multi-class problem and the data set is split into 31 data pairs for training and 31 pairs for testing. Each part has 21 samples belonging to DLCL, 4 belonging to FL and 6 belonging to CLL. The simulation results are shown in Fig. 4.5.

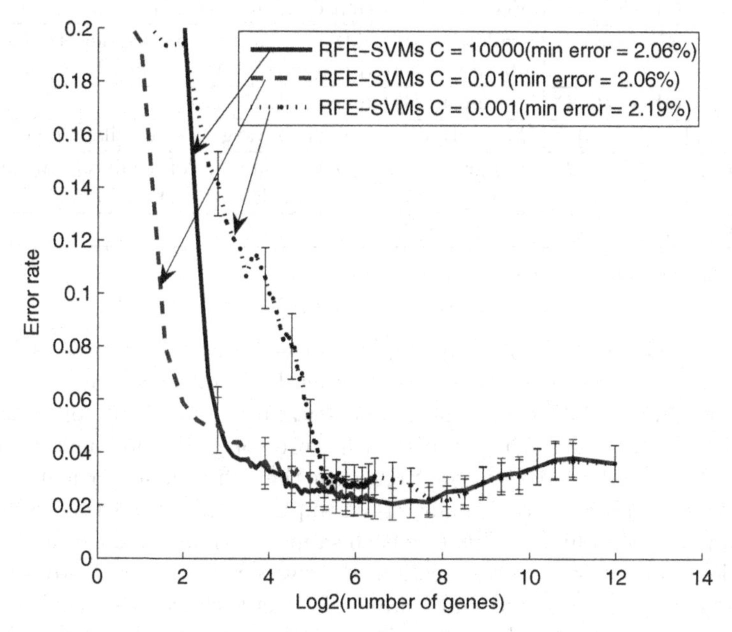

Fig. 4.5. Simulation results on the lymphoma data set with various C parameters.The error bar represents the 95% confidence interval.

As shown in Fig. 4.5, there is not too much difference in terms of the lowest error rate between the larger C values and the smaller ones, and all the models have approximately 2% error rate. In this case, the choice of C parameter does not influence the performance very much. This may be due to the fact that this data set is an easy one (meaning, well separated), and that it may be a relatively simple problem to perform the sorting out between different classes disregarding the true value of parameter C. The top 10 genes for the lymphoma data set selected by RFE-SVMs are listed in Table 4.2.

4.5.2 Simulation Results with Different Preprocessing Procedures

This section investigates the preprocessing of gene expressions after they have been obtained via procedure such as MAS5 in Affymetrix array and their

Table 4.2. Top 10 genes for lymphoma data obtained by RFE-SVM with C =0.01. Genes are ranked in order of decreasing importance.

Ranking	GAN[a]	Description
1	GENE1636X	osteonectin=SPARC=basement membrane protein; Clone = 487878
2	GENE1610X	Unknown; Clone=711756
3	GENE1637X	*Fibronectin 1; Clone=139009
4	GENE1635X	*Fibronectin 1; Clone=139009
5	GENE2328X	*FGR tyrosine kinase; Clone=347751
6	GENE263X	Similar to HuEMAP=homolog of echinoderm microtubule associated protein EMAP; Clone=1354294
7	GENE1648X	*cathepsin B; Clone=261517
8	GENE1609X	*Mig=Humig=chemokine targeting T cells; Clone=8
9	GENE3320X	Similar to HuEMAP=homolog of echinoderm microtubule associated protein EMAP; Clone=1354294
10	GENE1641X	*cathepsin B; Clone=261517

[a] Gene Accession Number

effects on classification performance are investigated. A very common preprocessing step before inputting the training data into various machine learning algorithms (or into various other statistical methods), is to normalize each feature vector. After such a step each feature (or gene here) has a mean of zero and a standard deviation of one. On the other hand, in a microarray analysis, it is common to normalize the sample vector so that the array effect is minimized, obtaining in this way each sample with a zero mean and a standard deviation of one. In this section, the two straightforward preprocessing procedures are investigated and compared using the colon cancer data set.

In the first one, all the expressions are converted to log expressions by taking a logarithm on all the expressions. Then all the sample vectors in the data set are normalized as follows: first the mean expression value of the sample is subtracted from each gene expression within the sample. Then each gene expression is divided by the standard deviation of the sample. After this preprocessing step, each sample will have a mean of zero and a standard deviation of one. This procedure is referred to as the sample normalization. Note that here one does not have a selection bias phenomenon because each sample is treated separately.

For the second preprocessing procedure, the same sample normalization is applied to the complete data first and then a feature normalization step to all the features in the data set follows. In order to perform a feature normalization without the selection bias for a given feature, the mean expression of the feature (calculated from the training data) is subtracted first from all the expression values of the feature in the complete data set. Then all the expression values of the feature in the complete data set are divided by the standard deviation of the feature which is also calculated from the training data. Consequently, after the feature normalization the mean of a feature will be zero and its standard deviation will be one. However, note that the mean and the standard deviation of the feature for the complete data set will not equal zero and one respectively. The second procedure, which will be referred to as the sample and feature normalization, is very similar to the one in [61] except that the data are not passed through a squashing function here. To test these two procedures, the same setting is used as in the previous section, i.e. 50 random splits of 50% training data and 50% testing ones. Various C parameters are tested again and the best results are presented for each procedure.

Example 4.1. The example below shows how the two simple preprocessing steps work. Consider a DNA microarray data set with 3 samples and 4 gene expressions as follows:

$$\mathbf{X} = [\mathbf{x}_1, \mathbf{x}_2, \mathbf{x}_3]^T = \begin{bmatrix} 1 & 3 & 2.5 & 7 \\ 2 & 0.8 & 9 & 32 \\ 4 & 0.1 & 6 & 18 \end{bmatrix} \overset{\log}{\rightarrow} \begin{bmatrix} 0 & 1.098 & 0.91 & 1.94 \\ 0.69 & -0.22 & 2.19 & 3.46 \\ 1.38 & -2.3 & 1.79 & 2.89 \end{bmatrix}.$$

To perform the sample normalization, the mean value of each sample is subtracted from each sample and then in this way obtained each gene expression is divided by the standard deviation of the sample. The mean expression and standard deviation of each sample are given as follows,

$$\begin{aligned} [\bar{x}_1, \bar{x}_2, \bar{x}_3]^T &= [0.987, 1.53, 0.94]^T \\ [s_{x1}, s_{x2}, s_{x3}]^T &= [0.7959, 1.62, 2.25]^T \end{aligned} \tag{4.1}$$

After the sample normalization, the input matrix $\mathbf{X}$ is now given as follows,

$$\mathbf{X}_{SN} = \begin{bmatrix} -1.24 & 0.37 & 1.14 & 2.43 \\ -0.51 & -1.13 & 1.34 & 2.12 \\ 0.19 & -2.026 & 0.79 & 1.28 \end{bmatrix}. \tag{4.2}$$

While performing feature normalization one averages over all training data samples and consequently splitting a data set into the training and test part causes an bias dubbed as selection bias. Suppose sample $\mathbf{x}_3$ is the test data. In order to perform feature normalization without selection bias, the mean and the standard deviation of the feature are evaluated using training data set $\mathbf{x}_1$ and $\mathbf{x}_2$ ones . As a result, the mean and the standard deviation of each feature is as follows:

$$\begin{aligned} [\bar{f}_1, \bar{f}_2, \bar{f}_3, \bar{f}_4] &= [-0.87, -0.377, 1.24, 2.28] \\ [s_{f1}, s_{f2}, s_{f3}, s_{f4}] &= [0.511, 1.07, 0.14, 0.21]. \end{aligned}$$

Now these values are used for the normalization of all three samples, namely, $\mathbf{x}_1, \mathbf{x}_2, \mathbf{x}_3$. After feature and sample normalization, the input matrix $\mathbf{X}$ is now given as follows,

$$\mathbf{X}_{SN\&FN} = \begin{bmatrix} -0.707 & 0.70 & -0.707 & 0.710 \\ 0.707 & -0.707 & 0.707 & -0.707 \\ 2.09 & -1.53 & -3.12 & -4.57 \end{bmatrix} \quad (4.3)$$

Note that the mean of each feature is zero for the training data $\mathbf{x}_1$ and $\mathbf{x}_2$. However, the mean over the feature is not equal to zero if calculating with $\mathbf{x}_3$.

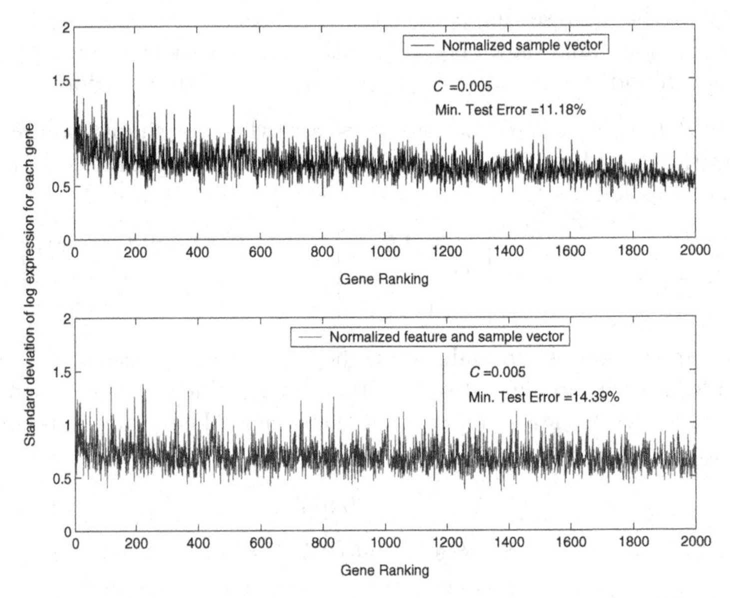

Fig. 4.6. Effect of different preprocessing procedures on the gene ranking for colon cancer data set. The genes are ranked in the order of decreasing importance. The gene with rank 1 is the most relevant gene.

In Fig. 4.6 [72], the gene rankings from RFE-SVMs with two different preprocessing procedures are compared. In the figure, the standard deviation of the log expression for each gene (calculated from the complete data set without sample or feature normalization) is plotted on the vertical axis. Their respective gene rankings from RFE-SVMs are plotted on the horizontal axis.

The top graph shows the result for the sample preprocessing procedure. It is interesting to observe that the gene with the higher standard deviation tends to have higher ranking. This trend suggests that RFE-SVMs with the sample normalization are likely to pick up genes with expression that vary more across the samples. This fits well with the assumption that a gene is less relevant if its expression does not vary much across the complete data set. Such a general trend cannot be observed in the bottom graph (where both the sample and the feature normalization are applied). Note that there is no connection between the standard deviation of the gene and gene ranking now. This phenomenon may be due to the fact that the feature normalization

Table 4.3. Colon cancer data. RFE-SVMs' top 10 genes for $C = 0.005$ for both the sample and the feature vectors normalization.

Ranking	GAN	Description
1	R39681	EUKARYOTIC INITIATION FACTOR 4 GAMMA (Homo sapiens)
2	R87126	MYOSIN HEAVY CHAIN, NONMUSCLE (Gallus gallus)
3	H20709	MYOSIN LIGHT CHAIN ALKALI, SMOOTH-MUSCLE ISOFORM (HUMAN);
4	H06524	GELSOLIN PRECURSOR, PLASMA (HUMAN);.
5	H49870	MAD PROTEIN (Homo sapiens)
6	R88740	ATP SYNTHASE COUPLING FACTOR 6, MITOCHONDRIAL PRECURSOR (HUMAN);
7	J02854	MYOSIN REGULATORY LIGHT CHAIN 2, SMOOTH MUSCLE ISOFORM (HUMAN); contains element TAR1 repetitive element ;.
8	M63391	Human desmin gene, complete cds
9	H09273	PUTATIVE 118.2 KD TRANSCRIPTIONAL REGULATORY PROTEIN IN ACS1-PTA1 INTERGENIC REGION (Saccharomyces cerevisiae)
10	X12369	TROPOMYOSIN ALPHA CHAIN, SMOOTH MUSCLE (HUMAN);.

step in the second preprocessing procedure will make each gene have the same standard deviation. Hence, a gene with higher standard deviation originally will no longer be advantageous over a gene having a smaller standard deviation. In Table 4.3 [72], the top 10 genes selected by $C = 0.005$ for both sample and feature vectors normalization are presented. By comparing Table 4.3 and Table 4.1, it is clear that the two preprocessing steps discussed here produced two different rankings and only five of the top 10 genes are selected by both preprocessing steps. This supports the trend that is observed in Fig. 4.6.

A general practice for producing good results with SVMs is to normalize each input (feature) to the one with a mean of zero and a standard deviation of one, as in the feature normalization step. However, in this case, this simple rule does not perform as well as expected: the error rate of applying both sample and feature normalization is higher than only the sample normalization. This phenomenon may be due to the fact that the feature normalization step in the second preprocessing procedure filters out the information about the spread of the expression for each gene as discussed previously and this information is helpful for selecting the relevant gene and classification. It is beyond the scope of this book to find out the optimal preprocessing procedure for RFE-SVMs. However, the simulation results presented here suggest that different preprocessing procedures may influence the performance of RFE-SVMs. Therefore, for a given problem, it may be wise to try different preprocessing procedures in order to obtain an optimal performance while using RFE-SVMs [72]. However, note that sample normalization only as shown in Fig. 4.6 produced slightly better result.

4.6 Comparison between RFE-SVMs and the Nearest Shrunken Centroid Method

4.6.1 Basic Concept of Nearest Shrunken Centroid Method

The nearest shrunken centroids method was first developed in [137, 138] and it is currently one of the most popular methods for class prediction of DNA microarrays data. This algorithm uses the nearest centroid method as a basis for classification and it incorporates the innovative idea of incrementally shrinking the class centroid towards the overall centroid for feature reduction. As a result, the algorithm can achieve equal or better performance than the classical nearest centroid method with fewer genes. In order to understand how the nearest shrunken centroid works, it is essential to review the nearest centroid algorithm for classification.

A nearest centroid classification first uses the training data to calculate the centroid of each class (or class centroid) as shown in Fig. 4.7. Then a new sample will be classified to the class whose centroid has the shortest square distance to the sample. In Fig. 4.7, L_1 is smaller than L_2, so the sample

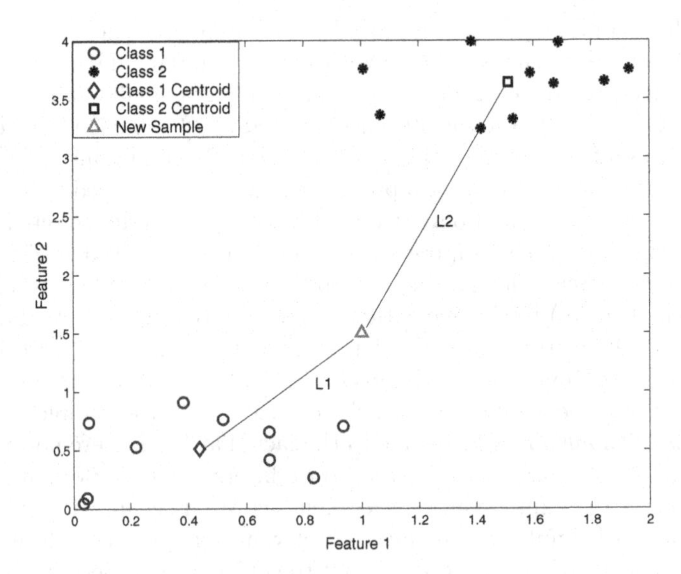

Fig. 4.7. A simple illustration of the nearest centroid classification. Centroids of the circle and the asterisk classes shown here as diamond and square respectively are used to classify new samples.

is classified to the circle class. This simple procedure can yield good results as demonstrated in [137]. However a major drawback is that it uses all the features available.

To overcome this drawback, in a nearest shrunken centroid method all the class centroids are shrunk towards the overall centroid as shown in Fig. 4.8(a). In other words, the components of the class centroids in the feature are increased or decreased in order to get closer to the component of the overall centroid in the feature. For example, to move class 1's centroid towards the overall centroid in Fig. 4.8(a), both components of the class 1's centroid have to increase. Once all class centroids' components in the feature are equal to the overall centroid's component in the feature (i.e., all of the class centroids reach the overall centroid of the feature), this feature no longer contributes to the classification. As an example, Fig. 4.8(b) shows the result of applying an overall shrinkage of 0.5374 to both feature 1 and 2 of the original problem in Fig. 4.7. In Fig. 4.8(b), the centroids of both classes in feature 1 direction are equal to each other and it implies that feature 1 can no longer contribute in differentiating the two classes after the shrinkage of 0.5374. The fact that feature 2 still can contribute to the classification after the shrinkage means that the centroids of the two classes are further apart from the overall centroid and more separable in this direction. Therefore, feature 2 will have a higher ranking than feature 1.

In a real world problem with thousand of genes, the algorithm uses all the genes available to calculate the class centroids first, then it shrinks the class centroids towards the overall centroid by an amount Δ to form shrunken class centroids. The shrunken class centroids are then used as nearest centroid classifiers on the test set to determine the cross-validation errors. These classifiers are referred to as the nearest shrunken centroid classifiers. This process of generating nearest shrunken centroid classifiers is repeated until all the class centroids reach the overall centroid. As a result of this process, a set of nearest shrunken centroid classifiers is generated and the optimal shrinkage is determined by the cross-validation with the test set. The more shrinkage Δ is applied on the class centroids, the less the number of genes will be used in the nearest shrunken centroid classification, hence a set of optimal genes may be selected from the process of shrinking the class centroids.

A crucial step to ensure good performance from the nearest shrunken centroid classifier is to apply Δ on the standardized distances between the class centroid and the overall centroid as in [138]. For a given problem with classes $1 \ldots c$, the kth component of the standardized distance d_{kj} for class j can be computed as follows [138]:

$$d_{kj} = \frac{\bar{x}_{kj} - \bar{x}_k}{m_j \cdot (s_k + s_0)} \tag{4.4}$$

where $\bar{x}_{kj}$ is the kth component of the centroid for class j, $\bar{x}_k$ is the kth component of the overall centroid, the value s_0 is a positive constant and m_j makes the $m_j \cdot s_k$ equal to the estimated standard error of the numerator

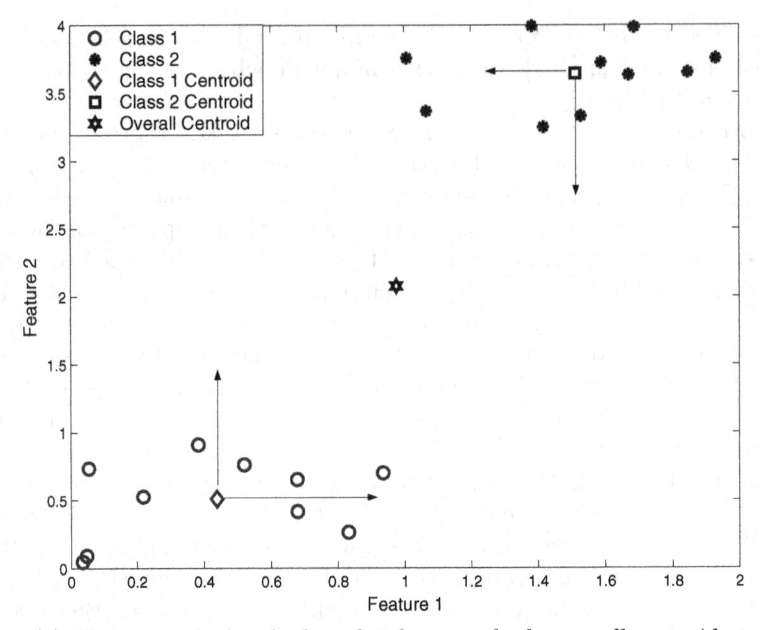

(a) The centroid of each class shrinks towards the overall centroid.

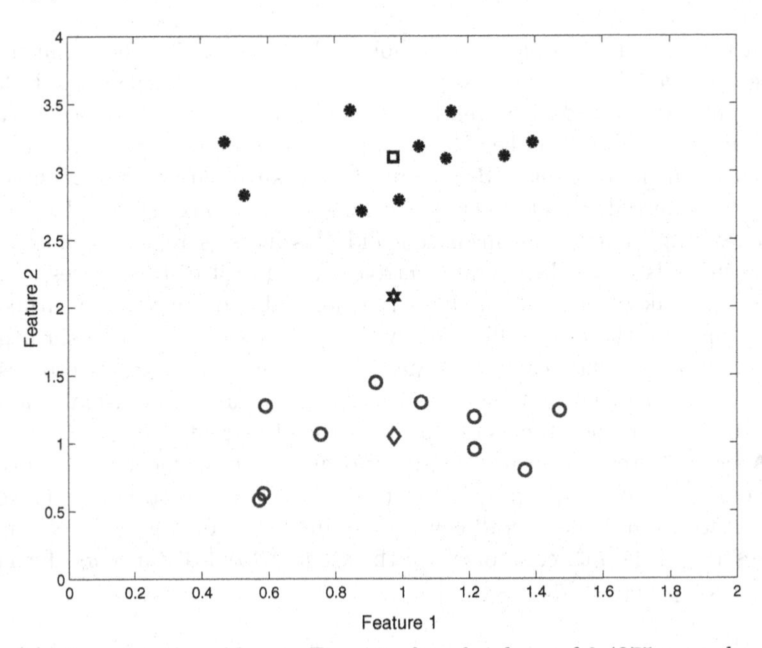

(b) The original problem in Fig. 4.7 after shrinkage of 0.4375 towards overall centroid. The feature 1 no longer contributes to the nearest centroid classification.

Fig. 4.8. The class centroids are shrunk to extract useful features.

in d_{kj}. (More details about nearest shrunken centroid can be found in [137]). The purpose of this standardization is to give higher weight to the gene whose expression is stable within samples of the same class. This means that a gene will be ranked highly if its class centroids deviate more from the overall centroid. At the same time it must also have a smaller standard deviation within the samples of the same class.

4.6.2 Results on the Colon Cancer Data Set and the Lymphoma Data Set

In this section, the results of comparisons between the nearest shrunken centroid method and the RFE-SVMs are presented.

Figure 4.9 shows the result of applying the nearest shrunken centroid to the colon cancer data set using the package Prediction Analysis for Microarrays in R (Pamr) [63]. As the shrinkage (or threshold) Δ increases, the number of genes used for classification is reduced. As a result, the leave-one-out errors

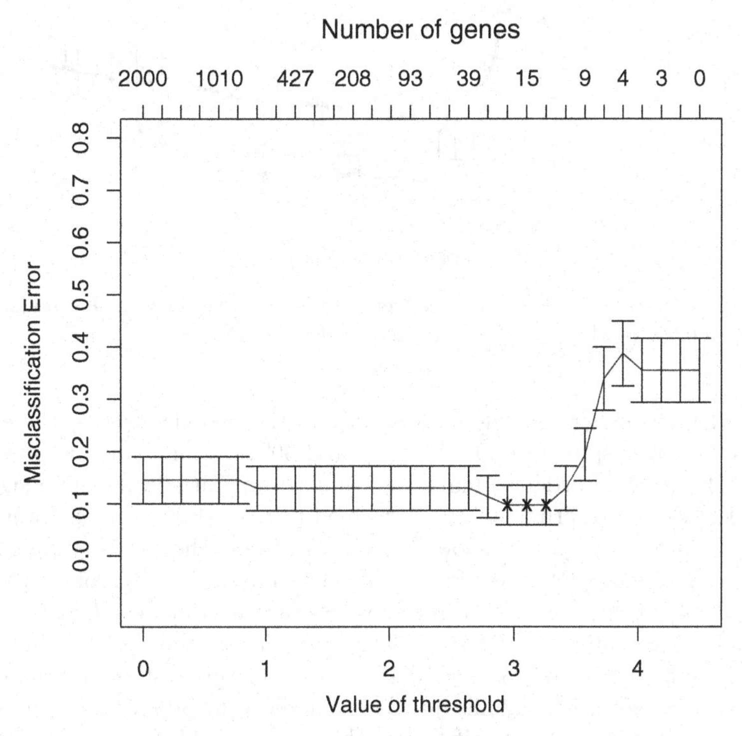

Fig. 4.9. The leave-one-out error of the nearest shrunken centroid on the colon cancer data set. Note that there are two different horizontal axis in the plot. The top horizontal axis shows the number of genes that are used for classification, whereas the bottom axis shows the size of the threshold Δ. When a larger threshold Δ is applied, a smaller number of genes is left to be eliminated.

are also changed. The minimum leave-one-out error is 9.68% in this case when approximately 15 genes are used. This error rate is only slightly higher than RFE-SVMs' leave-one-out error rate of 8.06%.

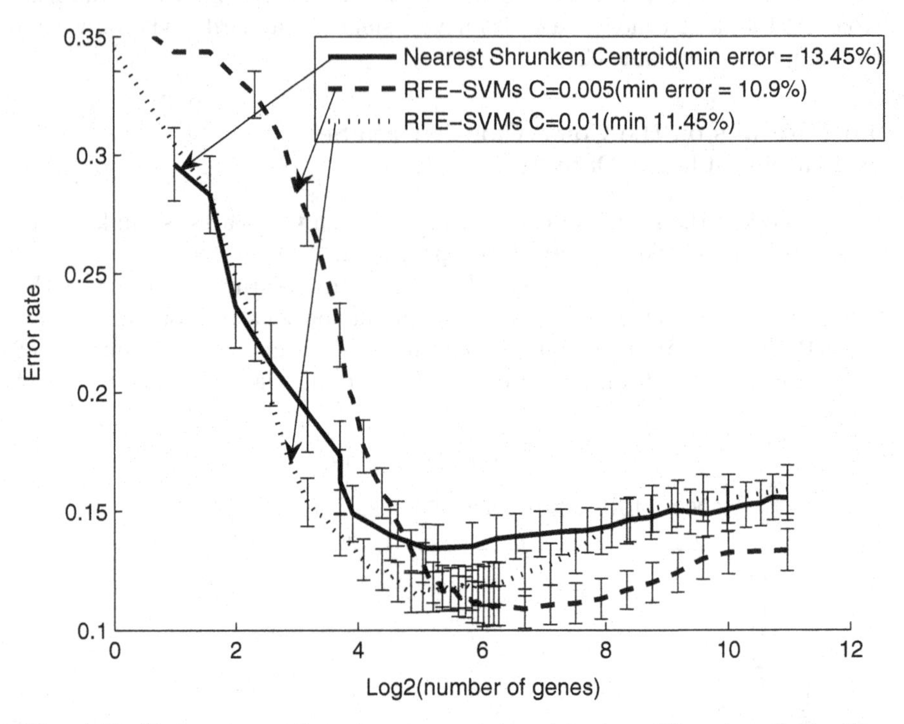

Fig. 4.10. Test errors on the colon cancer data set for two different methods. The error bar represents the 95% confidence interval.

In order to further test the performance of the two algorithms, the colon cancer data set is split into 50% training and 50% testing as in the previous section. However, in order to make the comparison statistically more significant, 100 random experiments are performed instead of 50 as in the previous section. Figure 4.10 from [71] shows the test errors and the corresponding 95% confidence interval of RFE-SVMs and the nearest shrunken centroid with various number of genes. As shown in the figure, the performance of RFE-SVMs (minimum error rate = 10.9%$\pm$ 0.7%) is superior to the nearest shrunken centroid (minimum error rate = 13.45%$\pm$ 1%) in this test setting and the difference is statistically significant. The difference in performance between the two algorithms is more significant in this more difficult setting than in the leave-one-out setting. This indicates that RFE-SVMs may have more superior performance when the number of samples is low.

The results for the lymphoma data set are shown in Fig. 4.11. Again, in order to make the comparison statistically more significant, 100 random ex-

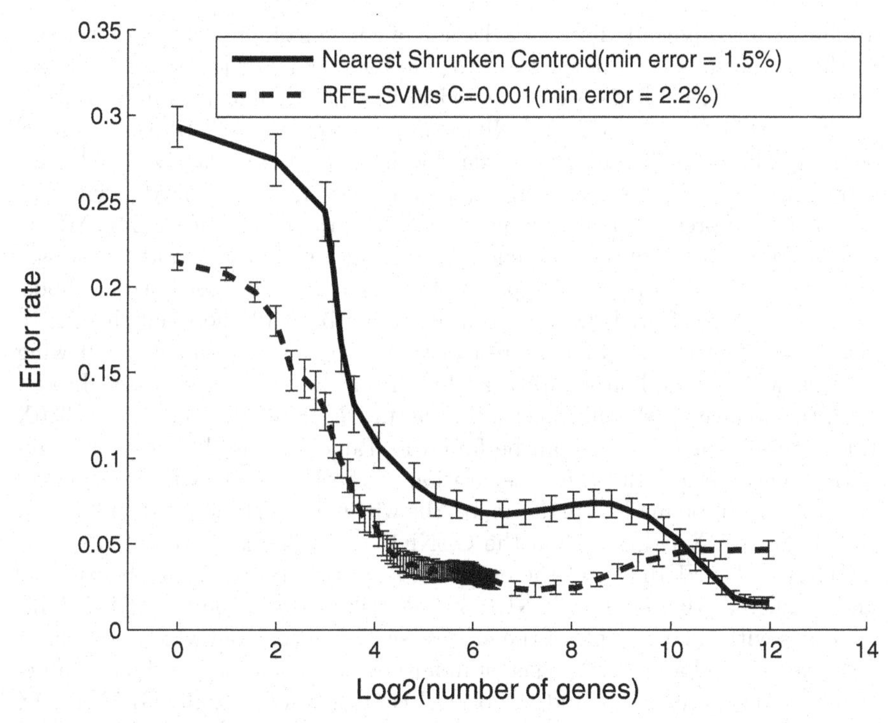

Fig. 4.11. Test errors on the lymphoma data set for two different methods. The error bars refer to 95% confidence interval.

periments are performed instead of 50 as in the previous section. The training is performed on 50% (31 samples) of the data set and the error rates are calculated on the other half of the data. The adaptive thresholding method is used for the nearest shrunken centroid, because it improves the performance on this data set as reported in [138]. The nearest shrunken centroid method achieves the lowest error rate of 1.5%±0.3%, whereas the RFE-SVM has the lowest error rate of 2.2%±0.4%. As a result, the difference between the two methods does not reach a statistically significant level. Furthermore, the nearest shrunken centroid method requires almost all 4026 genes to achieve the minimum error rate and its performance deteriorates significantly as the number of genes is reduced. In contrast, RFE-SVMs' best performance occurs when only 155 genes are used. This makes the model from RFE-SVMs more favourable to use, because it is developed in a lower dimensional space. Furthermore, the performances of RFE-SVMs are much better when fewer genes are used. Note that the results obtained here are different from the ones shown in [138]. In [138], the nearest shrunken method can achieve an error rate close to 0 % when the number of genes used is equal to 81. However, [138] used a 10-fold cross-validation for calculating the error rates instead of the harder test setting used here (50% for training and 50% for testing).

To illustrate graphically how the selection of genes is different between the two methods, the subspaces of the top two genes selected by the nearest shrunken centroid and RFE-SVMs are presented in Fig. 4.12. First, it is important to point out that the top gene (GENE1636X) selected by RFE-SVMs is ranked as the second most discriminative gene by the nearest shrunken centroid. This shows strong consensus between the two algorithms and coincides with the finding in [72] presented later in Sect. 4.7, where the results of RFE-SVMs are compared against 8 other different methods. Second, the FL and CL classes cannot be separated in the subspaces of the top two genes for both methods, as shown in Fig. 4.12. However, separation can be made between the DLCL class (the biggest class) and the rest of the classes. As a result, the following analysis will be based on separating DLCL from the rest of the classes. In Fig. 4.12(a), the nearest shrunken centroid selects GENE1622X over GENE1636X (more details on this gene can be found in Table 4.2) as the most discriminative genes because a perfect separation of the DLCL class from the other two classes can be made. Furthermore, the within class standard deviation for the GENE1622X is lower than the GENE1636X. This is represented by the intensity of GENE1622X of the entire data set which distributes between -2 and 2, whereas the one for GENE1636X distributes between 6.5 and -2. This coincides with the principle of the nearest shrunken centroid which is to select genes that have lower within-class standard deviation and larger deviation between the class centroids and the overall centroid. As a result, GENE1622X is regarded as a more discriminative feature than GENE1636X. In terms of RFE-SVMs, the DLCL class cannot be separated from the rest of the classes in either of the directions along (either GENE1636X or GENE1610X) as shown in Fig. 4.12(b). Only when the top two genes are combined together, can the separation be made (as shown by the decision boundary in Fig. 4.12(b)). This demonstrates an important property of RFE-SVMs: to identify the "combination" of genes that can give separation with maximal margin. The comparison presented above shows the major difference between the two approaches. RFE-SVMs make no assumption about the underlying distribution of the data set and it looks for a set of genes that can give best performance. In contrast, the nearest shrunken centroid assumes data are normally distributed around the centroids and it is more focused on an identification of the individual differentially expressed genes. As a result, the nearest shrunken centroid cannot pick up combinations of genes that are not differentially expressed but can still make good separation when combined together. In this section, two of the latest approaches for feature reduction and classification from the field of machine learning and statistics, and the comparisons of their performance in DNA microarray analysis are presented. In both of the benchmarking data sets tested, RFE-SVMs have noticeably better performance than the nearest shrunken centroid method when fewer genes and samples are used. The difference in the performance is more significant as the number of data available for training is smaller. This may be due to the fact that the position of the centroid and the spread of the classes are harder to estimate accurately as

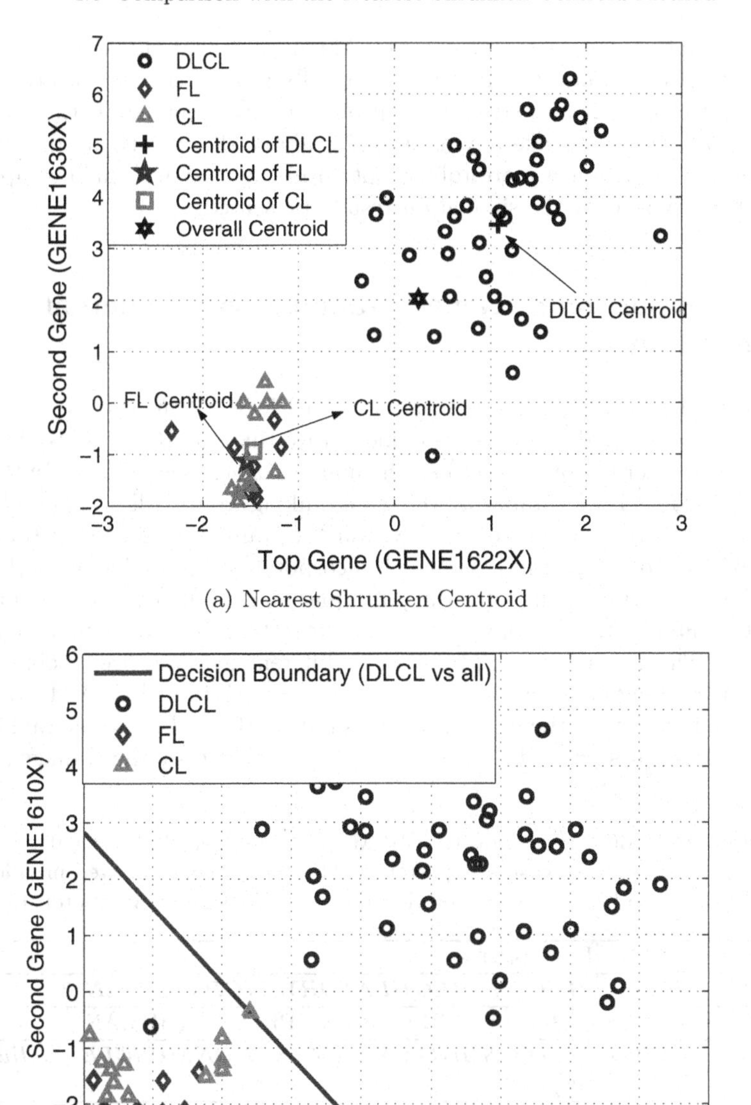

(a) Nearest Shrunken Centroid

(b) RFE-SVMs

Fig. 4.12. The subspace of the top two genes that are selected by (a) the nearest shrunken centroid and (b) RFE-SVMs for the lymphoma data set. The values for the top gene are plotted on the horizontal axis, whereas the values for the second gene are plotted on the vertical axis. The corresponding gene associated numbers (GAN) for selected genes are in brackets. Note that the nearest shrunken centroid's second-most discriminative gene is selected as the most discriminative one by RFE-SVMs.

the amount of data is extremely low. This means that the assumption that samples of the same class distribute normally around the class centroid does not suffice in such an extreme situation. On the other hand, the fact that RFE-SVMs do not make any assumption about the underlying distribution of the data makes it a more robust option in the case of a small sample and this is supported by the simulation results presented.

4.7 Comparison of Genes' Ranking with Different Algorithms

In Sects. 4.5.1 and 4.6, the performance of RFE-SVMs in two real world data sets were presented. The results demonstrate that the RFE-SVMs classifiers can have better performance when the number of genes is reduced. In this section, the focus is on comparing the gene ranking from RFE-SVMs with eight different algorithms implemented within the Rankgene software. (Rankgene is a package developed in [133] which incorporates eight different methods to produce a genes' ranking, including information gain, twoing rule, sum minority, max minority, gini index, sum of variances, t-statistic and one dimensional SVMs). The genes' rankings from eight different methods are combined into a single ranking and that ranking is compared with the RFE-SVMs' ranking. Table 4.4 [72] shows the top ten genes from the RFE-SVMs genes' ranking for $C = 0.01$. Genes printed in italics have been selected by the Rankgene too.

Table 4.4. Colon cancer data. RFE-SVMs' top 10 genes with $C = 0.01$. Gene are ranked in order of decreasing importance. The genes' names printed in bold italic letters have also been picked up by the Rankgene software within its top 10 genes.

Ranking	GAN	Description
1	***J02854***	***MYOSIN REGULATORY LIGHT CHAIN 2***
2	H06524	GELSOLIN PRECURSOR, PLASMA (HUMAN);.
3	***R87126***	***MYOSIN HEAVY CHAIN, NONMUSCLE (Gallus gallus)***
4	***M63391***	***Human desmin gene, complete cds.***
5	X86693	H.sapiens mRNA for hevin like protein.
6	***M76378***	***Human cysteine-rich protein (CRP) gene, exons 5 and 6.***
7	***T92451***	***TROPOMYOSIN, FIBROBLAST AND EPITHELIAL MUSCLE-TYPE (HUMAN)***
8	***Z50753***	***H.sapiens mRNA for GCAP-II/uroguanylin precursor.***
9	M31994	Human cytosolic aldehyde dehydrogenase (ALDH1) gene, exon 13.
10	M36634	Human vasoactive intestinal peptide (VIP) mRNA, complete cds.

Table 4.5. Colon cancer data. Top seven genes listed in [61]. Only the first gene, printed in bold italic letters, has also been picked up by the Rankgene software within its top 10 genes.

Ranking	GAN	Description
1	***H08393***	***COLLAGEN ALPHA 2(XI) CHAIN (Homo sapiens)***
2	M59040	Human cell adhesion molecule (CD44) mRNA, complete cds.
3	T94579	Human chitotriosidase precursor mRNA, complete cds.
4	H81558	PROCYCLIC FORM SPECIFIC POLYPEPTIDE B1-ALPHA PRECURSOR(Trypanosoma brucei brucei)
5	R88740	ATP SYNTHASE COUPLING FACTOR 6, MITOCHONDRIAL PRECURSOR (human)
6	T62947	60S RIBOSOMAL PROTEIN L24 (Arabidopsis thaliana)
7	H64807	PLACENTAL FOLATE TRANSPORTER (Homo sapiens)

Table 4.6. The 10 genes selected by RFE-SVMs and eight other different methods implemented in Rankgene software within their respective top 100 genes. (Avg Ranking = Average ranking of genes in nine different methods).

Avg Ranking[a]	GAN	Description
2.3	M76378	Human cysteine-rich protein (CRP) gene, exons 5 and 6.
13.5	M63391	Human desmin gene, complete cds.
27.2	Z50753	H.sapiens mRNA for GCAP-II/uroguanylin precursor.
28.1	T60155	ACTIN, AORTIC SMOOTH MUSCLE (HUMAN).
5.7	R87126	MYOSIN HEAVY CHAIN, NONMUSCLE (Gallus gallus)
7.5	M76378	Human cysteine-rich protein (CRP) gene, exons 5 and 6.
15.7	T92451	TROPOMYOSIN, FIBROBLAST AND EPITHELIAL MUSCLE-TYPE (HUMAN);.
21.3	H43887	COMPLEMENT FACTOR D PRECURSOR (Homo sapiens)
6.9	J02854	MYOSIN REGULATORY LIGHT CHAIN 2, SMOOTH MUSCLE ISOFORM (HUMAN).
31.3	M36634	Human vasoactive intestinal peptide (VIP) mRNA, complete cds.

[a] Average Ranking of genes in nine different methods

Although the genes have been ranked differently, six out of the top ten genes selected by RFE-SVMs have also been selected within the top ten genes by the Rankgene package as shown in Table 4.4 . This means that there is still a great deal of consensus on the genes' relevance obtained by different ranking methods. This may help in narrowing down the scope of the search for the most relevant set of genes.

On the other hand, in Table 4.5 the top 7 genes listed in [61] are shown. There is only one gene overlapping with the top ten genes from the Rankgene package. Also, only the gene ATP SYNTHASE listed in [61] was selected by the RFE-SVMs method as shown in Table 4.3. Furthermore, there are only 10 genes which have been selected by all nine methods (namely by the RFE and by 8 different methods implemented in the Rankgene software) within their respective top 100 genes. They are listed in Table 4.6. The average ranking of these 10 genes shows that only the top ranked genes are overlapped and that they are more likely to be selected by all the different methods. This strongly suggests that the ten listed genes may be very relevant in an investigation of a colon cancer.

4.8 Conclusions

This chapter has presented the performance of the improved RFE-SVMs algorithm for genes extraction in diagnosing two different types of cancers. Why and how this improvement is achieved by using different values for the C parameter was discussed in details. With a properly chosen parameter C, the extracted genes and the constructed classifier will ensure less over-fitting of the training data, leading to an increased accuracy in selecting relevant genes. These effects are more remarkable in a more difficult data set such as the colon cancer data. The simulation results also suggest that the classifier performs better in the reduced gene spaces selected by RFE-SVMs than in the complete 2000 dimensional gene space in the colon data set. This is a good indication that RFE-SVMs can select relevant genes, which can help in the diagnosis and in the biological analysis of both the genes' relevance and their function.

In terms of the raw data preprocessing, it is clear that the performance of RFE-SVMs can also vary with different preprocessing steps. In the colon cancer data set, normalizing only the sample vector produces better result. The comparison of genes' rankings obtained by the RFE-SVMs and by the Rankgene software package (which implements 8 different methods for a gene selection) shows that there is a great deal of consensus on genes' relevance. This may help in narrowing down the scope of search for the set of 'optimal' genes using machine learning techniques. The comparison between the improved RFE-SVMs and the nearest shrunken centroid on the colon data set suggested that the improved RFE-SVMs performs better when the number of data used for training is reduced. Although the two methods have similar minimal errors on the lymphoma data set, RFE-SVMs has much better performance when fewer genes are available. This makes the classifier developed by RFE-SVMs more desirable and robust.

Finally, the results in this book were developed from a machine learning and data mining perspective, meaning they are unrelated to any valuable insight from a biological and medical perspective. Thus, there is a need for a tighter cooperation between the biologists and/or medical experts and data miners in all future investigations. The basic result of this synergy should gives the meaning to all the findings presented here and in this way ensure a more reliable guide to the future research.

5

Semi-supervised Learning and Applications

5.1 Introduction

So far, the discussion in the previous chapters are centered around the supervised learning algorithm SVMs which attempts to learn the input-output relationship (dependency or function) $f(x)$ by using a training data set $\{\mathcal{X} = [\mathbf{x}(i), y(i)] \in \Re^m \times \Re, i = 1, \ldots, n\}$ consisting of n pairs $(\mathbf{x}_1, y_1), (\mathbf{x}_2, y_2), \ldots (\mathbf{x}_n, y_n)$, where the inputs $\mathbf{x}$ are m-dimensional vectors $\mathbf{x} \in \Re^m$ and the labels (or system responses) $y \in \Re$ are continuous values for regression tasks and discrete (e.g., Boolean) for classification problems. Another large group of standard learning algorithms are the ones dubbed as unsupervised ones when there are only raw data $\mathbf{x}_i \in \Re^m$ without the corresponding labels y_i(i.e., there is a 'no-teacher' in a shape of labels). The most popular, representative, algorithms belonging to this group are various clustering and (principal or independent) component analysis routines. They will be introduced and discussed in Chap. 6.

Recently, however, there are more and more instances in which the learning problems are characterized by the presence of (usually) a small percentage of labeled data only. In this novel setting, we may have enough information to solve a particular problem, but not enough information to solve a general problem. This learning task belongs to the so-called semi-supervised or transductive inference problems. The cause for an appearance of the unlabeled data points is usually expensive, difficult and slow process of obtaining labeled data. Thus, labeling brings the additional costs and often it is not feasible. The typical areas where this happens are the speech processing (due to the slow transcription), text categorization (due to huge number of documents, slow reading by people and their general lack of a capacity for a concentrated reading activity), web categorization, and, finally, a bioinformatics area where it is usually both expensive and slow to label a huge number of data produced.

In a semi-supervised learning problem, the algorithm must be able to utilize both labeled and unlabeled data to predict the class labels of the unlabeled points. Many semi-supervised learning algorithms have been developed

T.-M. Huang et al.: *Kernel Based Algorithms for Mining Huge Data Sets*, Studies in Computational Intelligence (SCI) **17**, 125–173 (2006)
www.springerlink.com

by using the original supervised learning algorithm as a basic classifier and they have incorporated unlabeled data with different strategies or heuristics. For example, the pre-labeling approach shown in [55] designs a set of initial classifiers from all the labeled data in the first iteration, then the second set of classifiers is constructed from the original and the newly labeled data in the second iteration. Another example will be the popular transductive SVMs (TSVM) proposed in [144] which is developed by extending the idea of margin maximization (More details can be found in Sect. 5.5.1). However, these algorithms generally converge to some locally optimal solutions instead of the globally optimal ones. This is due to the fact that the underlying cost functions of these algorithms are normally non-convex and it is not possible to find the globally optimal solution for a large problem. For example, to find the globally optimal solution for TSVM, ideally, one requires to try all the possible combination of labeling on the unlabeled data [79], but this approach becomes intractable for a data set with more than 10 data points. As a result, most of the TSVM implementations involve some heuristics to find a good local minimum. In this chapter, two very popular graph based semi-supervised learning algorithms, namely, the Gaussian random fields model (GRFM) introduced in [160] and the consistency method (CM) for semi-supervised learning proposed in [155, 158] will be presented and improved. These two algorithms are based on the theory of graphical model and they explore the manifold structure of the data set which leads to their global convergence. Because of the fact that both algorithms try to explore the manifold structure of the data set, they are dubbed as the manifold approaches when discussed together in this chapter. The manifold approaches are also motivated by the novel idea [144] of not estimating the underlying function on the whole input space (i.e., not trying to find $f(x)$ as in the supervised learning), but to directly estimate the function value on the point of interest (trying to find y_u without finding $f(x)$ for every x). This means that in a classification problem only the label (a belonging to some class) of the unlabeled data point is estimated and there is no model developed as in the supervised learning. This idea may be more suitable in the case of semi-supervised learning problem, because the amount of information available (labeled points) may only be enough to solve a particular problem (labeling the unlabeled points present) well, but not enough to solve a general problem (labeling all the possible unlabeled points).

Because none of the original manifold approaches successfully analyzes the possible problem connected with the situation where the proportion of the labeled data is unbalanced (the number of labeled data in each class differs very much), the main scope of this chapter will be on presenting the novel decision strategy (first proposed in [70, 74]) for improving the performance of the two models in the cases of unbalanced labeled data. The novel decision strategy is dubbed as normalization, because it is based on normalizing the model output. The focus of this chapter will also be on the problem of text classification, because it is where the semi-supervised learning can help to speed up and reduce the cost of developing an automatic text categorization system. The software package SemiL for solving the semi-supervised problems by implementing the manifold approaches was developed as part of this work in [75] and it will be presented at the end of the chapter. Finally, this chapter will also present and discuss some of the important issues related to its implementation.

This chapter is organized as follows:

- Section 5.2: The basic forms and concepts of the two manifold approaches namely GRFM and CM are presented here.
- Section 5.3: The effects of the unbalanced labeled data on the performance of the manifold approaches are discussed.
- Section 5.4: The novel decision rule for improving the performance of the manifold approaches is presented and the simulation results on text classification problems are shown.
- Section 5.5: The improved manifold approaches are compared with other state of the art semi-supervised learning algorithms on five different benchmarking data sets.
- Section 5.6: The important implementation details of the software package SemiL which is developed within this book is presented [75]. SemiL is the first efficient large-scale implementation of the manifold approaches introduced in this chapter.
- Section 5.7: Most of the results in this part of the work are first obtained on the text classification problem which is one of the major application for the semi-supervised learning algorithm, this section provides an overview of the problem and the pre-processing steps used in this work for the readers who do not have much backgrounds in this area.
- Section 5.8: This section concludes the presentations here and possible avenues for the further research in this novel area of semi-supervised learning.

5.2 Gaussian Random Fields Model and Consistency Method

In this section, two basic types of semi-supervised learning algorithms will be introduced. These two algorithms can be understood from the theory of regularization or from the point of view of lazy random walks. More emphasis will be on the later approach, because it is more intuitive to understand. Furthermore, it also provides a theoretical explanation to the effect of the unbalanced labeled data and the normalization which are the main themes of this chapter. However, the important properties of the algorithms deduced from the theory of regularization will still be mentioned.

5.2.1 Gaussian Random Fields Model

To begin, it is easier to first consider a binary classification with output $y \in \{-1, 1\}$. The manifold approaches treat the learning problem as a connected

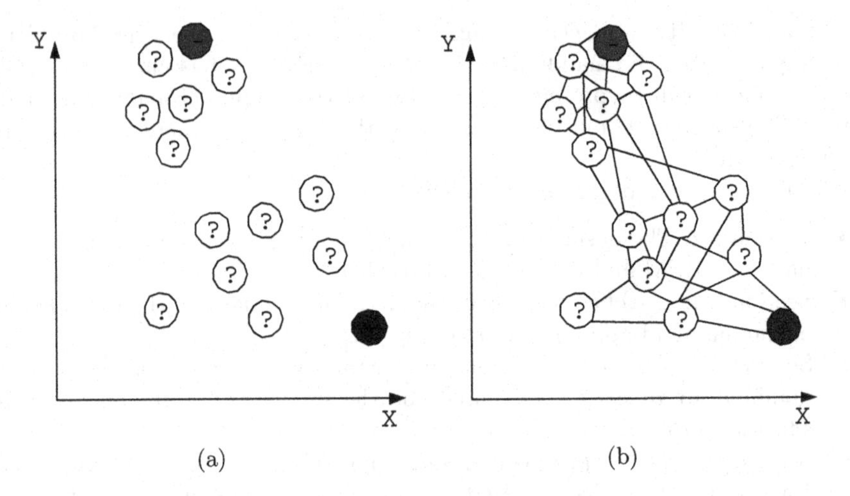

Fig. 5.1. Classification problem on a graph. Figure 5.1(a) shows the original classification problem with only two data labeled (one in positive class with blue color and another in the negative one with red color) and the rest of the points are unlabeled (with question mark inside). The data set in (a) is shown as a connected graph **G** in (b).

graph $\mathbf{G} = (\mathbf{V}, \mathbf{E})$ defined on the data set $\mathcal{X}$ (as shown in Fig. 5.1) with nodes **V** corresponding to the n data points and edges **E** (connection between the points) are weighted by an $n \times n$ symmetric affinity (or weight) matrix **W**. The first l points $\mathbf{x}_l$ are labeled, and the remaining points $\mathbf{x}_u (l + 1 \leq u \leq n)$ are unlabeled. The main idea behind GRFM is to find a real-valued function $f : \mathbf{V} \rightarrow \Re$ on **G** that does not vary much between the nearby unlabeled points (smoothness) and the value of the function on the initially labeled point f_l is equal to its initial label y_l, i.e., $f_i = y_i$ for $i = 1 \ldots l$. To ensure that f is smooth, the following quadratic energy function is used as the cost function of the algorithm [160].

$$E(f) = \frac{1}{2} \sum_{i,j=1}^{n} w_{ij}(f_i - f_j)^2 \tag{5.1}$$

It has been shown that the function f that minimizes the quadratic energy function is harmonic. This means that $(\mathbf{D} - \mathbf{W})\mathbf{f} = 0$ on the unlabeled data and $f_l = y_l$ on the labeled points [160], where $\mathbf{D} = diag(d_i)$ is the diagonal matrix with entries $d_i = \sum_j W_{ij}$. The matrix $(\mathbf{D} - \mathbf{W})$ is often referred to as the Laplacian matrix denoted as **L**.

The GRFM algorithm is shown in Algorithm 5.1 (the presentation here follows the basic model proposed in [160] tightly):

Example 5.1. The example below shows how the GRFM algorithms work. Consider a simple 1-D examples given in Fig. 5.2.

Algorithm 5.1 Gaussian Random Fields Model

1. Form the affinity matrix $\mathbf{W}$ defined by $W_{ij} = exp(-||x_i - x_j||^2/2\sigma^2)$ if $i \neq j$ and $W_{ii} = 0$.
2. Construct the diagonal matrix $\mathbf{D}$ with its (i,i)-element equal to the sum of the i-th row of $\mathbf{W}$ ($D_{ii} = \sum_{j=1}^{l} W_{ij}$, note $W_{ii} = 0$).
3. Form the following two matrices $\mathbf{W} = \begin{bmatrix} \mathbf{W}_{ll} & \mathbf{W}_{lu} \\ \mathbf{W}_{ul} & \mathbf{W}_{uu} \end{bmatrix}, \mathbf{D} = \begin{bmatrix} \mathbf{D}_{ll} & \mathbf{0} \\ \mathbf{0} & \mathbf{D}_{uu} \end{bmatrix}$ as well as the vector $\mathbf{f} = \begin{bmatrix} \mathbf{f}_l & \mathbf{f}_u \end{bmatrix}^T$, where l stands for the labeled data points and u for the unlabeled ones. Note $\mathbf{f}_l = \mathbf{y}_l$.
4. Solve for $\mathbf{f}_u$ as follows $\mathbf{f}_u = (\mathbf{D}_{uu} - \mathbf{W}_{uu})^{-1}\mathbf{W}_{ul}\mathbf{f}_l$, which is the solution for the unlabeled data points.
5. For $i = u$, if $f_i > 0$ then $y_i = +1$ otherwise $y_i = -1$.

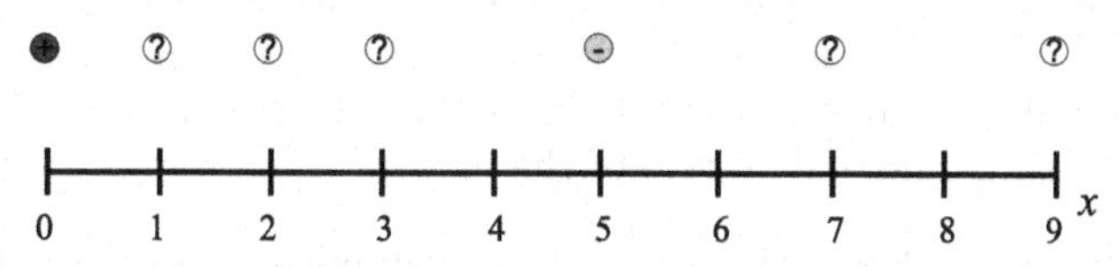

Fig. 5.2. A simple 1-D binary example. The task is to find out the labels of the unlabeled points having only one labeled point in each class.

The output $\mathbf{F}$ is calculated as follow with $\sigma = 0.8$ and $\alpha = 0.9$:

$$
\begin{aligned}
u &= \begin{bmatrix} 2\,3\,4\,6\,7 \end{bmatrix} \\
\mathcal{X} &= \begin{bmatrix} x_1\; x_2\; x_3\; x_4\; x_5\; x_6\; x_7 \end{bmatrix}^T = \begin{bmatrix} 0\,1\,2\,3\,5\,7\,9 \end{bmatrix}^T \\
\mathbf{y} &= \begin{bmatrix} 1\,0\,0\,0\,-1\,0\,0 \end{bmatrix}^T \\
\mathbf{f}_l &= \begin{bmatrix} 1\; -1 \end{bmatrix}^T \\
\mathbf{W}_{ul} &= \begin{bmatrix} 0.46 & 0.04 & 8.8e-04 & 0 & 0 \\ 0 & 8.8e-04 & 0.04 & 0.04 & 0 \end{bmatrix}^T \\
\mathbf{W}_{uu} &= \begin{bmatrix} 0 & 0.46 & 0.04 & 0 & 0 \\ 0.46 & 0 & 0.46 & 0 & 0 \\ 0.04 & 0.46 & 0 & 8.8e-04 & 0 \\ 0 & 0 & 8.8e-04 & 0 & 0.04 \\ 0 & 0 & 0 & 0.04 & 0 \end{bmatrix}
\end{aligned}
\tag{5.2}
$$

$$\mathbf{D}_{uu} = \begin{bmatrix} 0.96 & 0 & 0 & 0 & 0 \\ 0 & 0.969 & 0 & 0 & 0 \\ 0 & 0 & 0.5409 & 0 & 0 \\ 0 & 0 & 0 & 0.08 & 0 \\ 0 & 0 & 0 & 0 & 0.04 \end{bmatrix}$$

$$\mathbf{f}_u = (\mathbf{D} - \mathbf{W})^{-1}\mathbf{W}_{ul}\mathbf{f}_u = \begin{bmatrix} 0.87\ 0.76\ 0.64\ -0.99\ -0.99 \end{bmatrix}^T$$

$$\mathbf{y}_u = [+1 + 1 + 1 - 1 - 1]^T$$

The output $\mathbf{f}_u$ classifies the first three unlabeled points into the positive class and the rest into the negative one. It is interesting to point out that x_4 is closer (shorter distance) to the negative labeled point x_5, but it is still classified as belonging to the positive class. This phenomenon is due to the fact that the first four points are closer to each other, i.e., the average distance between the points is shorter and the density of data is higher from 0 to 3. Hence, they are considered by the GRFM algorithm as being in the same class. This phenomenon also demonstrates that the GRFM algorithm is concerned about finding out the manifold structure of the data.

This binary version of GRFM can be extended easily to a multi-class problem by performing the same binary classification on different classes (one-vs-all). In the next section, the multi-class version of CM will be presented.

5.2.2 Global Consistency Model

For a multi-class problem with c classes and n data points (both labeled and unlabeled), let $\mathcal{F}$ denote the set of $n \times c$ matrices with nonnegative entries. A matrix $\mathbf{F} = [\mathbf{F}_1^T, \ldots, \mathbf{F}_n^T]^T \in \mathcal{F}$ corresponds to a classification on the dataset $\mathcal{X}$ by labeling each point $\mathbf{x}_i$ as a label $y_i = j$ if $F_{ij} > F_{ik}$ $(k \neq j)$, $j = 1, \ldots, c$ (Note that $\mathbf{F}_i$ is a $1 \times c$ vector). We can understand $\mathbf{F}$ as a vectorial function $F : \mathcal{X} \to \Re^c$ which assigns a vector $\mathbf{F}_i$ of length c to each point $\mathbf{x}_i$. Define an $n \times c$ matrix $\mathbf{Y} \in \mathcal{F}$ as the initially labeled matrix with $Y_{ij} = 1$ if $\mathbf{x}_i$ is labeled as $y_i = j$ and $Y_{ij} = 0$ otherwise. Clearly, $\mathbf{Y}$ is consistent with the initial labels according the decision rule.

The complete CM algorithm is shown in Algorithm 5.2. First, one calculates a pairwise relationship $\mathbf{W}$ on the dataset $\mathcal{X}$ with the diagonal elements being zero. In doing this, one again can think of a graph $\mathbf{G} = (\mathbf{V}, \mathbf{E})$ defined on $\mathcal{X}$, where the vertex set $\mathbf{V}$ is just equal to $\mathcal{X}$ and the edges $\mathbf{E}$ are weighted by $\mathbf{W}$. In the second step, the weight matrix $\mathbf{W}$ of $\mathbf{G}$ is normalized symmetrically, which is necessary for the convergence of the iteration. The first two steps are exactly the same as in spectral clustering [155]. Note that self-reinforcement is avoided since the diagonal elements of the affinity matrix are set to zero in the first step ($\mathbf{W}_{ii} = 0$). The model labels each unlabeled point and assigns it to the class for which the corresponding $\mathbf{F}^*$ value is the biggest, as given in step 4 of Algorithm 5.2. The CM shown in Algorithm 5.2 is referred to as the one class labeling (or the original CM) version, because

Algorithm 5.2 Global Consistency Model

1. Form the affinity matrix $\mathbf{W}$ (same as in Algorithm 5.1.).
2. Construct the matrix $\mathbf{S} = \mathbf{D}^{-1/2}\mathbf{W}\mathbf{D}^{-1/2}$ in which $\mathbf{D}$ is a diagonal matrix with its (i,i)-element equals to the sum of the i-th row of $\mathbf{W}$.
3. Iterate $\mathbf{F}(t+1) = \alpha\mathbf{S}\mathbf{F}(t) + (1-\alpha)\mathbf{Y}$ until convergence, where α is a parameter in (0, 1) and t is the current step. Alternatively, one can solve the system of equation $(\mathbf{I} - \alpha\mathbf{S})\mathbf{F}^* = \mathbf{Y}$ for $\mathbf{F}$ without performing the iteration.
4. Let $\mathbf{F}^*$ denotes the limit of the sequence $\{F(t)\}$. Label each point $\mathbf{x}_i$ as a label $y_i = j$ if $F_{ij} > F_{ik}$ $(k \neq j)$, $j = 1, \ldots, c$, $i = 1, \ldots, n$.

in the j-th column of $\mathbf{Y}$ only the labeled points in the j-th class are different from zero. An alternative labeling method is to assign 1 for the labeled points in the j-th class, -1 for all the other labeled ones in other classes and 0 for all the unlabeled ones. CM using such a labeling method will be referred to as the two-class labeling CM in the rest of the chapter.

The iterative algorithm can be considered as trying to find a function $\mathbf{F}$ that minimizes the cost function derived in [157] as follows:

$$\min \quad \mathcal{Q}(F) = \frac{1}{2}\left(\sum_{i,j=1}^{n} W_{ij}\left\|\frac{1}{\sqrt{D_{ii}}}F_i - \frac{1}{\sqrt{D_{jj}}}F_j\right\|^2 + \lambda\sum_{i=1}^{n}\|F_i - Y_i\|^2\right). \tag{5.3a}$$

Note that in a matrix notation, the cost function is given as,

$$\min \quad \mathcal{Q}(\mathbf{F}) = \frac{1}{2}(\mathbf{F}^T\mathcal{L}\mathbf{F} - \mathbf{Y}^T\mathbf{F}). \tag{5.3b}$$

where $\mathcal{L}$ denotes the normalized Laplacian matrix and it is given as $\mathcal{L} = (\mathbf{I} - \alpha\mathbf{S})$. The first term on the right-hand side of (5.3a) is often referred to as the 'smoothness constraint' which ensures that function $\mathbf{F}$ does not vary much between nearby points. The second term which is often referred to as the 'fitting constraint' ensures that the classification function $\mathbf{F}$ will not disagree too much with the labels of the initially labeled data, i.e., it controls how well the classification fits the initially labeled data. The trade-off between these two terms is controlled by a positive parameter λ and it has been shown that the relationship between α and λ is $\alpha = \frac{1}{1+\lambda}$. Note that in such a formulation [157], both labeled and unlabeled data are within the fitting constraint.

The CM problems are solved in an iterative way only when the size of the problem is large (the amount of memory is not sufficient to store the $\mathbf{S}$ matrix). In general, the corresponding system of linear equations (5.5) which is derived by differentiating $\mathcal{Q}$ with respect to $\mathbf{F}$ as shown in (5.4) is solved by using conjugate gradient method which is a highly recommended approach for dealing with huge data sets. Also, instead of using the complete graph, the $\mathbf{W}$ matrix is approximated by using only the k-nearest neighbors. This step decreases the accuracy only slightly, but it increases the calculation speed significantly.

$$\left.\frac{\partial Q}{\partial \mathbf{F}}\right|_{\mathbf{F}=\mathbf{F}^*} = \mathbf{F}^* - \mathbf{S}\mathbf{F}^* + \lambda(\mathbf{F}^* - \mathbf{Y}) = 0 \quad (5.4)$$

$$\Rightarrow (\mathbf{I} - \alpha\mathbf{S})\mathbf{F}^* = \mathbf{Y} \quad (5.5)$$

$$\text{where } \alpha = \frac{1}{1+\lambda} \quad (5.6)$$

Example 5.2. The same 1-D problem in Example 5.1 is used again here to show how the CM algorithm work.

The output $\mathbf{F}$ is calculated as follow with $\sigma = 0.8$ and $\alpha = 0.9$:

$$\mathcal{X} = \begin{bmatrix} x_1 \, x_2 \, x_3 \, x_4 \, x_5 \, x_6 \, x_7 \end{bmatrix}^T = \begin{bmatrix} 0 \, 1 \, 2 \, 3 \, 5 \, 7 \, 9 \end{bmatrix}^T$$

$$\mathbf{Y} = \begin{bmatrix} 1\,0\,0\,0\,0\,0\,0 \\ 0\,0\,0\,0\,1\,0\,0 \end{bmatrix}^T$$

$$\mathbf{W} = \begin{bmatrix} 0 & 0.46 & 0.04 & 8.8e-04 & 0 & 0 & 0 \\ 0.46 & 0 & 0.46 & 0.04 & 4.0e-06 & 0 & 0 \\ 0.04 & 0.46 & 0 & 0.46 & 8.8e-04 & 0 & 0 \\ 8.8e-04 & 0.04 & 0.46 & 0 & 0.04 & 4.0e-06 & 0 \\ 0 & 4.0e-06 & 8.8e-04 & 0.04 & 0 & 0.04 & 4.0e-06 \\ 0 & 0 & 0 & 4.0e-06 & 0.04 & 0 & 0.04 \\ 0 & 0 & 0 & 0 & 4.0e-06 & 0.04 & 0 \end{bmatrix}$$

$$\mathbf{D} = \begin{bmatrix} 0.5 & 0 & 0 & 0 & 0 & 0 & 0 \\ 0 & 0.96 & 0 & 0 & 0 & 0 & 0 \\ 0 & 0 & 0.96 & 0 & 0 & 0 & 0 \\ 0 & 0 & 0 & 0.54 & 0 & 0 & 0 \\ 0 & 0 & 0 & 0 & 0.08 & 0 & 0 \\ 0 & 0 & 0 & 0 & 0 & 0.08 & 0 \\ 0 & 0 & 0 & 0 & 0 & 0 & 0.04 \end{bmatrix}$$

$$\mathbf{F} = (\mathbf{I} - \alpha\mathbf{D}^{-\frac{1}{2}}\mathbf{W}\mathbf{D}^{-\frac{1}{2}})^{-1}\mathbf{Y}$$

$$= \begin{bmatrix} 2.5182 \; 2.3764 \; 1.8791 \; 1.2593 \; 0.3348 \; 0.2519 \; 0.1603 \\ 0.3348 \; 0.5045 \; 0.6282 \; 0.6766 \; 1.6857 \; 1.2678 \; 0.8069 \end{bmatrix}^T$$

$$\mathbf{y} = [+1 + 1 + 1 + 1 - 1 - 1 - 1]^T$$

The output $\mathbf{F}$ classifies the first four points into the positive class and the rest into the negative one. It is interesting to point out that x_4 is closer (shorter distance) to the negative labeled point x_5, but it is still classified as belonging to the positive class. This phenomenon is due to the fact that the first four points are closer to each other, i.e., the average distance between the points is shorter and the density of data is higher from 0 to 3. Hence, they are considered by the CM algorithm as being in the same class. This phenomenon also demonstrates that the CM algorithm is concerned about finding out the manifold structure of the data as the GRFM algorithm.

To further demonstrate the properties of CM and the difference between the semi-supervised learning algorithms and the supervised ones, a more complicated toy example which is very similar to the one presented in [155] is shown in Fig. 5.3. This toy problem is known as the two moons problem and Fig. 5.3(a) shows the ideal solution for the problem. Figure 5.3(b) shows the same problem in a typical semi-supervised learning setting where the number of labeled points is much smaller than the number of unlabeled ones. In this particular example, only one point from each class is labeled and the rest of the data is unlabeled. The learning task is to find the labels for all the unlabeled points using the algorithm presented in this section. Figures 5.3(c) to 5.3(e) show how the labeling done by an iterative version of CM propagates as the number of iterations increases (where t denotes the time step). At the beginning of the iterations, the unlabeled data points are labeled by their closeness to the initially labeled point from each class. This can be observed from Fig. 5.3(d) where the separation between the two classes is in the middle of the two initially labeled points and perpendicular to the horizontal axis (at approximately 1). As the algorithm keeps converging towards the final solution, the influence of the unlabeled data which are labeled in the previous iterations (newly labeled data) becomes stronger. This ability of utilizing the information of the newly labeled data is due to the $\alpha\mathbf{SF}(t)$ term in the step 3 of Algorithm 5.2 and it helps the algorithm to explore the half moon shape (the manifold structure) of the problem. Therefore, it is clear that a perfect labeling (separation) can possibly be done. In contrast, the solution from a SVM which is based only on the initially labeled data (one from each class) is a decision boundary perpendicular to the horizontal axis and it fails to capture the half moon shape.

5.2.3 Random Walks on Graph

As mentioned previously, the graph-based approaches studied in this work can also be interpreted by random walks as shown in [156] for CM and [160] for GRFM. To help understanding the basic idea of random walks, a simple 1-D example (similar to the one in [6]) is shown in Fig. 5.4.

Example 5.3. Suppose one is interested in modeling the horizontal propagation of a bee between its hive and a flower. Assume there are 12 meters between the hive (x_1) and the flower (x_6). This distance is divided into four zones $x_2, \ldots, x_5$ where the bee is likely to stay before reaching the flower or the hive. Every minute the bee behaves in one of five ways, each with a given probability:

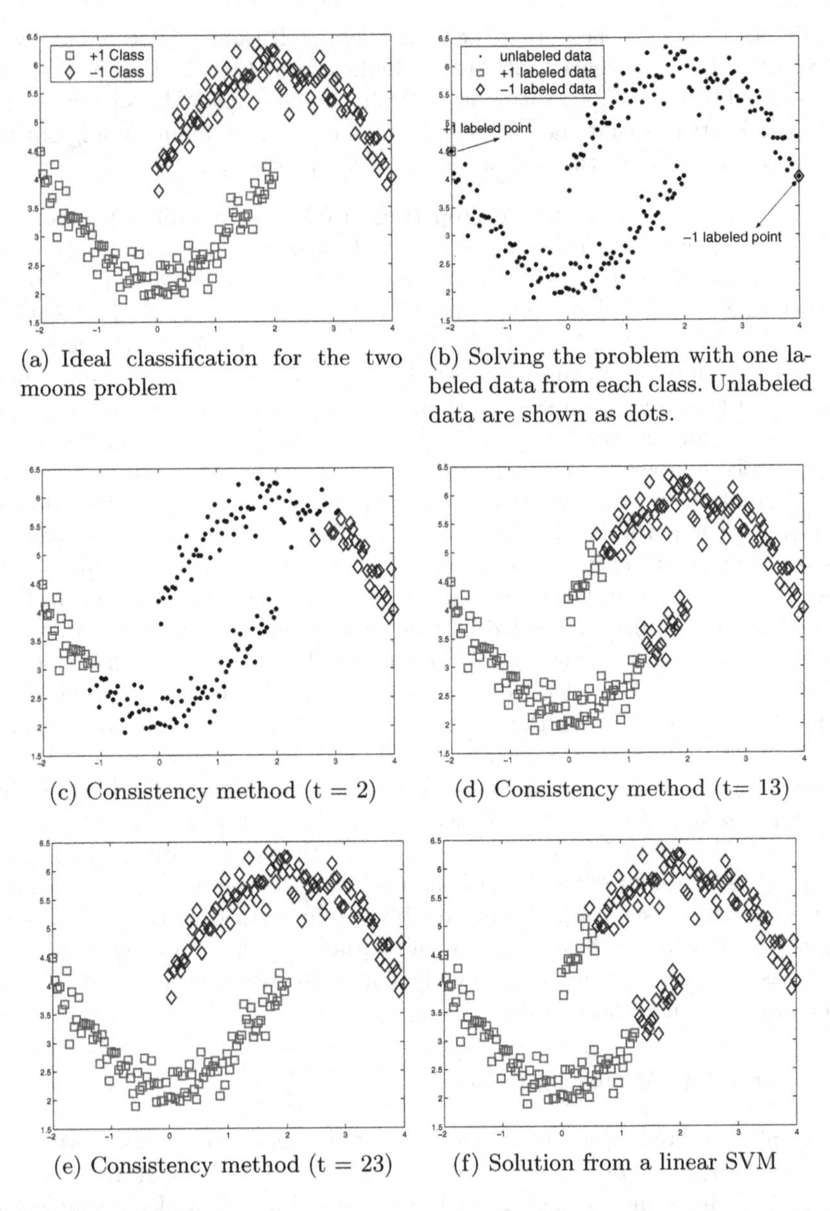

(a) Ideal classification for the two moons problem

(b) Solving the problem with one labeled data from each class. Unlabeled data are shown as dots.

(c) Consistency method (t = 2)

(d) Consistency method (t= 13)

(e) Consistency method (t = 23)

(f) Solution from a linear SVM

Fig. 5.3. Classification of the two moons problem. (a) Ideal classification. (b)-(e) Iteration of CM. (f) The solution of a linear SVM.

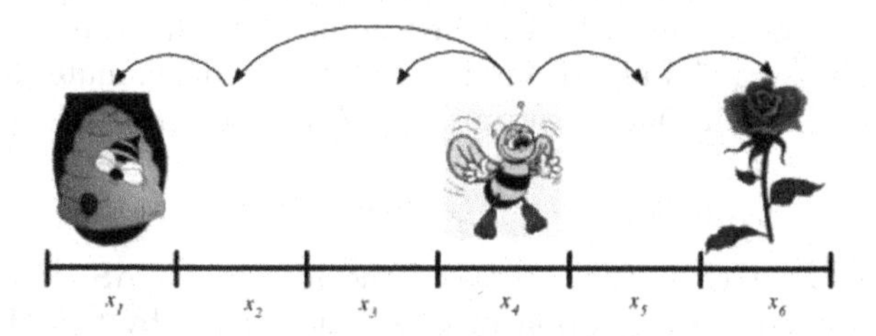

Fig. 5.4. A 1-D random walk

1. The bee flies toward the zone closer to the hive with probability of $\frac{1}{3}$.
2. The bee flies toward the zone closer to the flower probability of $\frac{1}{2}$.
3. The bee stay in the same zone with probability of $\frac{1}{6}$.
4. If the bee is in the flower or its hive, it will have probability of $\frac{1}{3}$ to leave.
5. If the bee is in the flower or its hive, it will have probability of $\frac{2}{3}$ of staying.

In this setting, it is possible to estimate where the bee is more likely to reach first given the initial position of the bee after certain amount of time. For example, if the bee has an initial position at x_4, then this initial condition can be represented by an output vector $\mathbf{y}_0 = [0, 0, 0, 1, 0, 0]$. As a result, the probabilities of its position after one minute are given by the vector $[0, 0, \frac{1}{2}, \frac{1}{6}, \frac{1}{3}, 0]$ and after two minutes by $[0, \frac{1}{4}, \frac{1}{6}, \frac{13}{36}, \frac{1}{9}, \frac{1}{9}]$. The probabilities for the position of the bee after s minutes can be computed by the following equation:

$$\mathbf{y}_s = \mathbf{y}_0 \mathbf{P}^s = \begin{bmatrix} 0 & 0 & 0 & 1 & 0 & 0 \end{bmatrix} \begin{bmatrix} \frac{2}{3} & \frac{1}{3} & 0 & 0 & 0 & 0 \\ \frac{1}{2} & \frac{1}{6} & \frac{1}{3} & 0 & 0 & 0 \\ 0 & \frac{1}{2} & \frac{1}{6} & \frac{1}{3} & 0 & 0 \\ 0 & 0 & \frac{1}{2} & \frac{1}{6} & \frac{1}{3} & 0 \\ 0 & 0 & 0 & \frac{1}{2} & \frac{1}{6} & \frac{1}{3} \\ 0 & 0 & 0 & 0 & \frac{1}{3} & \frac{2}{3} \end{bmatrix}^s . \tag{5.7}$$

The matrix $\mathbf{P}$ is known as the transition matrix for a random walk. Each row contains non-negative numbers, called transition probabilities. In this setting, one can find the expected number of steps for a random walk starting at some initial position x_i to reach x_j and to return, e.g, the expected number of minutes for a bee to reach flowers from its hive and to return. This expectation is often referred to as the commute time between the two positions and denoted by C_{ij}.

The CM presented in Sect. 5.2.2 can be interpreted as a random walk on graph with the following transition matrix as shown in [156]:

$$\mathbf{P} = (1 - \alpha)\mathbf{I} + \alpha \mathbf{D}^{-1}\mathbf{W}, \tag{5.8}$$

where α is a parameter in (0,1). But instead of using the commute time as a measure of closeness, CM algorithm uses a normalized commute time $\bar{C}_{ij}$ and it has also been shown that the normalized commute time is given as

$$\bar{C}_{ij} \propto \bar{G}_{ii} + \bar{G}_{jj} - \bar{G}_{ij} - \bar{G}_{ji} \quad if\ x_i \neq x_j, \tag{5.9}$$

where $\bar{G}_{ij} \in \bar{\mathbf{G}}$ which is the inverse of the matrix $\mathbf{I} - \alpha\mathbf{S}$. If we are now considering a binary classification in CM that is given by $\mathbf{f} = (\mathbf{I} - \alpha\mathbf{S})^{-1}\mathbf{y}$ where y is either 1 or -1, then the classification will be based on the comparison between $p_+(x_i) = \sum_{j|y_j=1} \bar{G}_{ij}$ and $p_-(x_i) = \sum_{j|y=-1} \bar{G}_{ij}$ [156]. This means that we are labeling an unlabeled point by summing up and comparing the normalized commute times of the point to all the positive labeled points and to all the negative labeled points. If an unlabeled point has a negative value of f, then it will mean that less steps are required to reach the unlabeled point from the negative labeled points. GRFM methods can be interpreted in a similar way, however it uses hitting times [160], instead of the normalized commute times between the labeled and the unlabeled points.

The use of random walk on a graph for semi-supervised learning has been successful and it is the fundamental idea behind the two algorithms discussed. However, classification based on comparing $p_-(x_i)$ and $p_+(x_i)$ for the unlabeled data has an obvious problem. Namely, the number of labeled data in each particular class can have an effect on the size of $p_-(x_i)$ or $p_+(x_i)$. The effect of unbalanced labeled data to CM and GRFM is quite significant and it will be discussed in the next section.

5.3 An Investigation of the Effect of Unbalanced labeled Data on CM and GRFM Algorithms

5.3.1 Background and Test Settings

The extensive simulations on various data sets (as presented in [155]) have indicated that both CM and GRFM behave similarly and according to the expectations that with an increase in the number of labeled data points l, the overall models' accuracies improve too. There was just a slightly more superior performance of CM from [155] with respect to GRFM, when faced with a small number of unbalanced labeled data. At the same time, the later model performs much better when only 1 labeled data in each class is available (note that labeled data is balanced in this particular case).

Such a behavior needed a correct explanation and it asked for further investigations during which several phenomena have been observed. While working with the balanced labeled data (meaning with the same number of labeled data per class) both models perform much better than in the case when the data is not balanced. Furthermore, the performance improvement by using the balanced labeled data can be as large as 50% [70, 74]. To demonstrate this

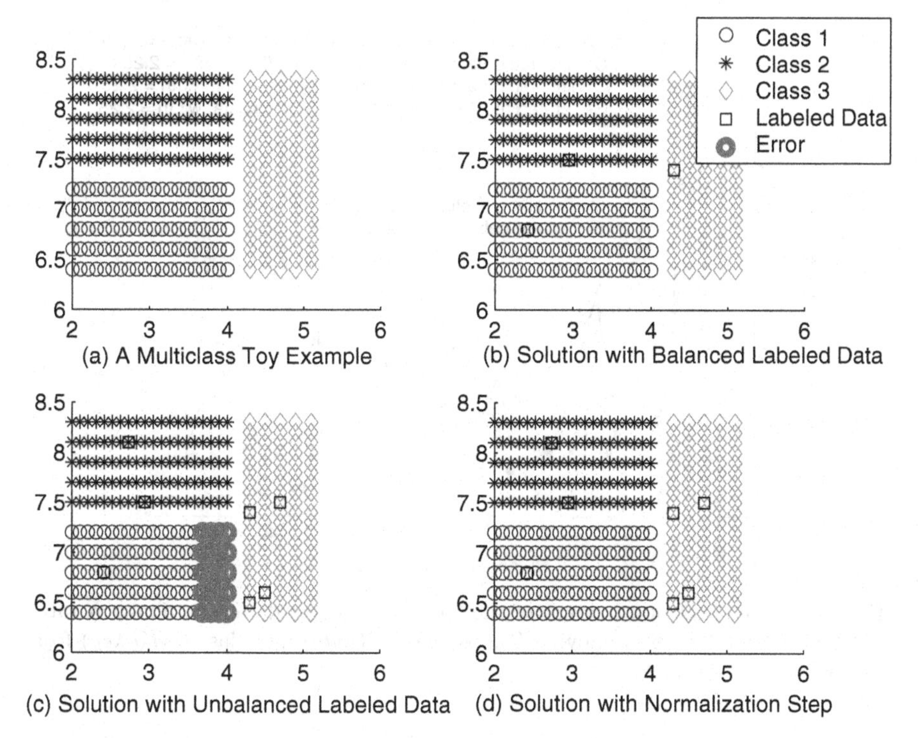

Fig. 5.5. (a) A multi-class toy example for demonstrating the effect of unbalanced labeled data. (b) With three balanced labeled data, the solution is perfect. (c) With seven unbalanced labeled data, the performance is worse than in (b). (d) The normalization step introduced in Sect. 5.4 corrects the problem of an unbalance in labeled data.

phenomenon, a simple 2D multi-class example is constructed and it is shown in Fig. 5.5. It is clear that with unbalanced labeled data, the performance of the algorithm can deteriorate significantly despite the fact that more labeled data is available. To further test this phenomenon on real world data sets, *rec* data set is used and the simulation results using the CM algorithm are shown in Fig. 5.6.

Rec data set is a subset of the popular 20-newsgroup (version 20-news-18828) text classification data set and it contains four topics, *autos*, *motorcycles*, *baseball*, and *hockey*. The articles were processed by the Rainbow software package with the following options as discussed in Sect. 5.7: (1) passing all words through the Porter stemmer before counting them; (2) tossing out any token which is on the stop list of the SMART system; (3) skipping any headers; (4) ignoring words that occur in 5 or fewer documents. No further pre-processing was done. After removing the empty documents, the data set consists of 3970 document vectors in a 8014-dimensional space. Finally the documents were normalized into TFIDF representation. The cosine distance

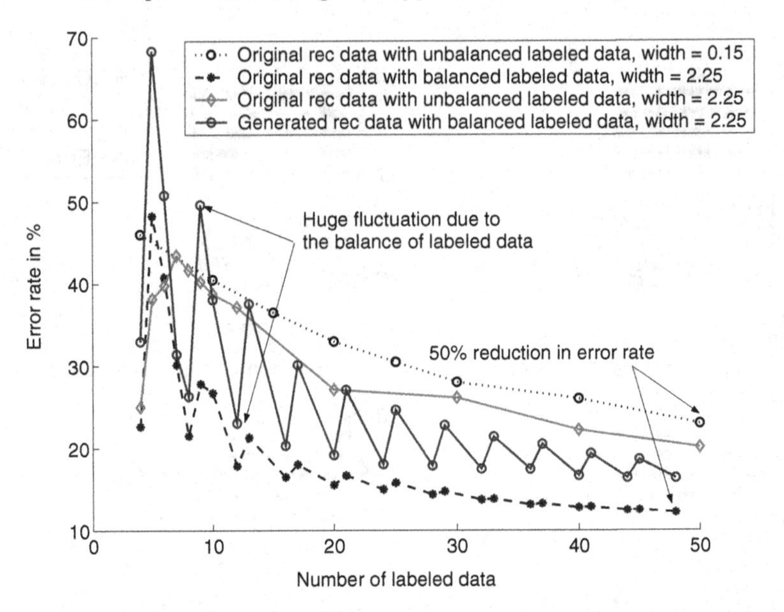

Fig. 5.6. The effect of a balance of the labeled data (*rec* data set). The generated *rec* data is produced by removing 50% of class 1 (*auto*) and class 4 (*hockey*) from the original *rec* data set.

between points was used as a measure of distance [20] here in forming matrix $\mathbf{W}$ (i.e., $W_{ij} = \exp(-d(\mathbf{x}_i, \mathbf{x}_j)/2\sigma^2)$ where $d(\mathbf{x}_i, \mathbf{x}_j) = 1-(\mathbf{x}_i^T\mathbf{x}_j/\|\mathbf{x}_i\|\,\|\mathbf{x}_j\|)$). The mentioned procedure is the same as in [155] just in order to ensure the same experiment's setting for the same data set.

In order to test the connection between the balance of the labeled data and the proportion of the unlabeled data in each class, another data set is generated from the original *rec* data set by removing 50% of class 1 (*auto*) and class 4 (*hockey*). As a result, the ratio of the unlabeled data in each class is 16.6% for class 1, 33.33% for class 2, 33.33% for class 3 and 16.6% for class 4. This proportion is different from the proportion of labeled data in each class when the labeled data is balanced (25% in each class).

Two different types of selection processes were used to select the labeled data. The balanced results shown in Fig. 5.6 were achieved by using a selection process which keeps the number of labeled data selected from each class as close to each other as possible. For examples, when five labeled points need to be selected, the selection process selects one from each class randomly first, then the fifth one is selected randomly from all the classes. As a result, the labeled data is still close to balance. The unbalanced results in Fig. 5.6 are obtained by randomly selecting labeled data from all classes. This selection procedure is the same as the one [155] and it will make the number of labeled data in each class being different (i.e., unbalanced).

5.3.2 Results on the *Rec* Data Set

The first curve (dotted-circles) in Fig. 5.6 is directly taken from [155] with a rather small σ value of 0.15. Working with such a small σ value, does not make the effect of the balance of the labeled data visible and the algorithm behaves according to the expectation that with more labeled points available, the error rate should decrease. A larger σ is used here to show the effect of the unbalanced labeled data. By comparing the lowest error rate of the first curve (dotted-circles) from [155] and the one with better balanced labeled data (the second curve (dashed-circles) in Fig. 5.6) using the original *rec*, it is clear that the minimum error rate reduces significantly from 25% (when 50 unbalanced labeled data are used) to 12% (when 48 balanced labeled data is available). The use of both balanced labeled data and a larger σ parameter improves the performance of the algorithm on this data set, but the former has stronger influence, because the performance of the algorithm with $\sigma = 2.25$ and the unbalanced labeling (the third curve (solid-diamonds) in Fig. 5.6) is similar to the one from [155].

An important observation is that the error rate for the CM algorithm increases sharply as soon as the labeled data become unbalanced. This phenomenon can be clearly observed from the increase in error rate between 4 labeled points (balanced) to 5 labeled points (unbalanced) in the second and the last curve in Fig. 5.6. It means that the method becomes very sensitive to the balance of the labeled data. Another important result is that this fluctuation also occurs on the generated *rec* data set where the four classes are not equal in size: the model with four balanced labeled data outperforms the model with five unbalanced labeled data as shown in the last curve of Fig. 5.6. This means that the improvement in the performance may occur even if the proportion of the labeled data in each class is not the same as the proportion of the unlabeled data in each class, i.e., it is the balance of the labeled data and the size of the σ parameter that produces improvements.

5.3.3 Possible Theoretical Explanations on the Effect of Unbalanced Labeled Data

Further investigation on the effect of the balance of the labeled data shows that a class with less labeled data will be more disadvantageous. It was found [70, 74] that a class j with more labeled data will be more likely to have a higher mean of $\mathbf{F}_j^*$ (where $\mathbf{F}_j^*$ is the j-th column vector of the output matrix $\mathbf{F}^*$ and its mean is given as $\bar{F}_j^* = \sum_{i=1}^{n} F_{ij}^*/n$) than the class with less labeled data when the σ parameter is relatively large. Recall that $\mathbf{F}_j^*$ is the output of class j's classifier as well as that the labeling of the i-th data is based on the biggest value in a row $\mathbf{F}_i^*$. This means that an unlabeled data will more likely be assigned to the class which has a higher $\bar{F}_j^*$ than to a disadvantageous class with lower $\bar{F}_j^*$. To demonstrate this phenomenon, the original *rec* data set is used again and a different number of labeled data is selected from each class.

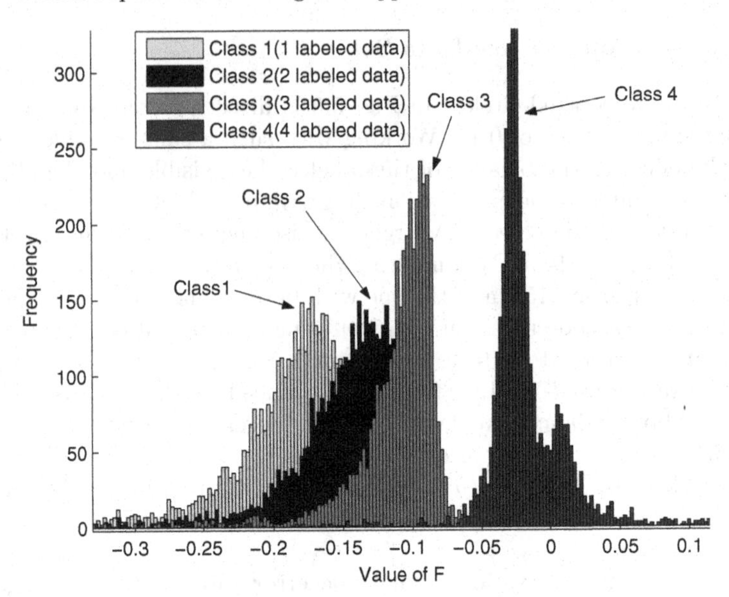

Fig. 5.7. Histogram of output $\mathbf{F}_j^*$ from CM for each class with unbalanced labeled data. Each $\mathbf{F}_j^*$ is represented by different colors. (*Rec* data set, $\sigma = 2.25$, Error rate $= 73.19\%$)

As shown in Fig. 5.7, the distribution of $\mathbf{F}_j^*$ is centered around its mean $\bar{F}_j^*$ and class 4 (with 4 labeled data) has a much higher mean $\bar{F}_4^*$ than class 1 (with 1 labeled data). This difference between the means of the distributions makes class 1 much more disadvantageous than class 4. This is due to the fact that a data point $\mathbf{x}_i$ will only be classified as class 1 if the value of F_{i1}^* is higher than F_{i4}^*. However this is unlikely to happen because $\bar{F}_1^*$ is much less than $\bar{F}_4^*$ and the values of $\mathbf{F}_1^*$ are distributed around its mean $\bar{F}_1^*$. It is this unfairness between the output of each classifier that makes the error rate 73.19% and almost all of the points are classified as class 4. Similar phenomenon can also be found in GRFM as shown in Fig. 5.8.

Figure 5.8 shows the output of GRFM on the same problem with the same points being labeled as in Fig. 5.7. The effect of unbalanced labeled points is even more significant in GRFM. This can be easily seen comparing the distances in Figs. 5.8 and 5.7. Namely, the distance between the class 1 and the class 4's centers of distributions is approximately 7 times larger in Fig. 5.8 ($d \cong 1$) than in Fig. 5.7 ($d \cong 0.15$).

The effect of the unbalanced labeled data on the mean of the classifier can be explained by interpreting CM algorithms as random walks on graph and GRFM as absorbing probability of standard random walk. As discussed in Sect. 5.2.3, a binary classification using $\mathbf{f} = (\mathbf{I} - \alpha\mathbf{S})^{-1}\mathbf{y}$ is based on the comparison between $p_+(x_i) = \sum_{j|y_j=1} \bar{G}_{ij}$ and $p_-(x_i) = \sum_{j|y=-1} \bar{G}_{ij}$ [156], i.e., on the total normalized commute times between positive and negative

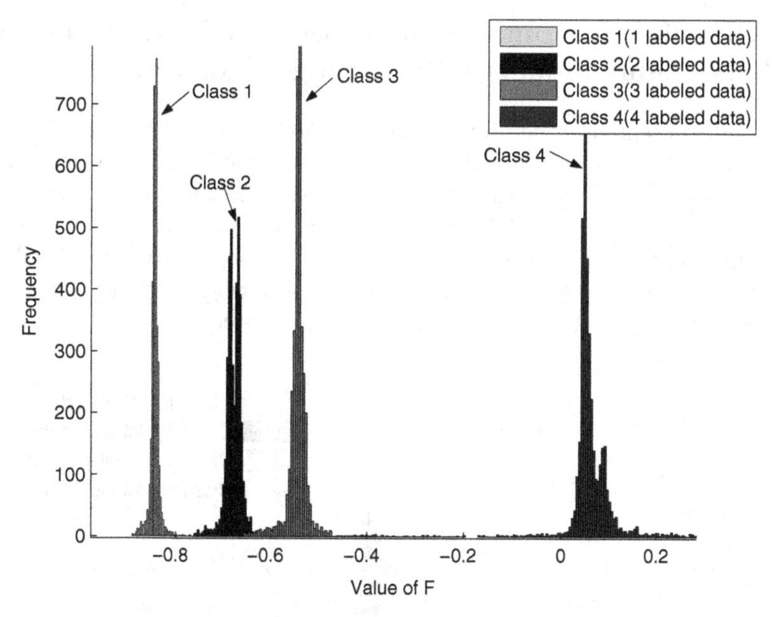

Fig. 5.8. Histogram of output $\mathbf{F}_j^*$ from the GRFM of different classes with unbalanced labeled data. (*Rec* data set, $\sigma = 2.25$, Error rate = 74.73%)

labeled data. With more positive labeled points than the negative ones, the mean of $\mathbf{f}$ will more likely be greater than zero and vice verse. Hence, it is more likely to assign an unlabeled point to the class with more labeled points. If we now consider solving a multi-class problem using several binary classifiers, then a binary classifier with less number of positive labeled points will be more disadvantaged than others, because the mean of its output $\mathbf{F}_j^*$ will be lower. Thus, the class with less labeled points will be handicapped. The GRFM algorithms use similar principle for labeling unlabeled data. However, one of the key differences between GRFM and CM is that the former uses the weighted hitting times between the point of interest and the rest of the unlabeled points. These hitting times are weighted and summed together to compare with outputs from other classifiers. Because the weights applied to the hitting times are controlled by the labeled points (the $\mathbf{W}_{ul}\mathbf{f}_l$ term), the effect of unbalanced labeled points will be magnified through the summation over all the unlabeled points.

In summary, with balanced labeled data and larger σ parameter, the performance of the manifold approaches can be much better than the results presented in [155] on the *rec* data set. However, the performance of the algorithm deteriorates as the labeled data become unbalanced. This led to the introduction of the normalization step proposed in [70, 74] as a novel decision rule to reduce the effect of unbalanced labeled data.

5.4 Classifier Output Normalization: A Novel Decision Rule for Semi-supervised Learning Algorithm

The result presented in the previous section shows that CM can perform much better if the labeled data is balanced. However, such a condition means that the algorithm may not utilize all the information (labeled data) available for a given problem, i.e., some labeled points may be eliminated in order to make the labeled data balanced. Looking from another perspective, the result also suggests that it is possible to achieve better performance if the effect of the unbalanced labeled data can be reduced or eliminated.

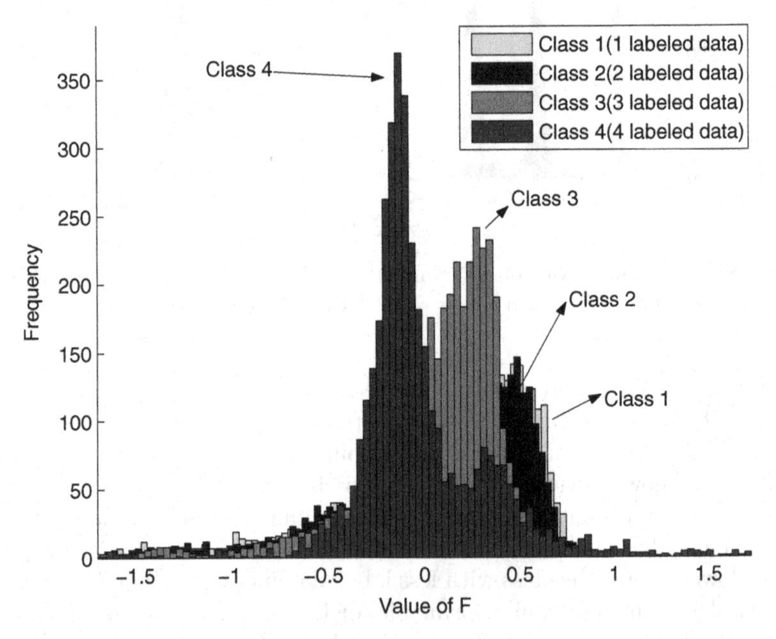

Fig. 5.9. Histogram of output F from CM of different classes with unbalanced labeled data after normalization. (Rec data set, $\sigma = 2.25$, Error rate = 25.86%)

To reduce the effect of unbalanced labeled data, a normalization step is introduced in [70, 74] for the elements of the column vectors $\mathbf{F}_j^*$ bringing them to the vectors $\hat{\mathbf{F}}_j^*$ with a mean = 0, and with a standard deviation = 1 as given below,

$$\hat{F}_{ij}^* = \frac{F_{ij}^* - \bar{F}_j^*}{s_j} \quad \text{where } s_j = \sqrt{\frac{\sum_{i=1}^{n}(F_{ij} - \bar{F}_j)^2}{n}} \tag{5.10}$$

is the standard deviation of $\mathbf{F}_j$.

Only now, after the normalization is performed, the algorithm searches for the maximal value along the rows of the standardized output matrix $\hat{\mathbf{F}}^*$ and labels the unlabeled i-th data to the class j if $\hat{F}^*_{ij} > \hat{F}^*_{ik}, k = 1, \ldots c,\ k \neq j$. The philosophy of adding this extra step is to make sure the classifier output $\mathbf{F}^*_j$ from each class is treated equally. The introduction of this normalization shifts the distribution of all the column vectors of $\mathbf{F}^*_j$ closer together, so the classes with more labeled points will no longer be advantageous. This can be shown by comparing Fig. 5.9 and Fig. 5.7. The normalization step reduces the error rate of the CM algorithm significantly from previously 73.19% (in Fig. 5.7) to 25.86% (in Fig. 5.9) in the example where 50 data are randomly labeled, i.e., the labeled data is (extremely) unbalanced. To test the normalization step in more general cases, the same simulations are conducted on the normalised CM and GRFM as in Sect. 5.3 on the *rec* data set and the results are presented in Fig. 5.10.

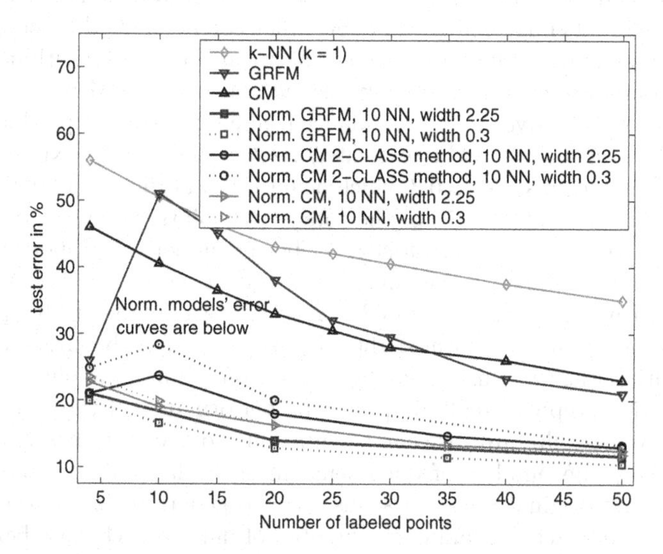

Fig. 5.10. The error rates of text classification with 3970 document vectors in an 8014-dimensional space for *rec* data sets from version 20-news-18828. At least one data in each class must be labeled. Thus the smallest number of labeled data here is 4. The normalized version of CM and GRFM algorithms outperforms the algorithms without normalization. 10 NN indicates that only 10 nearest neighbors are used for calculations of a matrix $\mathbf{W}$ making the computing time feasible.

Several interesting phenomena can be observed in Fig. 5.10. First, the normalization improves the performance of both methods very significantly. This can be observed easily by comparing the error rate between the model with and without normalization. The error rate of CM for four labeled points drops from 46% to 22%. When 50 points are labeled, the error rate drops from around 22% to about 13% and similar improvements can be found on the GRFM.

The only exception is in the case of the later method when there are only four labeled points available. In this situation, the error rate of GRFM is already much lower than the CM's one, even without the normalization, therefore the improvement by the normalization is not as significant as in other cases. This is a consequence of having balanced labeled data points from each class (1 in each class) as discussed in the previous section. Without the normalization, GRFM needs approximately forty unbalanced labeled points to match its very performance when having four balanced labeled points only. In contrast, the performance of the normalized model with ten unbalanced labeled data outperforms the result for the four balanced points. With a normalization step, GRFM seems to be slightly better than CM. This is not the case while working without the normalization as shown in [155]. Also, the normalized two-class labeling CM performs worse than the normalized original (one class labeling) CM (See the classification of the CM labeling in Sect. 5.2.2). By comparing the performances between the two algorithms when 10 labeled points are available, it looks as the normalized CM with original labeling, is less sensitive to the unbalanced labeled data and produces better result. The best model for the text categorization data in our experiments is a GRFM with width equal to 0.3 which achieves an accuracy of 90% with only 50 labeled points out of 3970 of the total data points. However, it should be said that it seems as CM has a slightly better numerical properties because the conditional numbers of the matrices involved are better. For both methods, with a normalization step of $\mathbf{F}^*$, models with smaller width parameter of a Gaussian kernels perform slightly better than with the larger widths.

At initial stage of this work, there were some doubts about whether the normalization step can really improve the performance of the algorithms, or the improved results are just coincidental due to the fact that the *rec* data set has the same number of documents in each class. To investigate these problems, the original *rec* data set is used to generate three more different unbalanced data sets (meaning the number of data in each class being different) and the simulation results on these three unbalanced data sets are shown in Fig. 5.11.

As shown in Fig. 5.11, the normalized models perform better than the models without the normalization and the overall trend is very similar to Fig. 5.10. The only exception is that the performance gap between the normalized GRFM and the normalized original CM is slightly larger. It is because the performance of the normalized GRFM is virtually unchanged between the balanced and the unbalanced data sets, whereas the normalized CM has a slight increase in the error rate.

In terms of the computational time, for a 3970 data, the learning run based on a conjugate gradient algorithm takes only about 25 seconds of CPU time on a 2GHz laptop machine for 100 random tests runs with width of RBF equal to 0.3 using the software package SemiL developed in [75] as part of this work. However, it had been found that the computational time required for GRFM is more sensitive to the width of the Gaussian kernel and it is usually bigger

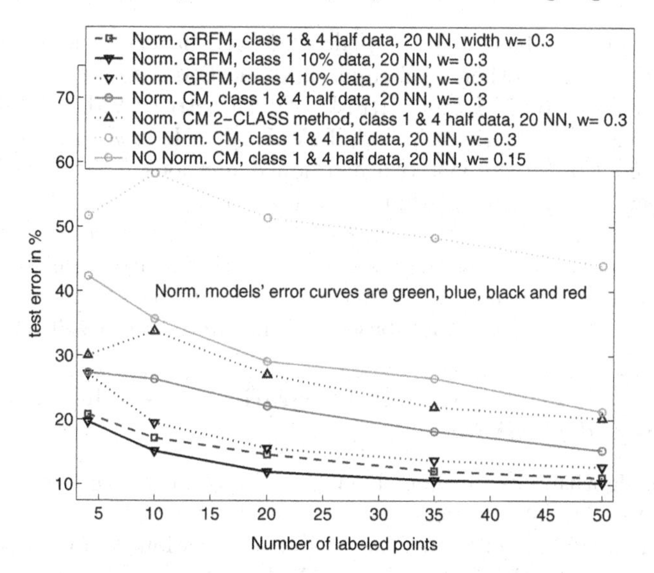

Fig. 5.11. The error rates of various generated and unbalanced *rec* data sets. The first unbalanced data set is generated by removing half of the first class and half of the fourth class. The rest are generated by removing either 90% of the first class or 90% of the fourth class. The widths of the RBFs equal 0.3 in all experiments but in one. Twenty nearest neighbors only are used to construct the **W** matrix.

than the time needed for CM. With a small width, the condition number of the Laplacian matrix can be quite high, hence more computational time is required. As a result, for huge data sets, CM may be more efficient (more details in Sect. 5.6.4).

In order to further test the idea of normalization and the performance of the two manifold approaches discussed in this work in relation with other known approaches, extensive simulations are carried out on five additional benchmarking data sets and the results obtained are presented and compared with other semi-supervised learning algorithms in the next section.

5.5 Performance Comparison of Semi-supervised Learning Algorithms

In the previous section, the main focus of a presentation is on the CM and GRFM which are often referred to as the manifold approaches in the field of semi-supervised learning. This is because both algorithms tried to explore the manifold structure of the data set. Recently several other approaches to the semi-supervised learning were also proposed. This section compares the Low Density Separation (LDS) algorithm as given in [29], with CM and GRFM as introduced in the previous sections. Benchmarking LDS is challenged and it follows from the fact that Chapelle and Zien have shown its superiority with respect to five other semi-supervised methods, namely to the SVMs, manifold (similar to CM and GRFM), Transductive SVM (TSVM), graph and Gradient Transductive SVM (∇TSVM). The LDS algorithm is an efficient combination of the last two mentioned methods namely, of the graph approach and ∇TSVM algorithm. In such a combination, one first calculates the graph based distances that emphasize low density regions between clusters, and then a novel Chapelle-Zien's ∇TSVM algorithm which places the decision boundary in the low density regions is applied. In the next section, the main idea behind the graph-based distance and TSVM will be presented.

5.5.1 Low Density Separation: Integration of Graph-Based Distances and ∇TSVM

The LDS algorithm is developed on the strong belief that the cluster assumption for the data is necessary for a development of the successful semi-supervised learning algorithm. The cluster assumption is also present in the foundation of both the CM and GRFM algorithms, and in this respect all three methods are similar. The LDS algorithm is a two steps procedure - in the first step one calculates the graph-based distances that emphasize low density regions between clusters and, in the second part, by using the gradient descent, one optimizes the transductive SVM which places the decision boundary in low density regions. The latter is the property of the algorithm that gives its name.

The graph-based distance used in the LDS algorithm is based on the connectivity kernel developed in [52] for clustering. The basic idea of this kernel is to assign higher similarity between the two points if the paths joining these two points do not cross region with low data density. As shown in Fig. 5.12(a), a path is likely to go through a low density region if the longest link of the path is larger. Let $\mathcal{P}_{ij}$ denote all paths from $\mathbf{x}_i$ to $\mathbf{x}_j$. In order to make points which are connected by "bridges" of other points more similar, [52] define for each path $p \in \mathcal{P}_{ij}$ the effective similarity d_{ij}^p (in this case will be the Euclidean distance) between $\mathbf{x}_i$ and $\mathbf{x}_j$ connected by p as the maximum weight on this path, i.e. the longest and weakest link on this path. The total dissimilarity between $\mathbf{x}_i$ and $\mathbf{x}_j$ is then defined as the minimum of all path-specific effective dissimilarities:

$$d(i,j) = \min_{p \in \mathcal{P}_{ij}} \Big(\max_{1 \le k \le |p|-1} d'(p_k, p_{k+1}) \Big), \tag{5.11}$$

and the connectivity kernel is given as

$$k(\mathbf{x}_i, \mathbf{x}_j) = c \cdot \exp\left[-\frac{1}{2\sigma^2} \left(\min_{p \in \mathcal{P}_{ij}} \Big(\max_{1 \le k \le |p|-1} d'(p_k, p_{k+1}) \Big) \right)^2 \right]. \tag{5.12}$$

Because the kernel value does not depend on the length of a path, it is possible to have a group of outliers to act as a 'bridge' between two separated clusters

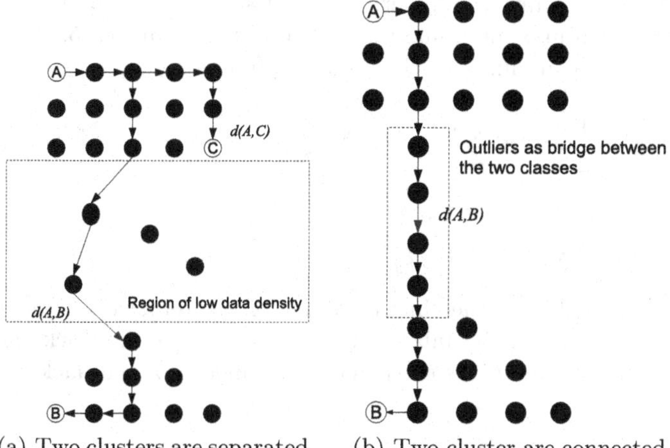

(a) Two clusters are separated by region of low data density.

(b) Two cluster are connected by bridge of outliers.

Fig. 5.12. The basic idea of a connectivity kernel. (a) Two points A and B from different clusters are separated by a region of low data density. This region can be detected by the fact that the longest link $d(A, B)$ in the path p_{AB} is longer than the longest link $d(A, C)$ in the path connecting A and C. (b) Two clusters are now connected by a bridge of outliers. This makes $d(A, B)$ having the same length as $d(A, C)$ as in (a). As a result, point A and B will be considered as in the same class. However, by taking the path length into account, the path length between A and B is longer than the one between A and C. Hence, it is possible to tell that A and B belong to different clusters.

as shown in Fig. 5.12(b). To prevent this problem, [29] soften the 'max' in (5.12) by replacing it with

$$\text{smax}^{\rho}(p) = \frac{1}{\rho} \ln \left(1 + \sum_{k=1}^{|p|-1} \left(e^{\rho d(p_k, p_{k+1})} - 1 \right) \right). \tag{5.13}$$

Thus the ρ-path distance matrix D_{ij}^{ρ} [29] is given as:

$$\mathbf{D}_{ij}^{\rho} = \frac{1}{\rho^2} \ln \left(1 + \min_{p \in \mathcal{P}_{ij}} \sum_{k=1}^{|p|-1} \left(e^{\rho d(p_k, p_{k+1})} - 1 \right) \right)^2. \tag{5.14}$$

The resulting distance matrix $\mathbf{D}^{\rho}$ is in general not positive definite, therefore [29] applied classical Multidimensional Scaling (MDS) to matrix $\mathbf{D}^{\rho}$ in order to form a new representation of the input vectors $\mathbf{x}_i$. More details on this part of the algorithm can be found in [29]. Once a new representation of $\mathbf{x}_i$ is obtained from MDS, it is used as the input to ∇TSVM or TSVM for training.

TSVM was first proposed in [144] and implemented in [79] and [19]. It is an extension of SVMs to solve semi-supervised learning problem. The main idea behind TSVM is to find the hyperplane that separates both labeled and unlabeled data with maximum margin as demonstrated in Fig. 5.13. Therefore, TSVM is aimed at minimizing the following functional,

$$\min \frac{1}{2}\mathbf{w}^2 + C\sum_{i=1}^{l} \xi_i + C^* \sum_{j=1}^{u} \xi_j^* \tag{5.15a}$$

$$\text{s.t.} \quad y_i(\mathbf{w} \cdot \mathbf{x}_i + b) \leq 1 - \xi_i \qquad 1 \leq i \leq l \tag{5.15b}$$

$$y_j^*(\mathbf{w} \cdot \mathbf{x}_i + b) \leq 1 - \xi_j^* \qquad l+1 \leq i \leq n\ , 1 \leq j \leq u, \tag{5.15c}$$

where $y_j^* \in \{-1, +1\}$ is the (in the course of an iterative solving) assigned label of the j-th originally unlabeled data point, ξ_j^* is the slack variable for the unlabeled data and C^* is the penalty parameter for the slack variable ξ^*.

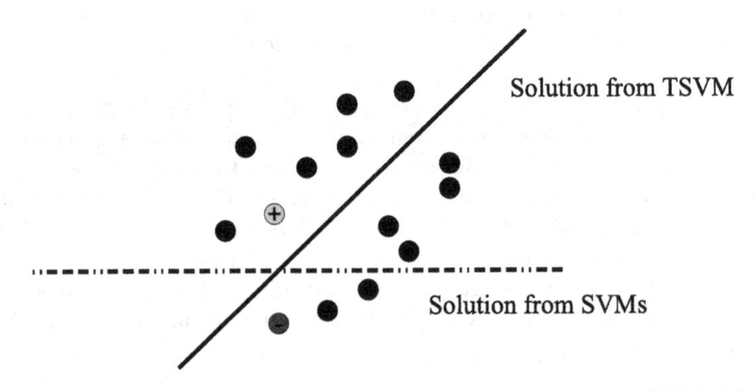

Fig. 5.13. The solution of TSVM. The separating hyperplane from SVMs (dash line) is based only on the two labeled points (points with + and -) in the figure, whereas the solution from TSVM (solid line) is based on all the labeled and the unlabeled ones (solid points). The TSVM is aimed at finding the separating hyperplane which gives maximum margin on both labeled and unlabeled data.

The main difference between TSVM and the SVMs presented in the previous chapters is on the second inequality constraint (5.15c) which is applied to the unlabeled data. This constraint makes the cost function of TSVM being non-convex and difficult to solve [79, 31]. This is due to the fact that the label of the unlabeled data y_i^* is unknown. Thus, to find the globally optimal solution for TSVM, in theory, one needs to try out all the possible combinations of y_i^* and find the combination of y_i^* that gives the separating hyperplane with maximum margin as shown in Fig. 5.13. It is clear that the problem is NP hard and that it can not be solved when the size of the problem is too big. Therefore, various optimization algorithms have been developed for solving a TSVM optimization problem by using different heuristics. In [79] a version of the TSVM algorithm which first labels the unlabeled points based on the classification of a SVM designed from all the initially labeled data was proposed. Then the switching of labels of the initially unlabeled data with heuristics to reduce the objective function take place and so on. This approach is implemented in the software SVMlight [78] and it extends the limit of TSVM to more than 10000 data points. Previously, it was not possible to achieve this by using a branch-and-bound solver [79]. However, a major drawback of Joachims' algorithm is that the proportion of data that belongs to the positive and the negative classes needs to be specified for the algorithm in advance and this is usually an unknown parameter. Although the progressive TSVM proposed in [31] is aimed to resolve this problem, it is not clear how the solutions obtained by TSVM algorithms are close to the globally optimal solution. In summary, the second constraint of TSVM limits the application of TSVM algorithms to small and medium size problems only and it also makes the finding of the globally optimal solution intractable.

The ∇TSVM [29] is the latest variant of TSVM algorithm aiming at the solution of the TSVM optimization problem more effectively. The main idea is to perform a gradient descent on the cost function (5.15a) and approximate the non-differentiable part of the cost function (5.15c) with an exponential function (more details can be found in [29]). The simulation results in [29] show that the ∇TSVM algorithm outperforms the original TSVM algorithm proposed in [79] and it also indicates that the solution from ∇TSVM may be closer to the global solution. By combining the graph-based distance and ∇TSVM together to form the LDS algorithm, [29] shows that their LDS outperforms five other state of the art semi-supervised learning algorithms including a manifold approach which is similar to the two manifold approaches (CM and GRFM) mentioned in the early part of this chapter. However, the manifold approach used in [29] did not take into account the normalization step and also the fact that the connectivity kernel may be better than RBF kernel used in manifold approaches in some problems. In the next section, details on how to combine graph-based distance and the manifold approaches together to form a new class of algorithm will be presented and compared with the LDS.

5.5.2 Combining Graph-Based Distance with Manifold Approaches

Both manifold approaches are based on the belief that 'adjacent' points and/or the points in the same structure (group, cluster) should have similar labels. This can be seen as a form of regularization [130] pushing the class boundaries toward regions of low data density which is similar to LDS. This regularization is often implemented by associating the vertices of a graph to all the (labeled and unlabeled) samples, and then formulating the problem on the vertices of the graph [87]. Both algorithms have similar property of searching the class boundary in the low density region and in this respect they have similarity with the ∇TSVM method too. Thus, it is somehow natural to compare the different algorithms developed around the same principles. This leads to the use of both the CM and GRFM to the same data sets as in Chapelle-Zien's

Algorithm 5.3 Combining Graph and the Manifold methods

1. Build a fully connected graph with edge length $W_{ij} = exp(\rho d(i,j)) - 1)$.
2. Use Dijkstra's algorithm to compute the shortest path lengths $d_{SP}(i,j)$ for all pairs of points.
3. Form the matrix $\mathbf{D}^\rho$ of squared ρ-path distances by $\mathbf{D}^\rho_{ij} = (\frac{1}{\rho}\log(1 + d_{SP}(i,j)))^2$.
4. Apply Multidimensional scaling on the matrix $\mathbf{D}^\rho$ (step 4 to 6 of the LDS algorithm in [29]) to produce a new representation of $\mathbf{x}_i$.
5. Treat the new representation of $\mathbf{x}_i$ as the input to the manifold approaches (Algorithm 5.1 and 5.2) and predict the labels of the unlabeled data. Note that the affinity matrix $\mathbf{W}$ is still computed by using an RBF function.

paper. Similarly, it was a natural idea to replace the second part of the LDS (namely the ∇TSVM part) by both the CM and GRFM algorithm. Thus, the last two algorithms compared in this section (and dubbed here with a prefix Graph &) are the combinations of the graph-based distances with the CM and GRFM. (Recall that the LDS algorithm is the combination of the graph-based distances with the ∇TSVM method). More precisely, both CM and GRFM are applied to a new representation of $\mathbf{x}_i$ which is computed by performing multidimensional scaling to the matrix of squared ρ-path distances $\mathbf{D}^\rho$, i.e., steps 1 to 6 of the LDS algorithm in Chapelle-Zien's paper are used and then followed by CM or GRFM. The algorithm is shown in Algorithm 5.3.

5.5.3 Test Data Sets

In this work, the same five data sets used in [29] are used for comparing the performances of various semi-supervised learning algorithms. Data are available at [28]. An overview of the data sets can be found in Table 5.1.

The *coil20* data set consists of grey-scale images of 20 different objects taken from 72 different angles [103]. It is used for evaluation of classifiers that detect objects in images. *g50c* is an artificial data set and it is generated from two standard normal multi-variate Gaussians. The means of the two Gaussians are located in a 50 dimensional space such that the Bayes error is 5%.

Table 5.1. Summary of Test Data Sets

Data Set	Classes	Dims	Points	labeled
coil20	20	1024	1440	40
g50c	2	50	550	50
g10n	2	10	550	50
text	2	7511	1946	50
uspst	10	256	2007	50

In contrast, *g10n* is a deterministic problem in 10 dimensions where the cluster assumption does not hold. The *text* data set is part of the 20-newsgroup data set and it contains the classes *mac* and *ms windows*. It is pre-processed as in [135]. The *uspst* data set contains the test data part of the well-known *USPS* data on handwritten digit recognition.

Although the same data sets are used, the test setting used in this work is slightly different than the one implemented in [29]. In order to make the results statistically more significant, the mean error rates (for the manifold approaches here) were calculated using 50 random splits of labeled and unlabeled data. The only exception is for the *coil20* data. In this data set, the four manifold algorithms under the investigation were applied to the same 10 random splits used in [29](the indices of labeled data and unlabeled data used are made available with the data sets). The reason for such a test setting is due to the fact that [29] selected 2 labeled data from each class, i.e., the labeled data is balanced in each class and this may also alter the outcome of the simulations.

In terms of model's parameters selections, and in order to reduce the computational time, we fixed some of the parameters in the algorithms and only considered combinations of values on a finite grid for the rest of the parameters. For the original CM methods, we fixed the α parameter in Sect. 5.2.2 to 0.99 and we only varied the σ parameter which determines the width of the Gaussian functions used in calculation of the affinity matrix $\mathbf{W}$.

Because the $2\sigma^2$ value plays a major role to the performance of the manifold algorithms, we tried to find the optimal value of $2\sigma^2$ between 0.005 and 200,000. However, it is important to point out that in some problems, it is not possible to solve the problem for $2\sigma^2$ being a small value such as 0.005, because the conditional number of the Laplacian and the normalized Laplacian matrices used in the GRFM and the CM algorithms respectively, will be very high and, consequently there will be problems with their inversions. In terms of the graph-based distance approach used in the LDS methods, we only tested the approach with values of ρ equal to 1, 4 and 16.

Also, the Chapelle-Zien's parameter δ is fixed at 0.1 and a full graph was used to construct the matrix $\mathbf{D}^{\rho}$ of squared ρ path distances in (5.14). For all data sets, except for the *coil20* one, 10 and 100 nearest neighbors are used for a construction of the affinity matrix $\mathbf{W}$ which is needed for CM and GRFM. In the *coil20* data, the use of the 10 nearest neighbors would not produce a fully connected affinity matrix and only until the number of the nearest neighbors exceeds 80, a fully connected affinity matrix could have been generated. The normalized versions of the CM and the GRFM proposed in Sect. 5.4 [70, 74] have also been used in the simulations. All simulations in this work (except for the experiments on *coil20* where a MATLAB based code was used) are generated using the software package SemiL [75] which is a software package designed to solve large scale semi-supervised learning problems using CM and GRFM. For the simulations that require calculation of graph-based distances, MATLAB based code from Chapelle and Zien was used to generate a new representation of $\mathbf{x}_i$. The new representation of $\mathbf{x}_i$ is then the input for a SemiL routine. Note that the *rec* data set is not used for comparison, because the computational time for the LDS is quite high and the MATLAB implementation from [28] is not suitable for problem such as *rec* data set.

5.5.4 Performance Comparison Between the LDS and the Manifold Approaches

Table 5.2 shows the lowest error rates achieved by CM and GRFM based approaches for all the five data sets included in this study. The results for the LDS methods have been taken directly from [29] with 10 random splits and they are used as references. Basic observations are as follows. First, both CM and GRFM preceded by the calculation of the graph-based distances are better for the multi-class problems than LDS, while the latter one is (slightly) better for the two-class ones. Second, for the two-class problems, the performance of GRFM is close to the results of LDS, and taking the stricter testing criterion used in our experiments (50 random runs compared to 10 ones in Chapelle-Zien's paper) they may be even, or there might be some advantages for the GRFM method as long as the cluster assumption for the data is fulfilled. For the *g10n* data set, without the cluster structure, LDS performs much better than manifold methods as expected.

Table 5.2. Comparisons of the mean test error rates of five semi-supervised algorithms (n stands for normalized).

Data set	LDS	CM	Graph & CM	GRFM	Graph & GRFM
coil20	4.86	8.9	**1.5**	9.83^n	2.9
g50c	**5.62**	7.52^n	7.38^n	6.56^n	6.84^n
g10n	**9.72**	22.29^n	23.66^n	17.93^n	20.8^n
text	**5.13**	13.6^n	13.09^n	7.27^n	7.33^n
uspst	15.8	9.74^n	$\mathbf{8.75}^n$	10.69^n	9.3^n

For the *coil20* data set, the lowest error rate of only 1.5% is achieved by combining the graph-based distances and the CM. The improvement in performance as a result of using the graph-based distances for CM and GRFM is quite significant in this case from 8.9% to 1.5% (6 times better) and 9.83% to 2.9% (3.3 times better) respectively. In this data set, both manifold based algorithms outperform the LDS approach and this coincides with the fact that manifold method used in [29] performs better than ∇TSVM which is the backbone of the LDS method. The normalized model did not perform as well as the non-normalized model in this case. This can be attributed to the fact that the $2\sigma^2$ value used here is very small ($2\sigma^2 = 0.005$) as well as to the use of balanced labeled data

For the *g50c* data set, the LDS method performs the best. However, the difference in performance between the manifold methods and the LDS method is much closer (6.56% vs 5.62%) than the difference (17.3% vs 5.62%) shown in [29]. Similarly, in the *text* data set, the performance difference is also reduced from 11.71% vs 5.13% to 7.33% vs 5.13%. These changes are attributed mostly to the normalization step that lowered the error rate by significantly reducing the effect of the unbalanced data. Also, the error of 7.33% was obtained by the GRFM method that implements Laplacian matrix.

The use of the graph-based distance does not significantly alter the performance in all the data sets except for the *coil20* data and partly for the *uspst* data. The same phenomenon is presented in Chapelle-Zien's paper too. This may be due to the fact that the idea of the connectivity kernel is developed from the image grouping problem [51] and both the *coil20* and the *uspst* data sets contain the images of different objects.

For the *g10n* data set, the performance of the LDS is better than the results of the manifold methods. This particular data set is generated in such a way that the cluster assumption does not hold. Therefore, it is not surprising that the manifold methods, relying strictly on the cluster assumption, have higher error rate. In contrast, LDS which is based on the ∇TSVM performs much better than the manifold approaches. This may be due to the fact that ∇TSVM is based on the idea of SVMs' margin maximization which does not rely on the cluster assumption. Also, the incorporation of the graph-based distances does not help for the non-clustered data very much.

In the *uspst* data set, the normalized version of CM with graph-based distances achieved the lowest error rate of 8.75%. Also, the performances of all the manifold methods (with or without using the graph-based distances) are significantly better than the performance of the LDS method. This is again attributed to two causes; first manifold algorithms perform better for multi-class problems and second the normalization step helps in the case of the unbalanced labeled data. In this multi-class problem the use of graph-based distances also improves the performance of both CM and GRFM methods.

The simulation results in this work suggest that incorporating graph-based distances to semi-supervised learning methods can bring more or less substantial performance improvement in the multi-class problems only. These improvements can be found not just for the ∇TSVM as shown in [29], but also for the manifold approaches used here.

Another interesting phenomenon is that using the graph-based distance with the manifold methods works the best when the value of ρ is in the lower region, meaning either 1 or 4 for all data sets. More investigations are needed to explain this behavior.

The reason why the manifold approach is better in the multi-class problems may be due to the fact that the manifold approaches perform global optimization over all n classifiers, while the ∇TSVM designs separately n classifiers by maximizing the margin of each classifier. The cost function of ∇TSVM is non-convex [79], and it always finds some suboptimal solutions for each particular classifier. In addition to that, it is well known that the sum of suboptimal solutions can not and does not produce an overall optimum. Hence, the performance of ∇TSVM will not be as optimal as the manifold approach in multi-class problems.

5.5.5 Normalizatioin Steps and the Effect of σ

From Table 5.2 it is clear that the normalized models dominated in the most of the data sets. Thus, it is important to know when and how the normalization step should be applied to the manifold algorithms. In Sect. 5.4 and 5.3, it is shown that the effect of the unbalanced labeled data is more significant to the performance of the CM and the GRFM algorithm on the *rec* data set. During the extensive simulations on the five data sets tested in this section, a very clear relationship between the size of the σ parameters and the performance of the normalized model is observed across all the data sets. Figure 5.14 shows the performance of the normalized CM and non-normalized CM with various σ parameters on a 10 nearest neighbor graph for the *uspst* data set. The performances of both models stay relatively constants, as the size of σ gets larger than a certain value.

However, with a larger σ parameter, the performance of the normalized model (error rate is 12%) is far more superior to the one of the non-normalized model (error rate is 30%). In contrast, the performance of the non-normalized model is better than the one of the normalized model when σ is relatively small. This means that the effect of the unbalanced data discussed in the previous sections [74] is more noticeable as the size of σ gets larger. This phenomenon can be explained as follows: when the σ value is large, the influence of the distance between the data points becomes less important, because in such a setting even a distant pairs of point will have relatively large similarity in the affinity matrix $\mathbf{W}$. As a result, this will make the classification of the unlabeled point dominated by the number of labeled points in each class. The normalized procedure tries to remove this effect by standardizing the output of $\mathbf{F}$. This also explains why the non-normalized models perform better than the normalized ones in the *coil20* data sets. This is because the size of $2\sigma^2$ used is only 1.7% of the mean value of the non-zero element in the $\mathbf{W}$ matrix.

The result shown in this section provides some guidelines as to when the normalization step can be used in relation with the σ parameter in order to obtain better performance. It also shows two possible zones where the σ parameter is optimal. For a given problem, one needs to compare the performance of the normalized model with relatively large σ, to the non-normalized model with relatively small σ and the better model should be used.

5.6 Implementation of the Manifold Approaches

The results presented in this chapter can not be made possible without the use of software package SemiL which is developed as part of this work [75].

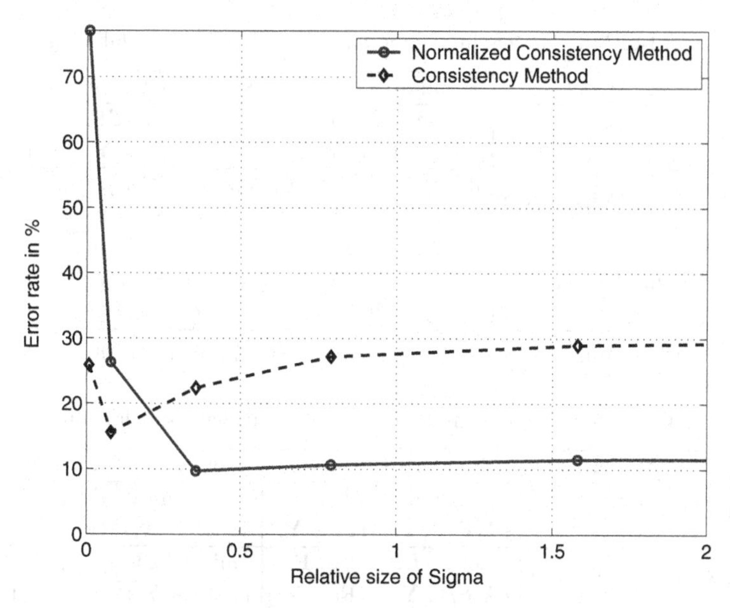

Fig. 5.14. The effect of a normalization step and the size of σ parameter on the *uspst* data set. The relative size of σ is calculated by finding out the ratio between $2\sigma^2$ and the mean value of all the non-zero elements in the affinity matrix $\mathbf{W}$.

SemiL is an efficient large-scale implementation of the manifold approaches discussed in this chapter. In this section, some of the important issues for implementing the manifold approaches will be discussed.

5.6.1 Variants of the Manifold Approaches Implemented in the Software Package SemiL

SemiL is capable of applying not only the two basic models presented in sections 5.2.1 and 5.2.2, but also ten other models [111] which are variants of the two basic models in real world problems. In this section, a brief overview of these models will be presented. Formulations that are derived and used in SemiL in order to accommodate all the models efficiently will also be given.

Tables 5.3 and 5.4 show formulations of all the models implemented in SemiL. The models are classified by the following attributes:

- Standard or Normalized Laplacian: As mentioned previously, the model that has the standard Laplacian matrix $\mathbf{L} = (\mathbf{D} - \mathbf{W})$ uses the hitting time as a measure of closeness between the labeled and the unlabeled data. In contrast, the model that has normalized Laplacian $\mathcal{L} = (\mathbf{I} - \alpha\mathbf{S})$ uses the normalized commute time as a measure of distance. Note that for all the models with normalized Laplacian in the Table 5.3 and 5.4 $\alpha = 1$, i.e. $\mathcal{L} = (\mathbf{I} - \mathbf{S})$.

Table 5.3. Models implemented in SemiL with hard labeling

Model	Standard Laplacian	Normalized Laplacian
Basic Model	$\min \mathbf{F}^T\mathbf{L}\mathbf{F}^*$ s.t. $\mathbf{F}_l = \mathbf{Y}_l$	$\min \mathbf{F}^T\mathcal{L}\mathbf{F}$ s.t. $\mathbf{F}_l = \mathbf{Y}_l$
Norm Constrained	$\min \mathbf{F}^T\mathbf{L}\mathbf{F} + \mu\mathbf{F}_u^T\mathbf{F}_u$ s.t. $\mathbf{F}_l = \mathbf{Y}_l$	$\min \mathbf{F}^T\mathcal{L}\mathbf{F} + \mu\mathbf{F}_u^T\mathbf{F}_u$ s.t. $\mathbf{F}_l = \mathbf{Y}_l$
Bound Constrained	$\min \mathbf{F}^T\mathbf{L}\mathbf{F}$ s.t. $\mathbf{F}_l = \mathbf{Y}_l$ $-C \leq \mathbf{F}_u \leq C$	$\min \mathbf{F}^T\mathcal{L}\mathbf{F}$ s.t. $\mathbf{F}_l = \mathbf{Y}_l$ $-C \leq \mathbf{F}_u \leq C$

* Harmonic Gaussian Random Field Model from [160].

Table 5.4. Models implemented in SemiL with soft labeling

Model	Standard Laplacian	Normalized Laplacian
Basic Model	$\min \mathbf{F}^T\mathbf{L}\mathbf{F}$ $+\lambda(\mathbf{F}_l - \mathbf{Y}_l)^T(\mathbf{F}_l - \mathbf{Y}_l)$	$\min \mathbf{F}^T\mathcal{L}\mathbf{F}$ $+\lambda(\mathbf{F}_l - \mathbf{Y}_l)^T(\mathbf{F}_l - \mathbf{Y}_l)$
Norm Constrained	$\min \mathbf{F}^T\mathbf{L}\mathbf{F} + \mu\mathbf{F}_u^T\mathbf{F}_u$ $+\lambda(\mathbf{F}_l - \mathbf{Y}_l)^T(\mathbf{F}_l - \mathbf{Y}_l)$	$\min \mathbf{F}^T\mathcal{L}\mathbf{F} + \mu\mathbf{F}_u^T\mathbf{F}_u$ $+\lambda(\mathbf{F}_l - \mathbf{Y}_l)^T(\mathbf{F}_l - \mathbf{Y}_l)^*$
Bound Constrained	$\min \mathbf{F}^T\mathbf{L}\mathbf{F}$ $+\lambda(\mathbf{F}_l - \mathbf{Y}_l)^T(\mathbf{F}_l - \mathbf{Y}_l)$ s.t. $-C \leq \mathbf{F} \leq C$	$\min \mathbf{F}^T\mathcal{L}\mathbf{F}$ $+\lambda(\mathbf{F}_l - \mathbf{Y}_l)^T(\mathbf{F}_l - \mathbf{Y}_l)$ s.t. $-C \leq \mathbf{F} \leq C$

* Consistency Method from [155].

- Hard Labeling or Soft Labeling approaches: Models with a hard labeling approach have to fulfill the equality constraint $\mathbf{F}_l = \mathbf{Y}_l$. This means that the initial labels of the labeled data must not be changed after the training, i.e., labels are fixed to either +1 or -1 for the initially labeled data. In contrast, the soft labeling approach allows the initial labels of the labeled data to be changed, i.e., in an extreme case, it is possible that the model produces a label which is different from the initial label of the labeled data. The size of the penalty parameter λ is used to control how close is the output of the labeled data $\mathbf{F}_l$ to the initial label $\mathbf{Y}_l$.
- Norm Constrained Model: The idea behind the norm constrained model is to make the size of the output $\mathbf{F}$ decreasing sharply beyond class boundaries [111]. This constraint is applied by adding an extra term '$\mu\mathbf{F}^T\mathbf{F}$' to the cost function and it is controlled by the penalty parameter μ. Currently, this constraint is applied only to the unlabeled data $\mu\mathbf{F}_u^T\mathbf{F}_u$.
- Bound Constrained Model: The bound constrained model enforces box constraints on the output $\mathbf{F}$, so that $\mathbf{F}$ can only be in certain range. It has been found [111] that using the box constraint may improve the performance of the semi-supervised learning algorithms slightly.

To accommodate all the models into SemiL efficiently, the formulations presented in Table 5.3 and 5.4 are slightly different from their original formulations. For example, CM in Sect. 5.2.2 has both initially labeled points and unlabeled points within the fitting constraint $(\lambda(\mathbf{F}-\mathbf{Y})^T(\mathbf{F}-\mathbf{Y}))$ of (5.3a). Given that $\mathbf{Y}_u = 0$ for the initially unlabeled points, the fitting constraint in (5.3a) is also equal to $\lambda(\mathbf{F}_l-\mathbf{Y}_l)^T(\mathbf{F}_l-\mathbf{Y}_l))+\lambda\mathbf{F}_u^T\mathbf{F}_u$, i.e., this fitting constraint can be viewed as a fitting constraint on the labeled data and a norm constraint on the unlabeled data. Instead of solving the linear system $(\mathbf{I}-\alpha\mathbf{S})\mathbf{F}=\mathbf{Y}$, the following linear system is solved in SemiL:

$$\begin{aligned} \left.\frac{\partial Q}{\partial \mathbf{F}}\right|_{\mathbf{F}=\mathbf{F}^*} &= \mathbf{F}^* - \mathbf{S}\mathbf{F}^* + \begin{bmatrix} \lambda(\mathbf{F}_l^* - \mathbf{Y}_l)) \\ \lambda\mathbf{F}_u^* \end{bmatrix} \\ &= (\mathbf{I}+\lambda\mathbf{I}-\mathbf{S})\mathbf{F}^* - \lambda\mathbf{Y} = 0 \\ &\Rightarrow (\mathbf{I}+\lambda\mathbf{I}-\mathbf{S})\mathbf{F}^* = \lambda\mathbf{Y}. \end{aligned} \tag{5.16}$$

The solution of the linear system (5.16) is the same as the solution of equation (5.5). For the model where λ and μ are different from each other, the following system is solved,

$$(\mathbf{I}+[\boldsymbol{\lambda}_l\ \boldsymbol{\mu}_u]^T\mathbf{I}-\mathbf{S})\mathbf{F}^* = \lambda\mathbf{Y}, \tag{5.17}$$

where $\boldsymbol{\lambda}_l$ is a row vector of length l and $\boldsymbol{\mu}_u$ is a row vector of length u. The reason for using equation (5.17) in SemiL instead of equation (5.5) is that the parameter selection on λ and μ can be implemented more efficiently. To change from one parameter setting to another, one only needs to update the diagonal of the Laplacian or the normalized Laplacian matrices. As a result, the diagonal of the Laplacian or normalized Laplacian matrices are stored separately in SemiL to fully utilize this advantage. In contrast, solving equation (5.5) with a different α parameter, one needs to update the complete $\mathbf{S}$ matrix. Therefore, solving equation (5.17) makes a huge saving in terms of computational time and the amount of memory required when performing model selection. *Note, however, that all the results presented previously are obtained by using two basic models only, namely, the Basic Model with Standard Laplacian for a GRFM method from Table 5.3 and the Norm Constrained Model with Normalized Laplacian for a CM method from Table 5.4 (these models are marked by star in the corresponding tables).* The rest of the 10 models are left for the interested readers to explore and study.

5.6.2 Implementation Details of SemiL

As mentioned previously, SemiL can apply all 12 different models listed in Tables 5.3 and 5.4. A detailed flow chart is shown in Fig. 5.15, which presents the flow of information and different processes in SemiL. The process of applying the manifold approaches in both tables (5.3 and 5.4) can be divided into two steps. The first step is to compute the graph that is needed for the model

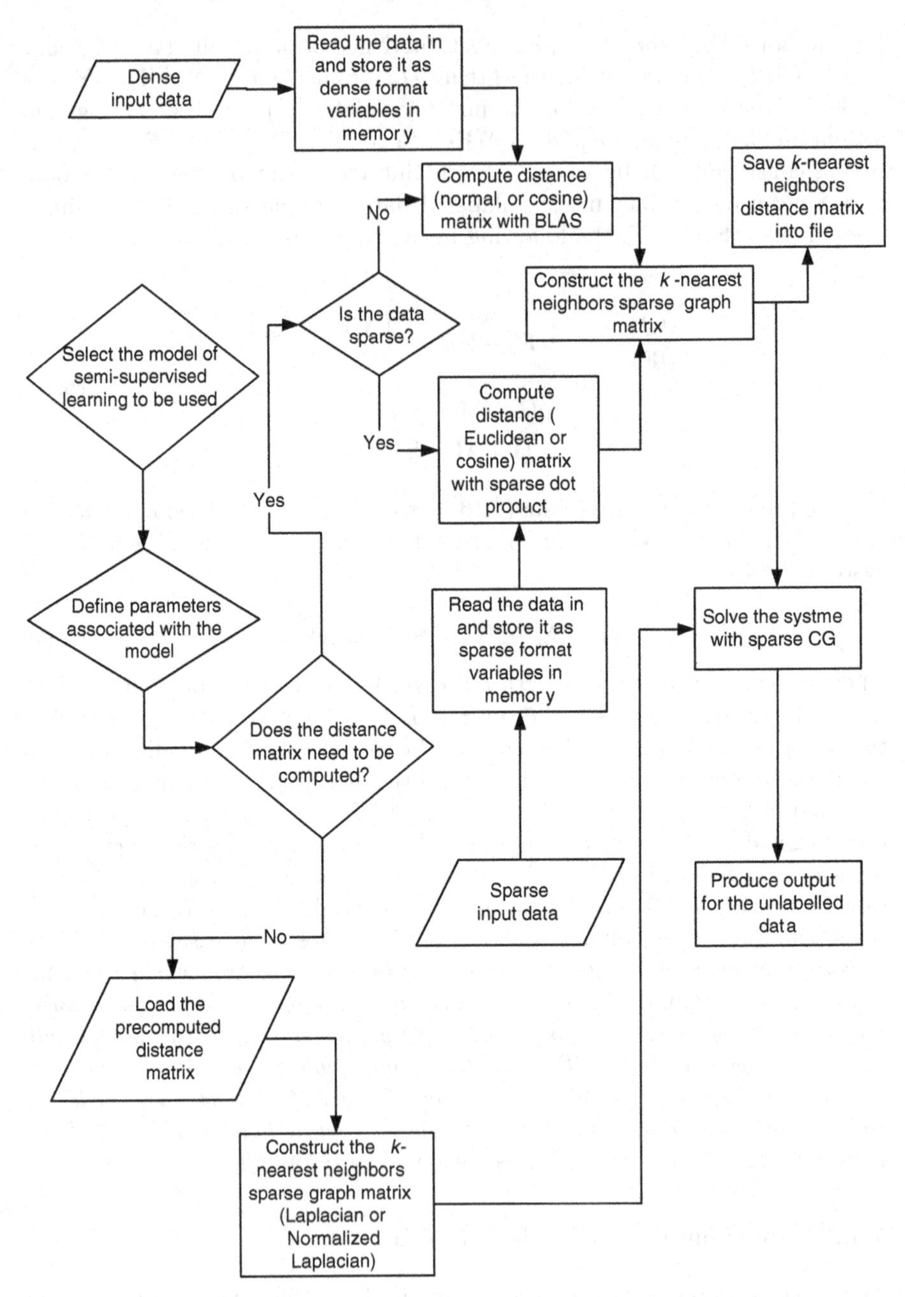

Fig. 5.15. Flow chart of software package SemiL.

selected by the user. There are two types of graphs implemented in SemiL, namely, the Laplacian and normalized Laplacian. In the second step, the corresponding system of linear equations is solved using a conjugate gradient (CG) solver.

Because SemiL is aimed to implement manifold approaches for solving large-scale problems on PCs, it is designed in such a way that the memory requirement is kept at a minimum, but at the same time the speed of obtaining a solution is maximized. To achieve this goal, SemiL is optimized to work on the k-nearest neighbors graph, i.e., when constructing the affinity matrix $\mathbf{W}$, only the weights $\mathbf{W}_{ij}$ of the k closest neighbors of each data point are included. This means that the full affinity matrix is approximated by the k closest neighbors graph. This approximation brings very significant amount of saving in terms of computer memory and the computational time. For example, the *MNIST* handwritten recognition problem has 60000 data points, therefore it will need 28.8 gigabytes of memory to store its complete graph. In contrast, for a 10 nearest neighbors (or 10 degrees) graph, it will require only 12 megabytes of memory to store. The use of complete graph required 2400 times more memory than the use of 10 nearest neighbors graph in this case, hence the amount of memory's saving is enormous and it allows the solving of relatively large-scale problems on a PC. To demonstrate how it works, consider the $\mathbf{W}$ matrix in Example 5.2. If only the 2 nearest neighbors are used, the $\mathbf{W}$ matrix is

$$\mathbf{W} = \begin{bmatrix} 0 & 0.46 & 0.04 & 0 & 0 & 0 & 0 \\ 0.46 & 0 & 0.46 & 0.04 & 0 & 0 & 0 \\ 0.04 & 0.46 & 0 & 0.46 & 0 & 0 & 0 \\ 0 & 0.04 & 0.46 & 0 & 0.04 & 0 & 0 \\ 0 & 0 & 0 & 0.04 & 0 & 0.04 & 4.0e-06 \\ 0 & 0 & 0 & 0 & 0.04 & 0 & 0.04 \\ 0 & 0 & 0 & 0 & 4.0e-06 & 0.04 & 0 \end{bmatrix}, \tag{5.18}$$

and the nonzero elements of the matrix are stored in a sparse format as [row index, column index,value] [15]. For a given problem, SemiL will first work out the k-nearest neighbors for each data points, then in order to keep the matrix $\mathbf{W}$ symmetric, some elements need to be copied from one side of the diagonal to another. After this step, SemiL will check whether the $\mathbf{W}$ matrix is a connected one or not. A connected graph has the property that any given vertex (point) can reach any other vertex (point) on the graph. This is a very crucial condition to check, because using an unconnected graph means that it is possible to have some parts of the unlabeled data still remaining unlabeled after the learning. Those unlabeled data are isolated and can not be reached by other points. In the case of (5.18), the matrix $\mathbf{W}$ will produce a connected graph. If the k-nearest neighbors graph is not connected, then one should increase the value of k so that the graph is connected.

Although the use of the k-nearest neighbors graph reduces the memory requirement significantly, one still has to compute the complete pairwise distance matrix $\mathbf{E}$ (n by n) first in order to construct the k-nearest neighbors graph. The distance measurement used in the matrix $\mathbf{E}$ can be Euclidean or cosine distances. This step can be very time consuming if the software is not optimized thoroughly. To overcome this problem, SemiL provides a sparse and a dense format for inputting data. Inputting data in the sparse format can save considerable amount of time when computing the pairwise distance matrix and it is particularly useful in the text classification problem, because each document vector $\mathbf{x}_i$ generally has small percentage of nonzero elements. Therefore, storing the data set in the sparse format may not only reduce the computational time but also reduces the amount of space to store the document vectors. To evaluate pairwise Euclidean distances between two data points stored in the sparse format, the following identity is used:

$$\left\|\mathbf{x}_i - \mathbf{x}_j\right\|^2 = \mathbf{x}_i \cdot \mathbf{x}_i - 2\mathbf{x}_i \cdot \mathbf{x}_j + \mathbf{x}_j \cdot \mathbf{x}_j. \tag{5.19}$$

The first term $\mathbf{x}_i \cdot \mathbf{x}_i$ and the last term $\mathbf{x}_j \cdot \mathbf{x}_j$ are pre-computed and stored at the beginning for each input vector, hence only the middle term $2\mathbf{x}_i \cdot \mathbf{x}_j$ is calculated during each iteration. To improve the efficiency of computing the middle term, a scatter-and-gather matrix-vector product is used to work out the middle term between $\mathbf{x}_i$ and the rest of the data points, i.e., for i=1, SemiL evaluates $\mathbf{X}^T\mathbf{x}_1$ where $\mathbf{X} = [\mathbf{x}_1\mathbf{x}_2 \ldots \mathbf{x}_j]$ and $\mathbf{X}^T\mathbf{x}_1$ is equivalent of evaluating $\mathbf{x}_1 \cdot \mathbf{x}_j$ for $j = 1$ to n. The scatter-and-gathering matrix-vector product is an efficient way of computing matrix-vector product for sparse vectors and matrices. The main idea is to first scatter the sparse vector into a full length vector, then looping through the non-zero element of the sparse matrix to evaluate the matrix-vector product. This strategy can explore the pipeline effect of the modern CPU to reduce the number of CPU cycles and save computational time. Similar strategy is used in evaluating cosine distance as well.

In terms of computing the pairwise distance matrix $\mathbf{E}$ with dense data format for Euclidean distance, Equation (5.19) is used again and Intel BLAS routines are used extensively. In order to use level 3 BLAS routines (the most efficient level of routines in all three levels), the pairwise dot products are computed in batch using level 3 BLAS routine *gemm*, which calculates the matrix-matrix product of two general matrices, with a predefined cache size c (in megabytes) by the user. The size of the batch n_b is determined by $n_b = (1048576 * c)/(n * 8)$ where 1048576 is the number of bytes in a megabyte and the number 8 is due to the double-precision (each element takes 8 bytes to be stored). The *gemm* is used to calculate the matrix-matrix product of the following equation

$$\mathbf{X}_{nb}\mathbf{X} = \begin{bmatrix} x_{k1} & \cdots & x_{km} \\ \vdots & \vdots & \vdots \\ x_{(k+n_b-1)1} & \cdots & x_{(k+n_b-1)m} \end{bmatrix} \begin{bmatrix} x_{11} & \cdots\cdots & x_{1n} \\ \vdots & \vdots \quad \vdots & \vdots \\ \vdots & \vdots \quad \vdots & \vdots \\ \vdots & \vdots \quad \vdots & \vdots \\ x_{m1} & \cdots\cdots & x_{mn} \end{bmatrix} = \mathbf{B}, \tag{5.20}$$

where B_{ij} is equal to the cross product $\mathbf{x}_k \cdot \mathbf{x}_j$ with $i = 1 \ldots n_b$, $j = 1 \ldots n$, $k = n_b * r + i$ and r is the current cycle. By dividing the total number of data n with the batch size n_b, SemiL can figure out how many cycles are needed to complete the calculation. For example, with *MNIST* problem, if 300 MB of memory is used as cache, then the batch size will be 655 and the number of cycles is 92.

This approach can save a lot of time by utilizing the efficiency of the Intel BLAS routine and the memory available in a computer. As an example, to work out the 10 nearest-neighbor graph for the MNIST problem with 300 megabytes of memory takes only about 2100 seconds on a Pentium 4 2.4 Ghz computer. In contrast, if no cache is used and the pairwise dot products are calculated with respect to one data point at a time using level 2 BLAS routines only, it will take about 12000 (approximately 3.33 hours) seconds to finish. Furthermore, if the BLAS routine was not used, it can take 33333 seconds (approximately 9.25 hours) to complete. It demonstrates why in the development stage of SemiL, a lot of efforts were focused on optimizing this part of the software, so the cost of using manifold algorithms can be as low as possible. It also makes the use of manifold algorithms on large-scale semi-supervised learning problems more feasible and attractive. Similarly, the cosine distance also benefits from this approach.

The core part of SemiL is an efficient sparse conjugate gradient (CG) solver which is designed to solve the following linear system of equations with box constraints.

$$\mathbf{A}\mathbf{f} = \mathbf{y} \tag{5.21a}$$

$$\text{s.t.} \; -C \leq f_i \leq C, \quad i = 1 \ldots n \tag{5.21b}$$

The matrix $\mathbf{A}$ above is either the k-nearest neighbors normalized Laplacian or Laplacian matrix and it is symmetric and positive definite. The reason for choosing a CG solver rather than ISDA, also developed in this work, is because the strength of CG is more distinct when the matrix $\mathbf{A}$ is sparse and the solution of $\mathbf{f}$ is dense. In contrast, ISDA developed in this work is highly optimized for SVMs where the solution $\mathbf{f}$ is sparse and the matrix $\mathbf{A}$ is dense. The popular Cholesky factorization for solving the system of linear equations is also considered during the design of SemiL, but it was found that the Cholesky factor $\mathbf{R}$ of the sparse matrix $\mathbf{A}$ can be a few dozen times denser than the matrix $\mathbf{A}$ [111]. This means that extra storage is needed when using Cholesky factorization and it is against the purpose of using k-nearest neighbor graph to reduce the memory footprint of the algorithm. In contrast, the use of the CG algorithm only needs a storage of matrix $\mathbf{A}$ in order to perform sparse matrix-vector product during each iteration. Therefore, CG is regarded as the most suitable solver for applying the manifold approaches discussed in this chapter to large-scale problems.

5.6.3 Conjugate Gradient Method with Box Constraints

The CG solver implemented in SemiL is based on a CG algorithm in [66] for finding a nonnegative minimum point of the quadratic function given below.

$$\text{Min} \quad \mathbf{Q}(\mathbf{f}) = \frac{1}{2}\mathbf{f}^T\mathbf{A}\mathbf{f} - \mathbf{y}\mathbf{f} + c \tag{5.22a}$$

$$\text{s.t.} \quad f_i \geq 0 \quad i = 1 \ldots n \tag{5.22b}$$

The matrix $\mathbf{A}$ must be a positive definite one in order to have a unique solution. The solving of (5.21) is similar to (5.22), because the Laplacian and the normalized Laplacian matrices introduced in Sect. 5.2 are positive definite. However, the only difference between the solving of (5.21) and (5.22) is the constraint on f_i. To solve (5.21), the algorithm in [66] must be extended to take upper and lower bound into account. To the best of author's knowledge, the extension presented here is original. The CG algorithm that can take box constraints into account is given in Algorithm 5.4 and the MATLAB code is given in Appendix D.

To solve the constrained optimization problem (5.21), the optimality conditions of the solution need to be established first. Similarly to the dual problem in SVMs, the following KKT conditions must be fulfilled at the optimal solution $\mathbf{f}^{b*}$ of the constraint optimization problem (5.21):

$$\left.\frac{\partial \mathbf{Q}}{\partial f_i}\right|_{f_i = f_i^{b*}} \geq 0 \ \text{ for } i \text{ in } I_{lo}, \tag{5.29a}$$

$$\left.\frac{\partial \mathbf{Q}}{\partial f_i}\right|_{f_i = f_i^{b*}} \leq 0 \ \text{ for } i \text{ in } I_{up}, \tag{5.29b}$$

$$\left.\frac{\partial \mathbf{Q}}{\partial f_i}\right|_{f_i = f_i^{b*}} = 0 \ \text{ otherwise}, \tag{5.29c}$$

where I_{lo} is the set of indices such that $f_i^{b*} = -C$, I_{up} is the set of indices such that $f_i^{b*} = C$. Note that the KKT conditions here need to be multiplied by -1 to be same as the ones in SVMs, because the task is to find the minimum of (5.22) not the maximum. These conditions can be illustrated graphically by considering the problem of minimizing a 1-dimensional quadratic function $Q(x) = ax^2 + bx + c$ $(a, b, c > 0)$ with a box constraint $-C \leq x \leq C$ as shown in Fig. 5.16. In Fig. 5.16, in both case 1 and case 3 the global minimum x_o

Algorithm 5.4 Conjugate Gradient Method with Box Constraints.

Consider solving the system of linear equation (5.21a) with the box constraints (5.21b).

1. Choose a point $\mathbf{f}^{(0)}$ that satisfies the box constraints (5.21b), e.g. $\mathbf{f}^{(0)} = \mathbf{0}$. The residue $\mathbf{r}^{(0)}$ is used as the first search direction, i.e. $\mathbf{d}^{(0)} = -\partial \mathbf{Q}/\partial \mathbf{f} = \mathbf{r}^{(0)} = \mathbf{y} - \mathbf{A}\mathbf{f}^{(0)}$.
2. Let I_{los} be the set of all indices $i \leq n$ such that
$$f_i^{(0)} = -C \quad \text{and} \quad r_i^{(0)} \geq 0. \tag{5.23}$$
In other words, I_{los} contains the indices of $f_i^{(0)}$ that fulfill the KKT conditions (5.29a).
3. Let I_{ups} be the set of all indices $i \leq n$ such that
$$f_i^{(0)} = C \quad \text{and} \quad r_i^{(0)} \leq 0. \tag{5.24}$$
In other words, I_{ups} contains the indices i of $f_i^{(0)}$ that fulfill the KKT conditions (5.29b). If $\left|r_i^{(0)}\right| < \tau$ for all indices i that are not in either I_{ups} or I_{los}, then $\mathbf{f}^{(0)}$ is the solution of the optimization problem (5.21). The algorithm terminates.
4. Set $\mathbf{d}^{(0)} = \bar{\mathbf{r}}^{(0)}$, where $\bar{\mathbf{r}}^{(0)}$ is the vector having
$$\bar{r}_i^{(0)} = 0 \quad \text{for } i \in (I_{los} \cup I_{ups}), \quad \bar{r}_i^{(0)} = r_i^{(0)} \quad \text{otherwise.} \tag{5.25}$$
5. Start with $t = 0$ and perform a standard CG step by computing the following:
$$a^{(t)} = \frac{\mathbf{d}^{(t)T}\mathbf{r}^{(t)}}{\mathbf{d}^{(t)T}\mathbf{A}\mathbf{d}^{(t)}}, \quad \mathbf{f}^{(t+1)} = \mathbf{f}^{(t)} + a^{(t)}\mathbf{d}^{(t)}, \quad \mathbf{r}^{(t+1)} = \mathbf{r}^{(t)} - a^{(t)}\mathbf{A}\mathbf{d}^{(t)}. \tag{5.26}$$
6. If $\mathbf{f}^{(t+1)}$ is outside the feasible region (i.e. some $f_i^{(t+1)}$ are either smaller than $-C$ or larger than C), go to step 7. Otherwise if $\left|r_i^{(t+1)}\right| < \tau$ for all i not in $(I_{los} \cup I_{ups})$, reset $\mathbf{f}^{(0)} = \mathbf{f}^{(t+1)}$, $\mathbf{r}^{(0)} = \mathbf{r}^{(t+1)}$ and go to step 2. Else compute $\bar{\mathbf{r}}^{(t+1)}$ and $\mathbf{d}^{(t+1)}$ as follows:
$$\bar{r}_i^{(t+1)} = 0 \quad \text{for } i \in (I_{los} \cup I_{ups}), \quad \bar{r}_i^{(t+1)} = r_i^{(t+1)} \quad \text{otherwise,} \tag{5.27}$$
$$\mathbf{d}^{(t+1)} = \mathbf{r}^{(t+1)} + \beta^{(t)}\mathbf{d}^{(t)}, \quad \beta^{(t)} = \frac{\left|\bar{\mathbf{r}}^{(t+1)}\right|^2}{\mathbf{d}^{(t)T}\mathbf{r}^{(t)}} \tag{5.28}$$
Replace t by $t+1$ and go to step 5.
7. Find $\eta^{(t)}$ using (5.30). Reset $\mathbf{f}^{(0)} = \mathbf{f}^{(t)} + \eta^{(t)}\mathbf{d}^{(t)}$ and $\mathbf{r}^{(0)} = h - A\mathbf{f}^{(0)}$. Redefine I_{los} to be the set of all indices $i \leq n$ such that $f_i^{(0)} = -C$. Redefine I_{ups} to be the set of all indices $i \leq n$ such that $f_i^{(0)} = C$. If $\left|r_i^{(0)}\right| < \tau$ for all i not in $(I_{los} \cup I_{ups})$, go to step 2. Else go to step 4.

of the function $Q(x) = ax^2 + bx + c$ $(a, b, c > 0)$ is outside or just on the boundary of the box constraints. As a result, the solution of the optimization problem with box constraints is on the boundary of the feasible region, i.e. $x = -C$ for case 1 and $x = C$ for case 3. The slope $\frac{dQ}{dx}$ of the quadratic function $Q(x) = ax^2 + bx + c$ in case 1 is greater or equal to zero at $x = -C$ which coincides with the first KKT conditions (5.29a). Similarly, the gradient of the quadratic function $Q(x) = ax^2 + bx + c$ in case 3 of Fig. 5.16 is less or equal to zero at point $x = C$ (KKT conditions (5.29b)). In the case 2, the solution is inside the feasible region. As a result, the gradient $\frac{dQ}{dx} = 0$ is equal to zero at the optimal solution. This simple example also can be extended to SVMs, but one needs to multiply the optimality conditions (5.29) by -1.

The active set approach is used here to extend CG for solving QP problems with box constraints. An active set method solves a sequence of subproblems of the same form as the original problem, but with some inequality constraints assumed to be fulfilled with equality. In other words, some f_i are assumed to be equal to C or $-C$ and they are excluded in the subproblem that are currently optimized by the solver. In Step 2 and 3 of the Algorithm 5.4, the variables that are on the boundary of the feasible region (either equal to C or $-C$) and at the same time also satisfied the KKT conditions (5.29a) or (5.29c) respectively are selected as the active set $(I_{los} \cup I_{ups})$. Consequently in Step 4, all search directions d_i whose indices i are in the active set are set to zero, i.e. the algorithm will not move the intermediate solution $\mathbf{f}^{(t)}$ in the directions of $f_i^{(t)}$ in the active set. As a result, $f_i^{(t+1)} = f_i^{(t)}$ in Step 5 of the Algorithm 5.4 for all indices i that are in the active set $I_{los} \cup I_{ups}$. The Step 5 of the Algorithm 5.4 is the standard CG step for solving the unconstrained

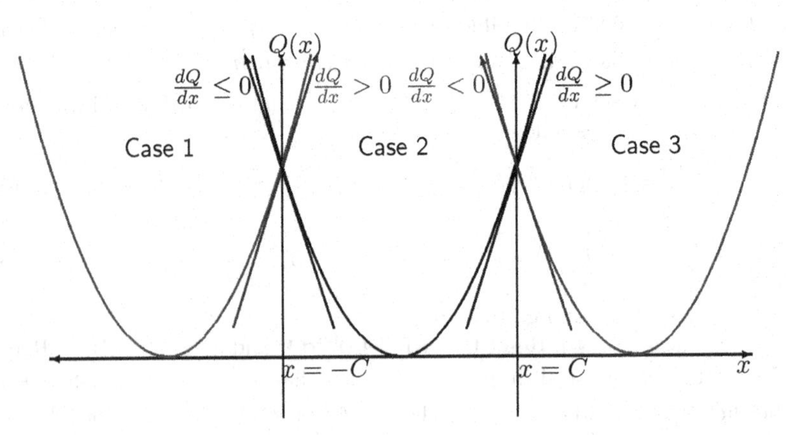

Fig. 5.16. A 1-D example of KKT conditions. There are three possible cases where the global minimum point of the function $Q(x)$ is located 1) the point is below or equal to the lower bound $-C$, 2) the global minimum point x_o is between $-C$ or C $(-C < x_o < C)$, 3) the global minimum point is above or equal to the upper bound C.

QP problem such as (5.22a). Because the focus of this section is on extending CG for solving QP with box constraints, the derivations of the equations used in Step 5 will not be presented here. Interesting readers can refer to [122] for an excellent introduction on the subject.

During the optimization process of the standard CG algorithm, it is highly likely that the updating of the intermediate solution $\mathbf{f}^{(t)}$ by (5.26) in Step 5 of Algorithm 5.4 results in another intermediate solution $\bar{\mathbf{f}}^{(t+1)}$ that is outside the feasible region, i.e. some elements of $\bar{\mathbf{f}}^{(t+1)}$ are either greater than C or smaller than $-C$. There are several ways to resolve this problem. The most common way is to find another intermediate solution $\mathbf{f}^{(t)}$ which is on the line joining $\mathbf{f}^{(t)}$ and $\bar{\mathbf{f}}^{(t+1)}$. The solution $\mathbf{f}^{(t+1)}$ should be inside the feasible region and at the same time it should give the greatest improvement on the objective function. This is often done by working out a ratio $\eta^{(t)}$ which controls how much of the solution should move from $\mathbf{f}^{(t)}$ towards the intermediate solution $\bar{\mathbf{f}}^{(t+1)}$ along the search direction $\mathbf{d}^{(t)}$, i.e. $\mathbf{f}^{(t+1)}$ is calculated as $\mathbf{f}^{(t+1)} = \mathbf{f}^{(t)} + \eta^{(t)}\mathbf{d}^{(t)}$ and $a^{(t)}$ in (5.26) is replaced by $\eta^{(t)}$. The optimal value of $\eta^{(t)}$ should bring $\mathbf{f}^{(t+1)}$ as close to $\bar{\mathbf{f}}^{(t+1)}$ as possible in order to have largest improvement, but at the same time $\mathbf{f}^{(t+1)}$ should stay within the feasible region. As a result, the optimal $\eta^{(t)}$ is calculated as follows:

$$u = k \quad \text{if} \quad \frac{C - f_u^{(t)}}{d_u^{(t)}} < \frac{C - f_k^{(t)}}{d_k^{(t)}} (u \neq k) \quad u, k \in U_v \tag{5.30a}$$

$$l = k \quad \text{if} \quad \frac{-C - f_l^{(t)}}{d_l^{(t)}} < \frac{-C - f_k^{(t)}}{d_k^{(t)}} (l \neq k) \quad l, k \in L_v \tag{5.30b}$$

$$\eta^{(t)} = \frac{-C - f_l^{(t)}}{d_l^{(t)}} \quad \text{if} \quad \frac{-C - f_l^{(t)}}{d_l^{(t)}} < \frac{C - f_k^{(t)}}{d_k^{(t)}} \quad \text{else} \quad \eta^{(t)} = \frac{C - f_k^{(t)}}{d_k^{(t)}} \tag{5.30c}$$

where U_v is the set of indices i such that $f_i^{(t)} > C$ and L_v is the set of indices i such that $f_i^{(t)} < -C$. The procedure above can be visualized by a simple 2D example in Fig. 5.17. In Fig. 5.17, the intermediate solution $\bar{\mathbf{f}}^{(t+1)}$ violates the box constraints on both f_1 and f_2 directions. In f_1 direction, $\bar{\mathbf{f}}^{(t+1)}$ exceeds the upper bound C, whereas in the f_2 direction, $\bar{\mathbf{f}}^{(t+1)}$ exceeds the lower bound $-C$. It is clear from the figure that the intermediate solution $\mathbf{f}^{(t+1)}$ in step $t+1$ should be the one that has $f_2^{(t+1)} = -C$, because it is the closest point to $\bar{\mathbf{f}}^{(t+1)}$ which is still inside the box constraints. As a result, the optimal $\eta^{(t)}$ in Fig. 5.17 is $(-C - f_2^{(t)})/d_2^{(t)}$, because it is smaller than $(C - f_1^{(t)})/d_1^{(t)}$. In other words, the solution $\bar{\mathbf{f}}^{(t+1)}$ is clipped to the $-C$ on the f_2 direction along $\mathbf{d}^{(t)}$ instead of C in the f_1 direction. The procedure above is used in step 7 of Algorithm 5.4 when $\mathbf{f}^{(t+1)}$ is outside the feasible region after the updating in step 5. Each time only one f_i will be clipped to either C or $-C$, as a result it has a similar role as the clipping operation (3.9) of the ISDA. The major difference is that more than one f_i are outside the feasible region

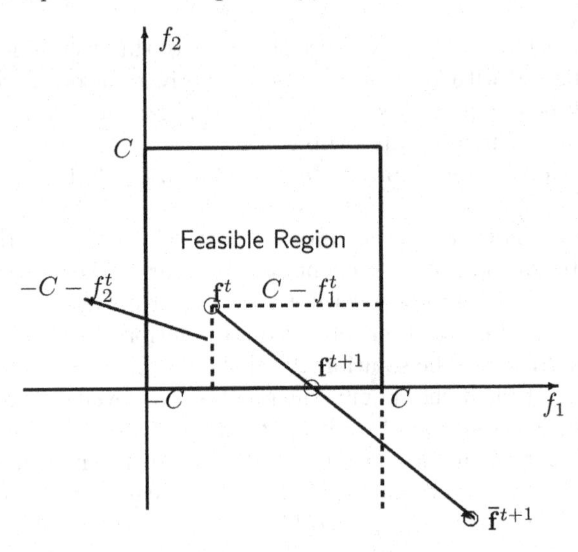

Fig. 5.17. A 2-D example to illustrate the procedure of enforcing the box constraints in CG.

in CG. This algorithm can also be used for the optimization of SVMs by changing the lower bound from $-C$ to 0. However, it has the same weakness as the active set method developed in [149] when the number of bounded SVs increases. This is because the procedure (5.30) can only clip one variable at a time and it is inefficient when many variables need to be clipped. In contrast, the ISDA developed in Chap. 3 is more suitable in such a case, because the clipping operation (3.9) is much cheaper. Finally, the algorithm terminates as the KKT conditions (5.29) of all the data points are fulfilled within the precision τ.

5.6.4 Simulation Results on the *MNIST* Data Set

In order to test the performance of SemiL and the manifold approaches, simulations are performed on the popular handwritten recognition data set MNIST used in the previous chapter. Because of the size of the data set, it is not feasible to operate on the complete Laplacian matrix and only the 10 nearest neighbors are used to approximate the graph. In Fig. 5.18, it is clear that the normalized models perform better than the models without normalization and the overall trend is very similar to Fig. 5.10. The normalized models achieve very high accuracy of 98% with only 0.1% of the data being labeled. This again shows the remarkable property of the semi-supervised learning algorithm to be able to utilize the unlabeled points as well as the labeled ones. The LDS method was not applied on this data set, because the Matlab implementation

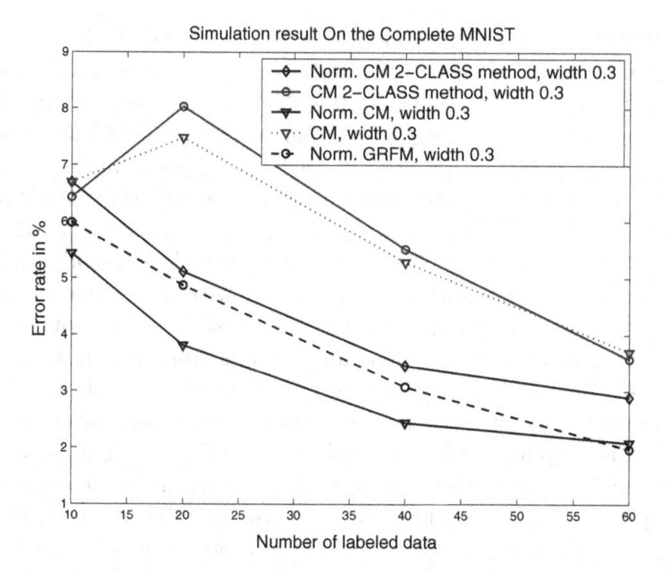

Fig. 5.18. Simulation result on the complete MNIST data set. Error rates for CM algorithm are based on 100 random trials, whereas the ones for GRFM are based on 50 random trials.

of LDS from [28] is not capable of solving such a large problem on a PC due to the time complexity and the memory requirement of the algorithm.

In terms of the computational time, the CM algorithm required much less time than the GRFM algorithm in this data set. This is why only fifty random experiments are performed on the GRFM. For each random experiment, CM algorithm requires on average only 17 seconds, whereas the GRFM algorithm requires on average 123 seconds. This is due to the fact that the condition number of the normalized Laplacian matrix $\mathcal{L}$ is much lower (better) than the Laplacian matrix $\mathbf{L}$. As a result, it takes less time for the CG solver to converge for CM than for GRFM. The reason for a better condition number is partly due to the fact that when applying the norm constraint and the fitting constraint in CM, the positive parameters λ and μ are added to the diagonal of the normalized Laplacian matrix. This improves the condition number of the normalized Laplacian matrix by making the diagonal element larger. The result in this section suggests that CM algorithm may be more desirable than GRFM when the size of the problem is large.

5.7 An Overview of Text Classification

In the last decade, there is an increase in interest on classification of documents into proper classes using computers. It is primarily due to the booming of documents available in electronic format on the Internet which allows us to access millions or even billions of documents at any time and any place. However, having so much information available on the net also means that finding the right information can be very difficult and time consuming. Therefore, the development of an automated system which can label documents with thematic categories from a predefined set has become a major research area. In the research community, the dominant approach to this problem is based on machine learning techniques such as support vector machines, artificial neural networks and more recently semi-supervised learning algorithms. This approach has several advantages over the knowledge engineering based approach which consists of manually building a set of rules for classification by domain experts. First, the machine learning approach is faster, cheaper and more reliable, because the computer will produce the classification rule instead of human experts. Second, once the machine learning system is developed, it can be applied easily to problems in different domains, whereas the knowledge engineering based approach will require different domain experts for developing the classification rules. To use machine learning approaches, the documents in electronic format need to be pre-processed first before submitting it into learning systems. Because a lot of the results in this part of the work is first obtained in text classification problems, the aim of this section is to explain the pre-processing steps used in this work to the readers who do not have any background in text classification. The pre-processing steps used here are the same as the ones in [155] and they are carried out using the Rainbow package which is developed in [96].

As mentioned previously, machine learning algorithms such as SVMs can not directly interpret texts, therefore the documents need to be converted into a compact representation which can be understood by the machine learning algorithm. This process is normally referred to as the indexing procedure. Through the indexing procedure, a document d_j will normally be represented as a vector of term weights $d_j = (v_{1j}, \ldots, v_{\mathcal{T}j})$, where $\mathcal{T}$ is the set of terms or features that occur at least once in at least one document of all the documents and the set $\mathcal{D}$ denotes all the n documents available for pre-processing. The term weights v_{kj} in each document can be regarded as how much term t_k contributes to the semantics of the document d_j. There are many document representations, and they are different in terms of how a term is defined in the representation and also how the term weights are computed [50]. In this work, each term is recognized as a unique word in the complete set of documents. This type of document representation is often referred to as the bag of words approach which is a typical choice in this area. The simplest bag of word approach is to represent each document by a vector of frequency count of each term in the document. A more sophisticated bag of word approach is used in this study for pre-processing of the text document and it is important to note that using this approach, semantic qualities such as grammatical structure of sentences and phrases will be neglected. Although this approach looks very primitive and a lot of important information of the document is filtered out during the indexing, in a number of experiments, it has been found that having a more complicated representations do not yield significantly better performance [50]. In particular, some researchers have used phrases as indexing terms rather than individual words [50]. Therefore, the following pre-processing steps are applied to the text data using Rainbow package.

To generate a good representation of the document, all the words occurred at least once in all the documents are first passed through the Porter stemmer before counting them. The basic idea of using the Porter stemmer [117] is to improve the performance of the classifiers by conflating words of similar meanings into a single term. This may be done by removing the various suffixes of words. For example, 'connect', 'connected, 'connecting', 'connection' and 'connections' can be conflated into the term 'connect' by removing the various suffixes such as '-ed', '-ing', '-ion' and '-ions' [117]. This step can be regarded as reducing the dimensionality of the document representation, hence the complexity of the problem is reduced and the classifier will be less likely to overfit.

The second steps involved in tossing out any word which is on the stop list of the SMART retrieval system. The stop list consists of 524 common words, like 'the' and 'of' [96]. Because these words appear in all of the documents, they will not have discriminative power. The headers of all the documents will be skipped, it is because for a data set such as 20 newsgroup [89], the header of each document consists of the class information, i.e., the label of the document. The word that does not occur very frequently in all the documents will also be removed to reduce the dimensionality of the document representation. The final stage of the pre-processing step is to normalize each document using Term Frequency Inverse Document-Frequency metric (TFIDF). The standard TFIDF function is defined as [50]

$$\mathrm{TFIDF}(t_k, d_j) = \#(t_k, d_j) \cdot \log \frac{n}{\#_{\mathcal{D}}(t_k)}, \tag{5.31}$$

where $\#(t_k, d_j)$ denotes the number of times the term t_k occurs in d_j, and $\#_{\mathcal{D}r}(t_k)$ denotes the document frequency of term t_k, that is, the number of documents in $\mathcal{D}$ where t_k occurs. This function tries to incorporate the intuitions that the more often a term occurs in a document, the more it is representative of its content, and also the more documents a term occurs in, the less discriminating it is. The term weights from TFIDF function are normalized into the range [0,1] by the cosine normalization [50], given as

$$v_{kj} = \frac{\mathrm{TFIDF}(t_k, d_j)}{\sqrt{\sum_{s=1}^{\mathcal{T}} (\mathrm{TFIDF}(t_s, d_j)^2)}}. \tag{5.32}$$

After these steps the documents are ready for the machine learning algorithm for training. Two paragraphs from two different topics (one related to SVMs and another related to machine learning from [151]) in Fig. 5.19 are used to demonstrate the preprocessing steps mentioned in this section.

Support Vector Machines are the latest development from statistical learning theory. SVMs deliver state of the art performance in real world applications such as text categorization, hand written character recognition, image classification, biosequencies analysis, etc., and are now established as one of the standard tools for machine learning and data mining.

Machine Learning is an area of artificial intelligence concerned with the development of techniques which allow computers to "learn". More specifically, machine learning is a method for creating computer programs by the analysis of data sets.

Fig. 5.19. Two paragraphs are used to demonstrate the preprocessing of text data. The paragraph on the left is related to SVMs and the one on the right is related to machine learning.

machine 1 learning 2 development 1 analysis 1 data 1 support 1 vector 1 machines 1 latest 1 statistical 1 theory 1 svms 1 deliver 1 state 1 art 1 performance 1 real 1 world 1 applications 1 text 1 categorization 1 hand 1 written 1 character 1 recognition 1 image 1 classification 1 biosequencies 1 established 1 standard 1 tools 1 mining 1 *are 2 the 3 from 1 of 2 such 1 as 2 and 2 in 1 etc 1 now 1 for 1 one 1*

machine 2 learning 2 area 1 artificial 1 intelligence 1 concerned 1 development 1 techniques 1 computers 1 learn 1 specifically 1 method 1 creating 1 computer 1 programs 1 analysis 1 data 1 sets 1 *is 2 an 1 of 3 with 1 the 2 which 1 more 1 a 1 for 1 by 1*

Fig. 5.20. All the unique terms and their corresponding frequencies of the two paragraphs in Fig. 5.19. The left text box corresponds to the paragraph related to SVMs and the one on the right corresponds to the paragraph about machine learning. Note that the common words are listed in italic and there are 44 unique terms in the paragraph about SVMs and 28 terms in the one about machine learning.

To visualize how the preprocessing steps are performed, the first step is to treat the paragraphs as bags of words as shown in Fig. 5.20. After passing all the terms through the Porter stemmer and removing all the common words that are in the stop list of SMART system, all the unique terms and their corresponding frequencies are shown in Fig. 5.21. Note that the terms "machine" and "machines" in the paragraph about SVMs are conflated into the term "machin" in Fig. 5.21 by the Porter stemmer, as a result the term "machin" has a frequency of 2. These two steps reduce the number of unique terms from 44 to 31 for SVMs' paragraph and from 28 to 16 for the one related to machine learning.

The calculations of TFIDF values and the normalized TFIDF (v_{kj}) for some of the unique terms are listed in Table 5.5. For the term "machin", its TFIDF values in both paragraphs are zero, because the term $\log \frac{n}{\#_{\mathcal{D}}(t_k)}$

machin 2 learn 2 develop 1 analysi 1 data 1 support 1 vector 1 latest 1 statist 1 theori 1 svm 1 deliv 1 state 1 art 1 perform 1 real 1 world 1 applic 1 text 1 categoriz 1 hand 1 written 1 charact 1 recognit 1 imag 1 classif 1 biosequ 1 establish 1 standard 1 tool 1 mine 1

machin 2 learn 3 area 1 artifici 1 intellig 1 concern 1 develop 1 techniqu 1 comput 2 specif 1 method 1 creat 1 program 1 analysi 1 data 1 set 1

Fig. 5.21. All unique terms and their corresponding frequencies after passing all the terms in 5.20 through the Porter stemmer and then removing all the common words from the output of the Porter stemmer.

Table 5.5. Calculation of TFIDF values for some of the terms in Fig. 5.21.

terms	t_k (SVMs)	t_k (ML)[a]	$\#_{\mathcal{D}}(t_k)$	TFIDF (SVMs)	TFIDF (ML)	v_{kj} (SVMs)	v_{kj} (ML)
machin	2	2	2	0	0	0	0
learn	2	3	2	0	0	0	0
area	0	1	1	0	0.30	0	0.27
artifici	0	1	1	0	0.30	0	0.27
intellig	0	1	1	0	0.30	0	0.27
develop	1	1	2	0	0.00	0	0
comput	0	2	1	0	0.60	0	0.53
support	1	0	1	0.30	0	0.45	0
vector	1	0	1	0.30	0	0.45	0
latest	1	0	1	0.30	0	0.45	0
statist	1	0	1	0.30	0	0.45	0

[a] Machine Learning

of TFIDF is equal to $\log \frac{2}{2} = 0$. As a result, this term does not have any role in the classification and this coincides with the intuition that the more documents a term occurs in, the less discriminating it is. In contrast, the term "support" has a v_{kj} value of 0.45 in the SVMs' paragraph and 0 in ML's paragraph, because it appears twice in the SVMs' one, but not in the one about machine learning.

5.8 Conclusions

The chapter presents the basic ideas of and the original contributions to the field of semi-supervised learning, which is an introduction of the normalization step into the manifold algorithms.

The extensive simulations on seven different data sets have shown that an introduction of a normalization step as proposed in [70, 74] improves the behavior of both manifold approaches (namely, CM and GRFM) very significantly in all of the data sets when the σ parameter is relatively large. In both methods, the normalization of $\mathbf{F}^*$ can improve the performance up to fifty percents. Furthermore, the normalized models with large σ parameters are significantly better than the unnormalized models with small σ parameters in six out of the seven data sets tested. This result provides a strong evidence to support the use of the normalization in the two manifold approaches discussed in this chapter. This improvement in performance from normalizing the column of $\mathbf{F}^*$ is achieved by removing the effects of unbalanced labeled data. In the situation where the labeled data is unbalanced, the class with less labeled data will be more disadvantageous: this is due to the fact that the class with less labeled data will have a lower mean value at its output $\mathbf{F}_j^*$. The normalization step corrects this problem by standardizing the column of $\mathbf{F}^*$ to have mean of zero and standard deviation of one, so the output from each classifier is treated equally.

Although the normalization step can improve the performance significantly, it does not mean that it will work in all situations. The σ parameter plays a crucial role on the effectiveness of the normalization step. During the extensive simulations on all seven data sets, a very clear relationship between the size of σ parameters and the performance of the normalized model is found across all the data sets. With a relatively large value of σ, the performance of the manifold approaches can be improved significantly. In contrast, when a relatively small value of σ is more appropriate for a given data set, the normalization procedure does not seem to provide significant improvements [73]. This also means that there are two possible zones where the σ parameter is optimal (small σ without normalization or large σ with normalization). This result gives some guidance when performing the model parameters selection for the manifold approaches.

In terms of comparison with other semi-supervised learning algorithms on the five benchmarking data sets, the results suggest that the manifold algorithms have much better performance in both multi-class data sets (*coil20* and *uspst*), whereas the LDS performs slightly better for the two-class data holding cluster assumption. This may be due to the fact that the cost function of the manifold approach is convex, whereas the one for ∇TSVM is non-convex. Thus, the solution of ∇TSVM is not as optimal as the ones from manifold approaches. For the two-class data set without cluster structure (*g10n*), the LDS method performs much better than the manifold algorithms. The result in this work suggests that the manifold algorithms may be more suitable for handling multi-class problems than the LDS and ∇TSVM methods. This result also sheds the new light on the possible strength and weakness of the two known approaches (manifold approaches and LDS) for future development.

By combining the graph-based distances and the manifold approach, the performance of the algorithms is greatly improved in multi-class data sets only. It seems that the use of the graph-based distances does not help the manifold approaches for the two-class problems. More investigations are still needed to determine the usefulness of graph-based distances to manifold approaches.

The very first efficient software implementation for large scale problems of the manifold approaches was also developed as part of this work. The software package SemiL has been successfully applied to many real world data sets in this work and it is the backbone for this part of the work. It also demonstrates that manifold approaches are cost-effective for solving large-scale problems. The efficiency of SemiL has also made it a popular software for working in this area at the point of writing (worldwide there are more than 100 downloads per month of SemiL from its web site). The additional 10 models implemented in SemiL provide a lot of potentials for future improvement on the manifold approaches by exploring the impacts of the λ and μ parameters.

6

Unsupervised Learning by Principal and Independent Component Analysis

Unsupervised learning is a very deep concept that can be approached from different perspectives, from psychology and cognitive science to engineering [106]. Very often it is called "learning without a teacher". This means that a learning human, animal or man-made system observes its surroundings and, based on observations adapts its behavior without being told to associate given observations to given desired responses, as opposed to *supervised learning.* A result of unsupervised learning is a new representation or explanation of the observed data.

Principal component analysis (PCA) and independent component analysis (ICA) are methods for unsupervised learning. In their simplest forms, they assume that observed data can be represented by a linear combination of some unknown hidden factors called sources and causes [91, 76, 35]. The model is mathematically described as:

$$\mathbf{X} = \mathbf{AS} \tag{6.1}$$

where $\mathbf{X} \in \Re^{N \times T}$ represents observed data matrix obtained by N sensors with T observations, $\mathbf{A} \in \Re^{N \times M}$ is an unknown mixing matrix and $\mathbf{S} \in \Re^{M \times T}$ is an unknown source matrix consisted of M sources with T observations. The mixing matrix $\mathbf{A}$ is due to the environment through which source signals propagate prior to being recorded and due to the recording system. In order to focus on the essence of the problem, we have omitted an additive noise term in (6.1). This linear model represents a useful description for many applications. A few examples include brain signals measured by electroencephalographic (EEG) recording [93], functional Magnetic Resonance Imaging (fMRI) [98], radio signals received by a multiantenna base station in wireless communication systems [121], multispectral astronomical and remotely sensed images [104, 46, 47], multiframe blind deconvolution with non-stationary blurring process [86], near infrared spectroscopy [30] and nuclear magnetic resonance spectroscopy [105]. Recovery of the unknown source signals from the data is often called *blind source separation* (BSS). PCA and ICA are methods used to accomplish a BSS task.

T.-M. Huang et al.: *Kernel Based Algorithms for Mining Huge Data Sets*, Studies in Computational Intelligence (SCI) **17**, 175–208 (2006)
www.springerlink.com

It is our intention throughout this chapter to present concepts in a way that will enable interested readers to understand essence of the PCA and ICA. Therefore, presented mathematical derivations are not rigorous and very often reader is referred to references for finding out additional details about some topics. MATLAB code is provided after presentation of the important topics in order to enable the reader to easily and quickly reproduce the presented results, and get a better understanding of the topics. A minimal knowledge of MATLAB is required. Although some concepts such as entropy and mutual information, are specific and may not be known to the wider audience, the material is presented in such a way that a reader with average mathematical background should be able to follow the derivations and reproduce results.

We shall now illustrate the BSS problem by two examples for which we are going to need computer-generated data in accordance with the model given by (6.1). This means that both mixing matrix **A** and source matrix **S** will be provided by us in order to simulate the mixing process. The PCA and ICA algorithms yet to be described will, however, reconstruct the sources **S** based on data **X** only, i.e., the characteristics of the mixing process contained in the mixing matrix **A** will not be known to PCA and ICA algorithms.

In the first example we shall assume that there are two source signals, namely images shown in Fig. 6.1.

Fig. 6.1. Two source images.

They are mixed with the artificial mixing matrix:

$$\mathbf{A} = \begin{bmatrix} 2 & 1 \\ 1 & 1 \end{bmatrix} \tag{6.2}$$

The mixed images represented by the data matrix **X** are shown in Fig. 6.2.

In order to establish the connection between mixed images shown in Fig. 6.2, the mixing matrix given by (6.2) and the data model given by (6.1) we shall re-write data model given by (6.1) on the component level basis:

Fig. 6.2. Two mixed images.

$$x_1(t) = 2s_1(t) + s_2(t)$$
$$x_2(t) = s_1(t) + s_2(t)$$

where t represents a pixel index running from 1 to T. It is evident that, although images are two-dimensional signals, they can be represented as one-dimensional signals. Mapping from two-dimensional to one-dimensional signal is obtained by the row stacking procedure implemented by the MATLAB command *reshape*. Assuming that our sources images S_1 and S_2 have $P \times Q$ pixels the mapping is implemented through:

$$\mathbf{s}_n = reshape(S_n, 1, P * Q) \;\; n = 1, 2$$

In our example, $P = 181$ and $Q = 250$ implying $T = 181 \times 250 = 45250$. Consequently, the size of our data matrix $\mathbf{X}$ and source matrix $\mathbf{S}$ is 2×45250. In order to display mixed images, we need to convert one-dimensional signals $\mathbf{x}_1$ and $\mathbf{x}_2$ back to two-dimensional format. This is implemented through:

$$X_n = reshape(\mathbf{x}_n, P, Q) \;\; n = 1, 2$$

Later on, we shall learn that images belong to the sub-Gaussian class of stochastic processes, which can be characterized by a formal measure called kurtosis. Sub-Gaussian processes have negative values of kurtosis. As opposed to the sub-Gaussian processes, there are also super-Gaussian processes for which the kurtosis parameter has positive value. Speech and music signals are super-Gaussian processes.

In order to support and illustrate the algorithms to be described, we shall need both type of signals. Therefore, we show in Fig. 6.3 three seconds of two source speech signals, and in Fig. 6.4, two mixed speech signals using again the matrix given by (6.2). One can imagine that mixed speech signals shown in Fig. 6.4 were obtained by two-microphones recording when two speakers were talking simultaneously. Although for audio signals like speech, real mixing process cannot be described with a matrix given by (6.2), it will serve the purpose of illustrating the BSS problem. Physically relevant description of the mixing process in the reverberant in-door acoustic environment is described with a mixing matrix whose elements are filters. If we assume that two speakers talking simultaneously were recorded by two microphones, the mathematical description of the process on the component equation level is given with the convolutional model:

$$x_1(t) = \sum_{k=1}^{K} a_{11}(k)s_1(t-k) + \sum_{k=1}^{K} a_{12}(k)s_2(t-k)$$
$$x_2(t) = \sum_{k=1}^{K} a_{21}(k)s_1(t-k) + \sum_{k=1}^{K} a_{22}(k)s_2(t-k)$$

K represents the filter order, which is assumed here to be equal for all the filters.The corresponding BSS problem is more difficult in such a realistic case. We refer interested readers to [140, 88], and we proceed with the mixing matrix of (6.2) for pedagogic purposes. The fundamentally important question here

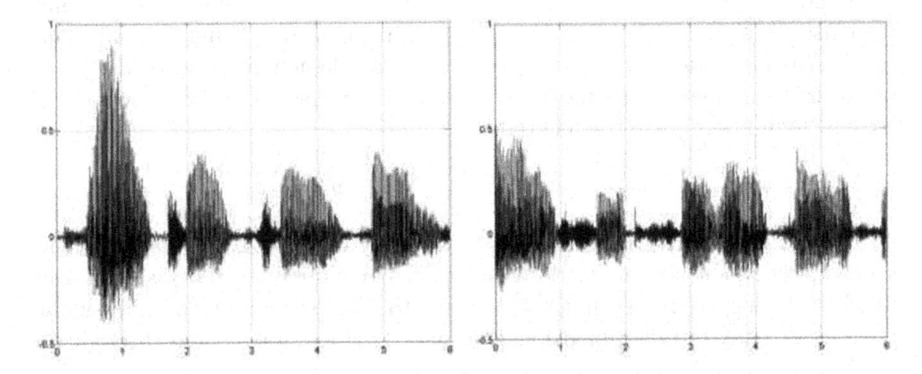

Fig. 6.3. Two source speech signals.

is how either PCA or ICA algorithms recover the source images shown in Fig. 6.1 from their mixtures shown in Fig. 6.2, or speech signals shown in Fig. 6.3 from their mixtures shown in Fig. 6.4, when only mixed signals are available to the algorithms? The "blindness" means that the mixing matrix $\mathbf{A}$ is unknown. If it were known, recovery of source signals $\mathbf{S} = \mathbf{A}^{-1}\mathbf{X}$ is straightforward. As noted in [67] any meaningful data are not really random, but are generated by physical processes. When the relevant physical processes are independent, the generated source signals are not related, i.e., they are independent. This is a fundamental assumption upon which all ICA algorithms are built. PCA also exploits the assumption that source signals are not related, but it only assumes that signals are not correlated. In order to derive mathematical equations for ICA algorithms built upon the independence assumption, cost functions that measure statistical (in)dependence must be formulated. Source signals are

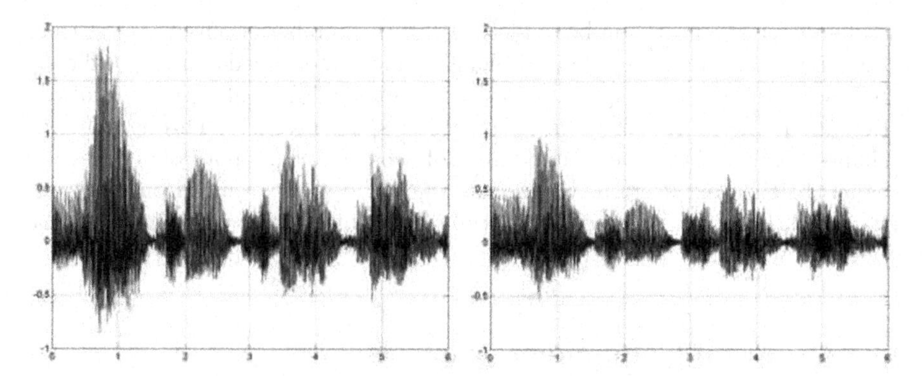

Fig. 6.4. Two mixed speech signals.

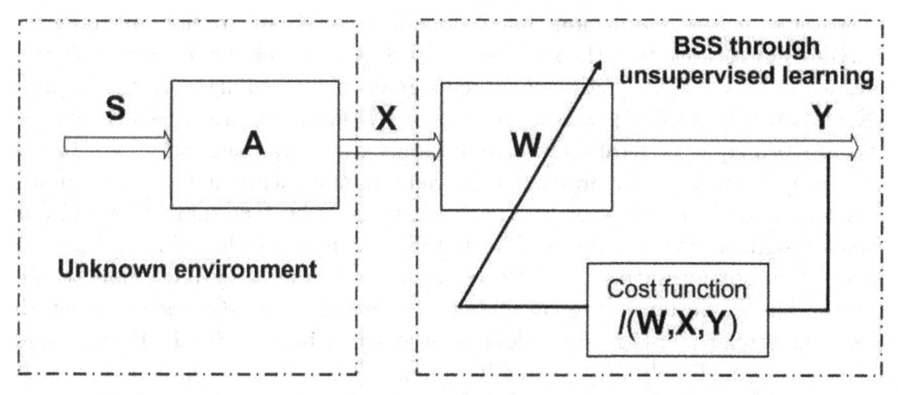

Fig. 6.5. Blind source separation through unsupervised learning. A cost function $I(\mathbf{W}, \mathbf{Y}, \mathbf{X})$ measures statistical (in)dependence between components of $\mathbf{Y}$. Demixing matrix $\mathbf{W} \simeq \mathbf{A}^{-1}$ is learnt by minimizing $I(\mathbf{W}, \mathbf{Y}, \mathbf{X})$.

then recovered through the unsupervised learning process by minimizing the measure of statistical dependence. The process is illustrated in Fig. 6.5 where $I(\mathbf{W}, \mathbf{X}, \mathbf{Y})$ denotes cost function.

A mathematical description of the BSS process is given by:

$$\mathbf{Y} = \mathbf{W}\mathbf{X} \tag{6.3}$$

where $\mathbf{Y} \in \Re^{M \times T}$ represents reconstructed source signals and $\mathbf{W} \in \Re^{M \times N}$ represents the unmixing matrix obtained through learning process as:

$$\mathbf{W} = \arg\min I(\mathbf{W}, \mathbf{Y}, \mathbf{X}) \tag{6.4}$$

It is obvious from (6.1) and (6.3) that $\mathbf{Y}$ will represent the source matrix $\mathbf{S}$ if $\mathbf{W} \simeq \mathbf{A}^{-1}$. In order for that to be possible, the number of sensors N and number of sources M should be the same. When number of sensors is greater than number of sources, the BSS problem is over-determined. By using PCA-based dimensionality reduction technique [76], the number of sources and number of sensors can be made equal. If the number of sensors is less than the number of sources the BSS problem is under-determined and very difficult. It has solutions only in special cases, when some additional *a priori* information about the sources are available, such as sparseness [92]. Throughout this chapter, we shall assume that number of sensors and number of sources are equal. In the next section, we discuss PCA and show results from using it. As noted, PCA can serve for the initial treatment of data, which will later be treated by ICA.

6.1 Principal Component Analysis

PCA and ICA are two types of methods used to solve the BSS problem. They exploit the assumption that hidden source signals are not mutually related. PCA reconstructs the source signals $\mathbf{S}$ by decorrelating observed signals $\mathbf{X}$, while ICA reconstructs the source signals $\mathbf{S}$ by making observed signals $\mathbf{X}$ statistically as independent as possible. Both methodologies are related to the principle of redundancy reduction, which was suggested in [14] as a coding strategy in neurons, where redundancy meant a level of statistical (in)dependence between extracted features. The fact that PCA obtains source signals through decorrelation process suggests that, in the general case, it is not an optimal transform from the redundancy reduction standpoint. The exception is the case when source signals are Gaussian, in which case uncorrelatedness is equivalent to statistical independence. Having said that, we immediately assert that ICA makes sense only when source signals are non-Gaussian. Therefore, PCA is quite often included as a first step in the implementation of the ICA algorithms through its special form known as whitening or sphering transform [76, 35]. Whitening transform is named in analogy with the name used for white stochastic processes. The stochastic process $z_n(t)$ is said to be white if its autocorrelation function equals delta function i.e. $E\left[z_n(t)z_n(t+\tau)\right] = \sigma_n^2\delta(\tau)$ where σ_n^2 represents variance of z_n and $\delta(\tau)$ is Kronecker delta equal to 1 for $\tau = 0$ and 0 otherwise.When PCA is applied on multivariate data set it makes it spatially uncorrelated i.e. $E\left[z_n(t)z_m(t)\right] = \sigma_n^2\delta_{nm}$. We say that PCA-transfomed data set $\mathbf{Z}$ is spatially white.

The statistical independence assumption is formally stated as:

$$p(\mathbf{S}) = \prod_{i=1}^{N} p_i(S_i) \tag{6.5}$$

that is, the joint probability density function $p(\mathbf{S})$ is a product of the marginal probability density functions $p_i(S_i)$. From now on we shall assume that source signals as well as the observation signals have zero mean. If signals do not have zero mean, it is easy to make them, to have a zero mean by subtracting the mean value from them. Under the zero mean assumption, the uncorelatedness assumption is formally expressed as:

$$\mathbf{R_S} = E\left[\mathbf{SS}^T\right] = \mathbf{\Lambda} \tag{6.6}$$

where $\mathbf{R_S}$ represents covariance matrix, $\mathbf{\Lambda}$ is some diagonal matrix and E denotes mathematical expectation. It is shown in Appendix E that (6.5) and (6.6) are equivalent for Gaussian processes. The PCA transform $\mathbf{W}$ is designed such that the transformed data matrix $\mathbf{Z} \in \Re^{M \times T}$:

$$\mathbf{Z} = \mathbf{WX} \tag{6.7}$$

has uncorrelated components, i.e., $\mathbf{R_Z} = \mathbf{\Lambda}$. Having this in mind, we derive the PCA transform from:

$$\mathbf{R_Z} = E\left[\mathbf{ZZ}^T\right] = \mathbf{W}E\left[\mathbf{XX}^T\right]\mathbf{W}^T = \mathbf{WR_XW}^T = \mathbf{\Lambda} \tag{6.8}$$

From (6.8), we can recognize that the PCA transform $\mathbf{W}$ is nothing else but matrix of eigenvectors $\mathbf{E}$ obtained through eigen-decomposition of the data covariance matrix $\mathbf{R_X}$, i.e.,

$$\mathbf{W} = \mathbf{E}^T \tag{6.9}$$

A special form of the PCA transform is a whitening or sphering transform that makes transformed signals uncorrelated with unit variance. This is formally expressed as $\mathbf{R_Z} = \mathbf{I}$, where $\mathbf{I}$ represents identity matrix. This is equivalent to writing (6.8) as:

$$\mathbf{R_Z} = E\left[\mathbf{ZZ}^T\right] = \mathbf{W}E\left[\mathbf{XX}^T\right]\mathbf{W}^T = \mathbf{\Lambda}^{-1/2}\mathbf{WR_XW}^T\mathbf{\Lambda}^{-1/2} = \mathbf{I} \tag{6.10}$$

Using (6.10), the whitening transform is obtained as:

$$\mathbf{W} = \mathbf{\Lambda}^{-1/2}\mathbf{E}^T \tag{6.11}$$

Because $\mathbf{W}$ has dimensionality $M \times N$, we see that $\mathbf{Z}$ will have dimensionality $M \times T$, i.e., by using the PCA transform we can reduce dimensionality of our original data set $\mathbf{X}$ if $N > M$. The problem is how to estimate the number of sources M . We use the fact that the covariance matrix $\mathbf{R_X}$, which is nonnegative real symmetric matrix, has real eigenvalues. Ordering of the first M eigenvectors used to construct PCA transforms given with (6.9) or (6.11) is done such that the corresponding eigenvalues $\lambda_1, \lambda_2, ..., \lambda_M, \lambda_{M+1}$ satisfy $\lambda_1 \geqslant \lambda_2 \geqslant ... \geqslant \lambda_M > \lambda_{M+1}$, where eigenvalue λ_{M+1} corresponds with noise. The concept is illustrated in Fig. 6.6. The difficulty in this approach to dimensionality reduction is that *a priori* knowledge about the noise level is required in order to determine which eigenvalue corresponds with the noise.

This knowledge is not often available and more robust approaches to number of sources detection must be used in practice. We refer interested readers to the two very often used approaches, the Akaikes's information criteria [5] and the minimum description length criteria [120].

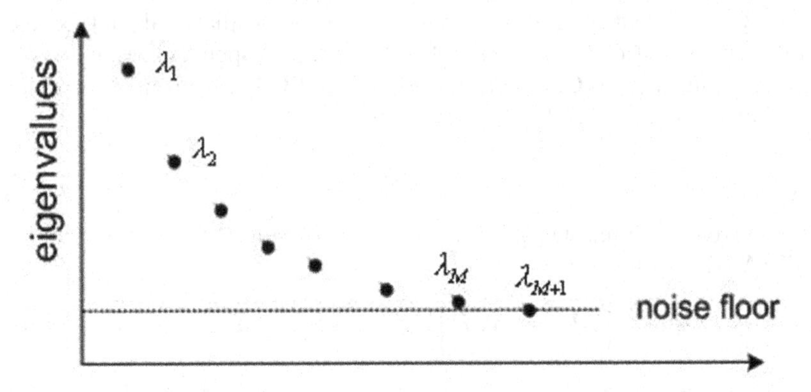

Fig. 6.6. Determination of number of sources by eigenvalue ordering.

We shall now illustrate the application of the PCA to the BSS problem of the mixed images and speech signals shown in Figs. 6.2 and 6.4, respectively. As we saw from our previous exposition, in order to derive the PCA transform, we need to have the data covariance matrix $\mathbf{R_x}$ available. In practice, we have to estimate it from the given number of T data samples according to:

$$\mathbf{R_X} = E\left[\mathbf{x}\mathbf{x}^T\right] \simeq \frac{1}{T}\sum_{t=1}^{T}\mathbf{x}(t)\mathbf{x}^T(t) \tag{6.12}$$

where $\mathbf{x}(t)$ is a column vector obtained from the $N \times T$ data matrix $\mathbf{X}$ for each particular observation $t = 1,...,T$. Equation (6.12) is written on the component level at the position (i,j) as:

$$\mathbf{R_X}(i,j) = E\left[x_i x_j\right] \simeq \frac{1}{T}\sum_{t=1}^{T}x_i(t)x_j(t) \tag{6.13}$$

i.e. elements of the data covariance matrix are cross-correlations between corresponding components of the data vector $\mathbf{x}(t)$. We want to recall that in the first example the number of sensors equals number of sources is $N = M = 2$ and the number of observations is $T = 181 \times 250 = 45250$. In the second example, that involves speech signals, $N = M = 2$ and $T = 50000$. Because speech signals were digitized with sampling frequency $F_s = 16kHz$ this implies data record length of slightly more than 3 seconds.

To implement the PCA transform given by (6.9), we need a matrix of eigenvectors which is obtained by the eigenvalue decomposition of the data covariance matrix $\mathbf{R_X}$ and implemented by MATLAB command $[\mathbf{E}, \mathbf{D}] = eig(\mathbf{R_X})$. $\mathbf{R_X}$ is estimated from data matrix $\mathbf{X}$ by the MATLAB command $\mathbf{R_X} = cov(\mathbf{X}^T)$. The PCA transform given by (6.7) and (6.9) is implemented by the following sequence of MATLAB commands:

```
Rx=cov(X');        % estimate data covariance matrix
[E,D]=eig(Rx);     % eigen-decomposition of the data covariance
                     matrix
Z=E'*X;            % PCA transform
```

In the given MATLAB code symbol '%' is used to denote comments. As we see only the data matrix $\mathbf{X}$ is used to implement the PCA transform. This is why we say that PCA is an unsupervised method. When the PCA transform is applied to the mixed images shown in Fig.6.2, the results shown in Fig. 6.7 are obtained:

Fig. 6.7. Source images reconstructed by PCA transform.

The reconstruction process was not fully successful. However, it can be observed that first reconstructed image is approximation of the *flowers* source image and second reconstructed image is approximation of the *cameraman* source image. Two PCA de-mixed images shown in Fig. 6.7 were displayed using the following sequence of MATLAB commands:

```
Z1=reshape(Z(1,:),P,Q); % transform vector into image
figure(1); imagesc(Z1)  % display first PCA image
Z2=reshape(Z(2,:),P,Q); % transform vector into image
figure(2);imagesc(Z2)   % display second PCA image
```

To understand better why PCA was not very successful in solving the BSS problem, we plot in Figs. 6.8, 6.9 and 6.10 histograms of the source images shown in Fig. 6.1, mixed images shown in Fig. 6.2 and PCA reconstructed images shown in Fig. 6.7.

Histograms are graphical representations of the number of times each signal or image amplitude occurs in a signal. The signal is divided into K bins

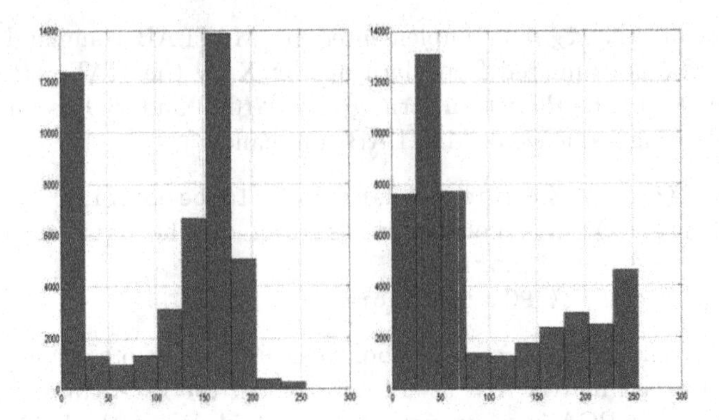

Fig. 6.8. Histograms of two source images shown in Fig. 6.1

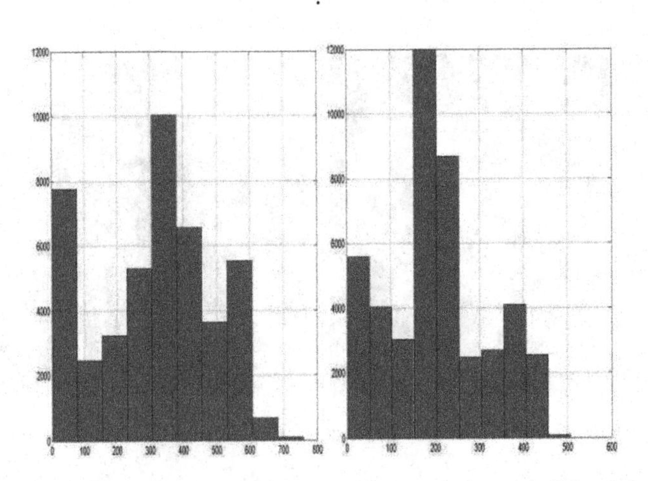

Fig. 6.9. Histograms of two mixed images shown in Fig. 6.2.

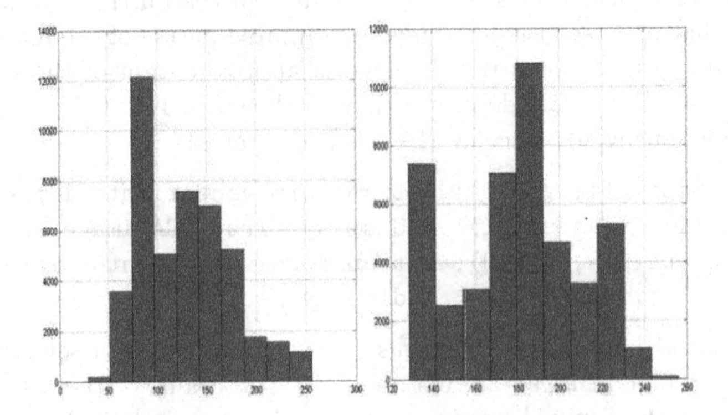

Fig. 6.10. Histograms of two PCA images shown in Fig. 6.7.

and frequency of occurrence is counted for each bin. If a histogram is scaled by the overall number of occurrences, an empirical estimate of the probability density function (pdf) is obtained. The histograms showed in Figs. 6.8, 6.9 and 6.10 are obtained by the MATLAB command *hist*. For example, the histogram of the first source image is obtained by $hist(\mathbf{S}(1,:))$. We see that histograms of the PCA reconstructed source images are more similar to those of the mixed images than to histograms of the source images. We can also see that histograms of the PCA reconstructed images are more like histograms that would be obtained from normal or Gaussian process than histograms of the original source images, which are very non-Gaussian. The histograms of the source images are bimodal, which means that values that occur at each side of the mean value are more probable than the mean value itself. Processes with bimodal pdf's are called sub-Gaussian processes. There are also processes whose pdf's are concentrated around the mean value, with probabilities that rapidly decrease as the distance from the mean value increases. Such processes are called super-Gaussian processes.

As we already mentioned, there is a measure called *kurtosis* that is used for classification of stochastic processes. It can also be used as a measure of non-Gaussianity. For some real zero mean process x, *kurtosis* $\kappa(x)$ is defined as:

$$\kappa(x) = \frac{E[x^4]}{\left(E\left[x^2\right]\right)^2} - 3 \tag{6.14}$$

where $E[x^4]$ and $E[x^2]$ denote fourth order and second order moments estimated according to:

$$E[x^p] \simeq \frac{1}{T}\sum_{t=1}^{T} x^p(t),\ p \in \{2,4\} \tag{6.15}$$

Empirical estimate of the $\kappa(x)$ can be obtained using the following sequence of MATLAB commands:

```
m4=sum(x.*x.*x.*x)/T; % estimate E[x^4]
m2=sum(x.*x)/T;       % estimate E[x^2]
kx=m4/(m2*m2)-3;      % estimate kurtosis
```

Kurtosis is negative for sub-Gaussian processes, zero for Gaussian process and positive for super-Gaussian processes. This is illustrated in Fig. 6.11.

The source images shown in Fig. 6.1 have the values of the kurtosis equal to $\kappa(s_1) = -1.36$ and $\kappa(s_2) = -0.88$. Mixed images have kurtosis value equal to $\kappa(x_1) = -0.93$ and $\kappa(x_2) = -0.64$. As expected, the kurtosis of the mixed images is closer to zero than the kurtosis of the source images. This is a consequence of the *central limit theorem* (CLT), which ensures that the pdf of mixed signals is always more gaussian than the pdf of its constituent source signals. The PCA reconstructed source images shown in Fig. 6.7 have kurtosis equal to $\kappa(z_1) = -0.4$ and $\kappa(z_2) = -0.82$. We have already noticed that

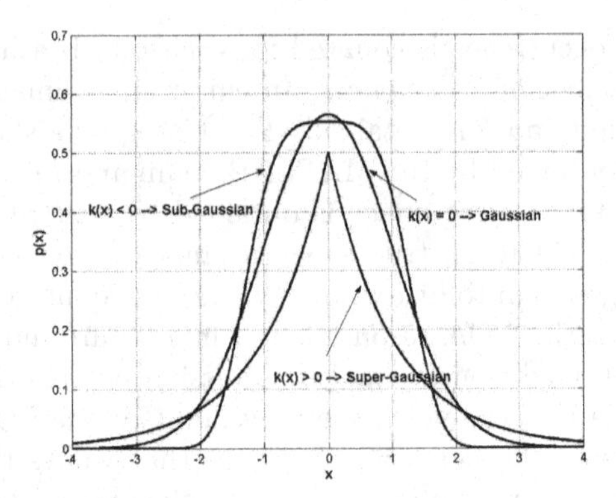

Fig. 6.11. Classification of distributions from stochastic processes by the value of the kurtosis parameter.

first PCA reconstructed source image shown in Fig. 6.7 actually corresponds with the second source image shown in Fig. 6.1 (*flowers*) and the second PCA reconstructed source image corresponds with the first source image shown in Fig. 6.1 (*cameraman*). Keeping that in mind, we verify that the PCA reconstructed source images are even more gaussian than mixed images. This is consequence of the fact that PCA forces uncorelatedness among the signals. This is evidently not optimal for non-Gaussian signals such as images. This explains why the PCA reconstructed source images do not approximate well the original source images, which is also evident from comparing the corresponding histograms.

We now present the application of PCA to separate the mixture of speech signals in the BSS problem shown as the second example in Fig. 6.4. We have used a whitening transform given by (6.11), which is a special form of PCA that produces decorrelated signals with unit variance. Again, it is constructed from eigen-decomposition of the data covariance matrix $\mathbf{R_X}$ and implemented by the MATLAB code:

```
Rx = cov(X');          % estimate data covariance matrix
[E,LAM] = eig(Rx);     % eigen-decomposition of Rx
lam = diag(LAM);       % extract eigen-values
lam = sqrt(1./lam);    % square root of inverse
LAM = diag(lam);       % back into matrix form
Z=LAM*E'*X;            % the whitening transform
```

The reconstructed speech signals are shown in Fig. 6.12.

If the extracted source signals are compared with the original speech source signals shown in Fig. 6.3 we see that the quality of a reconstruction is not good. This can be observed better if histograms of the source speech signals shown in Fig. 6.3, mixed speech signals shown in Fig. 6.4 and PCA reconstructed source signals shown in Fig. 6.12, are compared. The corresponding histograms

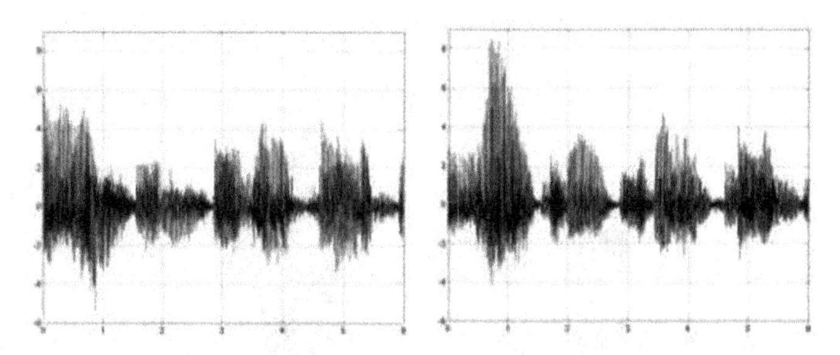

Fig. 6.12. Source speech signals reconstructed by whitening (PCA) transform.

are shown in Figs. 6.13, 6.14 and 6.15. From Fig. 6.13, we see that speech amplitudes are mainly concentrated around mean value of zero. Estimated kurtosis values of the corresponding speech source signals were $\kappa(s_1) = 9.52$ and $\kappa(s_2) = 2.42$. Estimated kurtosis values of the mixed speech signals were $\kappa(x_1) = 7.18$ and $\kappa(x_2) = 3.89$, while estimated kurtosis values of the PCA reconstructed speech signals were $\kappa(z_1) = 1.86$ and $\kappa(z_2) = 6.31$. Again, mixed and PCA recovered signals are more Gaussian than source signals. The reason why PCA failed to recover speech source signals is the same as in the first example, namely it forces decorrelation among the reconstructed signals, which is never optimal for non-Gaussian processes.

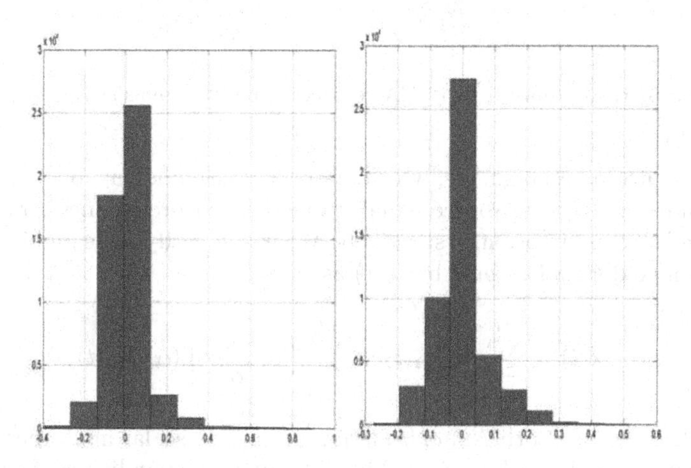

Fig. 6.13. Histograms of two speech signals shown in Fig. 6.3.

We have to notice that the second reconstructed speech signal shown in Fig. 6.12 corresponds with first source signal shown in Fig. 6.3 and the first reconstructed speech signal shown in Fig. 6.12 corresponds with the second source signal shown in Fig. 6.3. The same phenomenon occurred during reconstruction of the mixed image sources. It is a consequence of the scale

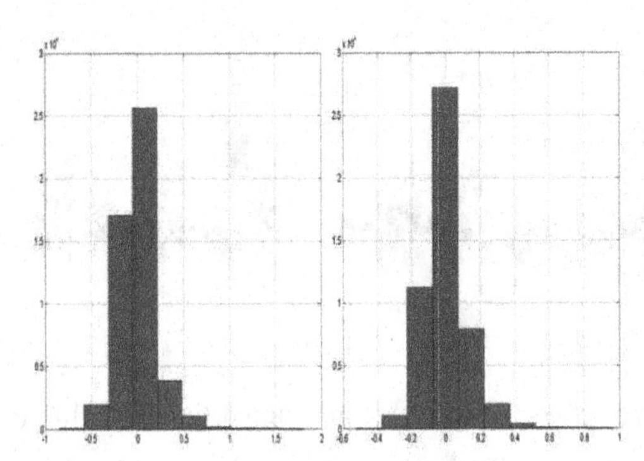

Fig. 6.14. Histograms of two mixed speech signals shown in Fig. 6.4.

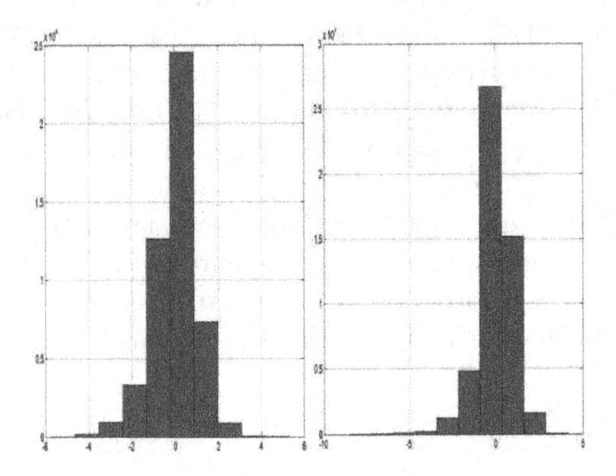

Fig. 6.15. Histograms of two PCA recovered speech signals shown in Fig. 6.12.

and permutation ambiguities, which are properties inherent to PCA and ICA algorithms that recover source signals using only uncorelatedness or statistical independence assumption, respectively. We easily verify scale ambiguity if we write a mixed signal defined by (6.1) as:

$$\mathbf{x}(t) = \sum_{m=1}^{M} \mathbf{a}_m s_m(t) = \sum_{m=1}^{M} \left(\mathbf{a}_m \frac{1}{\alpha_m} \right) (\alpha_m s_m(t)) \tag{6.16}$$

Evidently, because both $\mathbf{A}$ and $\mathbf{S}$ are unknown, any scalar multiplier in one of the sources can always be canceled by dividing corresponding column of $\mathbf{A}$ by the same multiplier. The statistical properties of the signals are not changed by scaling. The permutation ambiguity is verified by writing mixing model in (6.1) as:

$$\mathbf{X} = \mathbf{AS} = \mathbf{AP}^{-1}\mathbf{PS} \tag{6.17}$$

Components of the source matrix $\mathbf{S}$ and mixing matrix $\mathbf{A}$ can be freely exchanged such that $\mathbf{PS}$ is new source matrix and $\mathbf{AP}^{-1}$ is new mixing matrix. Matrix $\mathbf{P}$ is called permutation matrix having in each column and each row only one entry equal to one and all other entries equal to zero. We can also note that reordering of the source components does not change statistical (in)dependence among them.

We now want to provide further evidence why PCA is not good enough to reconstruct non-Gaussian source signals from their linear mixtures, which becomes the motivation to introduce ICA algorithms. In order to demonstrate PCA limitations, we generate two uniformly distributed source signals by using MATLAB command *rand.* Because the amplitudes are distributed randomly in the interval [0,1], the signals are made zero mean by subtracting mean value from each signal. We show in Fig. 6.16 scatter plots of two generated source signals for 100 samples (left) and 10,000 samples (right). The plot is obtained using MATLAB command $scatter(s_1, s_2)$. Equivalent plot would be obtained if MATLAB command $plot(s_1, s_2,' ko')$ was used. The scatter plot shows points of simultaneous occurrence in the amplitude space of the signals and gives nice visual illustration of the degree of statistical (in)dependence between the two signals. If two signals are independent, as they are in this example illustrated by Fig. 6.16, then for any particular value of signal s_1, signal s_2 can take any possible value in its domain of support. As we can see in Fig. 6.16-left, which shows scatter plot diagram for 100 samples, every time when signal s_1 took value 0.4 signal s_2 took different value in the interval $[-0.5, 0.5]$. This is even more visible in Fig. 6.16-right, which shows scatter plot diagram for 10,000 samples. If however signals $s_1$and s_2 would be statistically dependent, i.e., s_2 would be some function of s_1 given by $s_2 = f(s_1)$, then scatter diagram would be nothing else but plot of the function f. For example in a case of the linear relation $s_2 = as_1 + b$ the scatter plot diagram is a line. The process of plotting the scatter diagrams is implemented by the following sequence of MATLAB commands:

```
s1=rand(1,T);      % uniformly distributed source signal 1
s2=rand(1,T);      % uniformly distributed source signal 2
s1=s1-mean(s1);    % make source 1 zero mean
s2=s2-mean(s2);    % make source 2 zero mean
scatter(s1,s2);    % plot s1 vs. s2
```

In this example, we have chosen the number of samples to be $T = 100$ and $T = 10,000$. We have used $T = 10,000$ samples to generate mixed signals in accordance with (6.1) and to apply PCA and ICA algorithms. Consequently, our source signal matrix $\mathbf{S}$ will have dimensions $2 \times 10,000$. We observe that there is no structure present in the scatter plot i.e., no information about one source signal can be obtained based on information about other source signal. This means that if we knew the amplitude of the source signal s_1 at some time point, we cannot predict what would be the amplitude of source signal s_2 at the same time point. We can say that there is no redundancy between the source signals or that they are statistically independent.

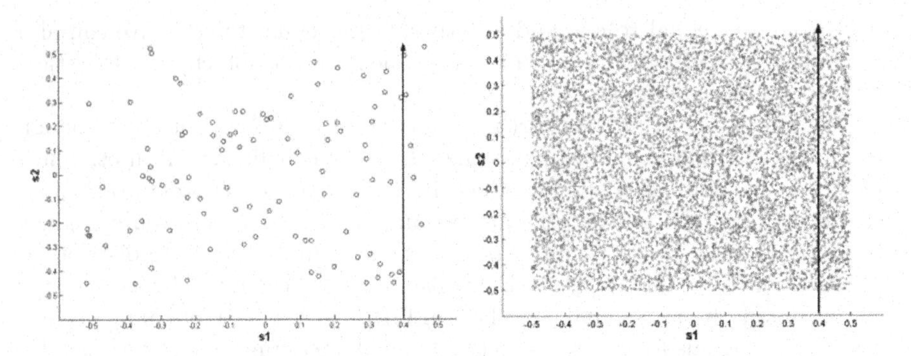

Fig. 6.16. Scatter plot of two uniformly distributed source signals: left-100 samples; right-10,000 samples.

We now generate mixed signals $\mathbf{X}$ according to (6.1) and mixing matrix $\mathbf{A}$ given by (6.2). Our data matrix $\mathbf{X}$ has dimensions 2×10000 and the process is implemented by the following sequence of MATLAB commands:

```
S=[s1;s2];                   % the matrix of source signals
X=A*S;                       % the matrix of mixed signals
scatter(X(1,:),X(2,:))       % plot the scatter plot x1 vs. x2
```

The corresponding scatter plot is shown in Fig. 6.17. Obviously, now there is a structure or redundancy present between the mixed signals. If we had knowledge about one mixed signal, we could guess the value of the other mixed signal with reasonably high probability. Next, we apply the PCA transform given by (6.7) and (6.9) to the mixed signals. The sequence of the MATLAB commands that implements PCA and plots the scatter plot is given with:

```
Rx=cov(X');                % estimate data covariance matrix
[E,D]=eig(Rx);             % eigen-decomposition of Rx
Z = E'*X;                  % PCA transform
scatter(Z(1,:),Z(2,:))     % plot z1 vs. z2
```

The scatter plot of PCA transformed signals $\mathbf{Z}$ is shown in Fig. 6.18.

Evidently, the decorrelation process realized through PCA transform reduced significantly the redundancy level between the transformed signals. However, it is still possible to gain the information about one signal having information about another, especially if a signal amplitude is at the borders of the range of values. If we compare scatter plots of source signals and PCA transformed signals, we can visually observe the similarity in the scatter plots, and also the fact that further rotation of the scatter plot of PCA de-mixed signals is necessary. This further rotation step can be implemented by an orthogonal matrix obtained by the ICA algorithms [76, 35].

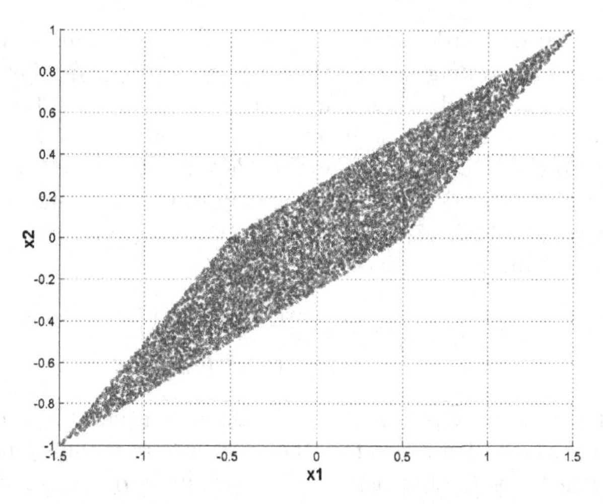

Fig. 6.17. Scatter plot of two mixed signals obtained by linear mixing of two uniformly distributed source signals.

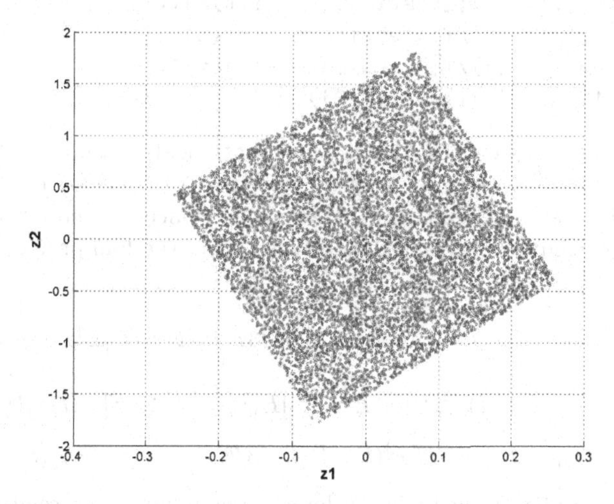

Fig. 6.18. Scatter plot of two PCA transformed signals.

The scatter plots in Figs. 6.16, 6.17 and 6.18 provide intuitive illustration of why the PCA is not good enough to de-mix non-Gaussian sources and why ICA is necessary. We want to provide an additional and more quantitative evidence for that statement based on some measures of statistical dependence between corresponding source, mixed and reconstructed signals. For that purpose, we shall use measures for the second order and fourth order statistical dependence between the signals given in the form of normalized cross-correlation and normalized fourth-order cross-cumulants, respectively. The reason why third order statistical dependence is not used is due to the fact that stochastic processes which are symmetrically distributed around the mean value have odd order statistics equal to zero and majority of signals used in practice (including signals that we use throughout this chapter) are symmetrically distributed. That is why the first statistics of the order higher than two that is used to measure statistical dependence is the fourth order statistics. We present in appendix a more detailed discussion between uncorrelatedness and independence, as well as definition of cumulants and some of their properties [126, 22, 97, 100]. The normalized cross-correlation between two processes x_i and x_j is defined by:

$$\overline{C_{11}}(x_i, x_j) = \frac{E[x_i(t)x_j(t)]}{\left(E[x_i^2(t)]\right)^{\frac{1}{2}} \left(E[x_j^2(t)]\right)^{\frac{1}{2}}} \tag{6.18}$$

which implies $-1 \leq \overline{C_{11}}(x_i, x_j) \leq 1$. This normalization makes cross-correlation invariant with respect to the fluctuations of the signal amplitude. $\overline{C_{11}}(x_i, x_j)$ can be estimated using MATLAB function $corrcoef(X')$ or the following MATLAB code may be used as well:

```
m11=sum(xi.*xj)/T;% estimate E[xi(t)xj(t)]
m2i = sum(xi.*xi)/T;% estimate E[xi(t)^2]
m2j = sum(xj.*xj)/T;% estimate E[xj(t)^2]
c11 = m11/sqrt(m2i)/sqrt(m2j);
```

In order to measure the fourth order statistical dependence between the stochastic processes we need to use the fourth order (FO) cross-cumulants [97, 100]. If we assume that shifts between sequences are equal to zero, then three FO cross-cumulants between the zero mean random processes x_i and x_j are defined as [97, 100]:

$$C_{13}(x_i, x_j) = E[x_i(t)x_j^3(t)] - 3E[x_i(t)x_j(t)]E[x_j^2(t)] \tag{6.19}$$

$$C_{22}(x_i, x_j) = E[x_i^2(t)x_j^2(t)] - E[x_i^2(t)]E[x_j^2(t)] - 2\left(E[x_i(t)x_j(t)]\right)^2 \tag{6.20}$$

$$C_{31}(x_i, x_j) = E[x_i^3(t)x_j(t)] - 3E[x_i(t)x_j(t)]E[x_i^2(t)] \tag{6.21}$$

We could proceed with the higher order cross-cumulants, but relations between them and moments become very complex and therefore computationally very expensive. In order to prevent fluctuations of the sample estimate of the FO cross-cumulants due to the fluctuations in the signal amplitude, we normalize them according to:

$$\overline{C_{pr}}(x_i, x_j) = \frac{C_{pr}(x_i, x_j)}{\left(E[x_i^2(t)]\right)^{\frac{p}{2}} \left(E[x_j^2(t)]\right)^{\frac{r}{2}}} \quad p, r \in \{1, 2, 3\} \; and \; p + r = 4 \tag{6.22}$$

We use the property satisfied by independent random variables x_i and x_j and some integrable functions $g(x_i)$ and $h(x_j)$:

$$E\left[g(x_i)h(x_j)\right] = E[g(x_i)]E\left[g(x_j)\right] \tag{6.23}$$

This follows from the definition of statistical independence given by (6.5). We choose $g(x_i)$ and $h(x_j)$ accordingly. Then it is easy to verify that $\overline{C_{11}}(x_i, x_j)$, $\overline{C_{22}}(x_i, x_j)$, $\overline{C_{13}}(x_i, x_j)$ and $\overline{C_{31}}(x_i, x_j)$ are equal to zero if x_i and x_j are statistically independent. Therefore they can be used as a measure of the second order and fourth order statistical (in)dependence. We provide here a MATLAB code that can be used to estimate normalized FO cross-cumulant $\overline{C_{22}}(x_i, x_j)$. It is straightforward to write a code for estimation of the other two FO cross-cumulants $\overline{C_{13}}(x_i, x_j)$ and $\overline{C_{31}}(x_i, x_j)$.

```
m22 = sum(xi.*xi.*xj.*xj)/T;% estimate E[xi(t)^2.*xj(t)^2]
m2i = sum(xi.*xi)/T;  % estimate E[xi(t)^2]
m2j = sum(xj.*xj)/T;% estimate E[xj(t)^2]
m11 = sum(xi.*xj)/T;% estimate E[xi(t)xj(t)]
c22 = (m22 - m2i*m2j - 2*m11*m11)/m2i/m2j;
```

Fig. 6.19-left shows the logarithm of the absolute values of the normalized cross-correlation $\overline{C_{11}}$ for two uniformly distributed source signals (with the scatter plot shown in Fig. 6.16, two mixed signals (with the scatter plot shown in Fig. 6.17), two PCA de-mixed signals (with scatter plot shown in Fig. 6.18) and two, yet to be derived, ICA de-mixed signals (with the scatter plot shown in Fig. 6.25). Fig. 6.19-right shows the logarithms of the absolute value of the normalized FO cross-cumulant $\overline{C_{22}}$ between corresponding signals in the same order as for the cross-correlation. Fig. 6.20-left and 6.20-right shows logarithms of the absolute values of the FO cross-cumulants $\overline{C_{13}}$ and $\overline{C_{31}}$ for the same signals. (Note that low values of log10 imply low level of the cross-correlation (Fig. 6.19, left) and low level of the fourth order statistical dependence (Fig. 6.19, right and Fig. 6.20), respectively).

It is evident, that while PCA perfectly decorrelated the mixed signals, it did not significantly reduce the level of the FO statistical dependence between transformed signals $z_1$and z_2. On the contrary, the minimum mutual information ICA algorithm to be described later managed to reduce equally well both second order and fourth order statistical dependence between transformed signals y_1 and y_2. We present values of the estimated cross-correlations and FO cross-cumulants in Table 6.1.

	Sources s_1, s_2	Mixtures x_1, x_2	PCA z_1, z_2	ICA y_1, y_2
$\overline{C_{11}}$	-0.0128	0.9471	1.5e-14	-1.4e-3
$\overline{C_{22}}$	-0.0364	-0.6045	-0.4775	-0.0015
$\overline{C_{13}}$	0.0016	-1.04	0.2573	0.0061
$\overline{C_{31}}$	0.014	-0.6939	-0.2351	0.0072

Table 6.1. Estimated values of the cross-correlations and fourth order cross-cumulants for source signals, mixed signals, PCA de-mixed signals and ICA de-mixed signals.

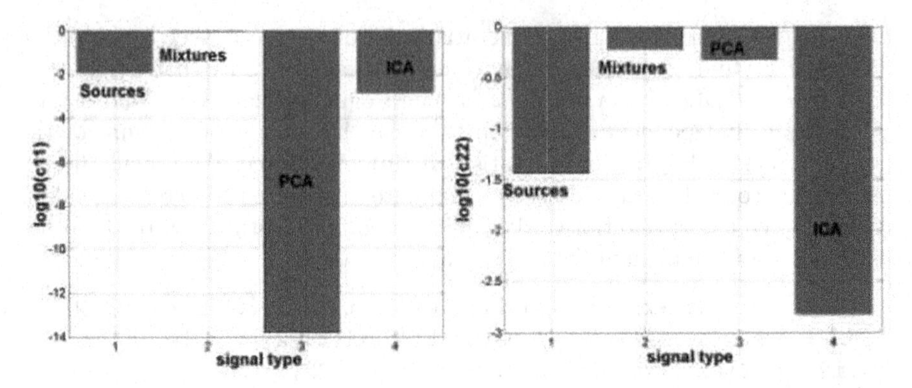

Fig. 6.19. (left) cross-correlations $\overline{C_{11}}$; (right) FO cross-cumulants $\overline{C_{22}}$. From left to right: the source signals, the mixed signals, the PCA de-mixed signals, the ICA de-mixed signals.

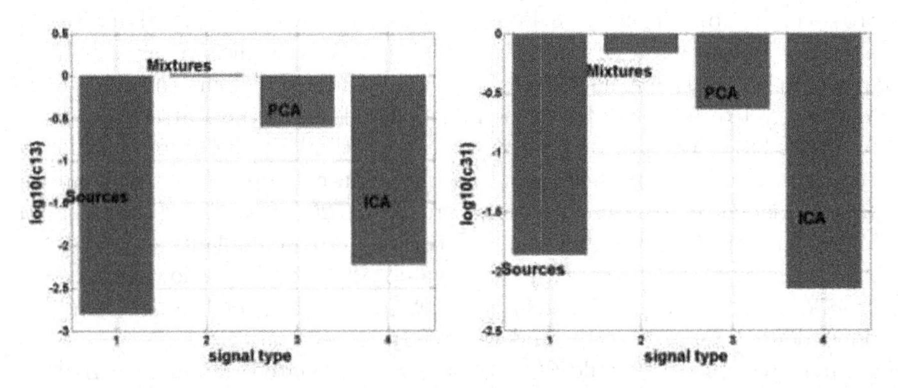

Fig. 6.20. (left) FO cross-cumulants $\overline{C_{13}}$; (right) FO cross-cumulants $\overline{C_{31}}$. From left to right: the source signals, the mixed signals, the PCA de-mixed signals, the ICA de-mixed signals.

We also want to illustrate why the PCA transform is optimal for Gaussian or normally distributed source signals, and why ICA is not needed for Gaussian sources. In order to do that we generate two Gaussian source signals using MATLAB command *randn.* Such obtained signals have zero mean and unit variance. We use the following MATLAB code to generate source signals and plot Fig. 6.19:

```
s1=randn(1,T);  % normally distributed source 1
s2=randn(1,T);  % normally distributed source 2
scatter(s1,s2); % plot s1. vs. s2
```

As in the previous example, the number of samples was chosen to be $T = 10,000$ which implies that source signal matrix $\mathbf{S}$ has dimensions $2 \times 10,000$. We show scatter plot of two normally distributed source signals in Fig. 6.21.

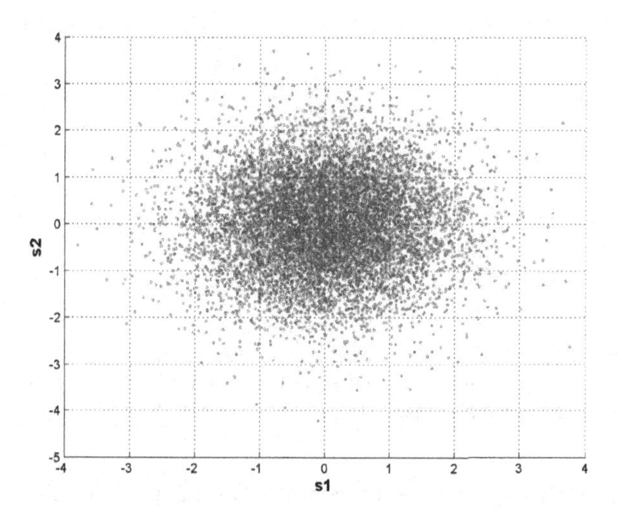

Fig. 6.21. Scatter plot of two normally distributed source signals.

No redundancy can be observed in the scatter plot, i.e., having information about one source signal does not help us to predict the value of another source signal. This is obviously not the case with the mixed signals $\mathbf{X}$, the scatter plot which is shown in Fig. 6.22. They are obtained by linear mixing model (6.1) and mixing matrix $\mathbf{A}$ given by (6.2). Dimensions of the data matrix $\mathbf{X}$ are 2×10000. The following MATLAB code is used to generate mixed signals and plot Fig. 6.22:

```
S = [s1; s2];                  % form source matrix
X = A*S;                       % generate data matrix
scatter(X(1,:),X(2,:));  % plot x1 vs. x2
```

When the whitening transform given by (6.11) is applied to mixed signals with the scatter plot shown in Fig. 6.22 we obtain signals $\mathbf{Z}$, the scatter plot which is shown in Fig. 6.23. The following MATLAB code is used to implement whitening transform and plot Fig. 6.23.

```
Rx = cov(X');                  % estimate data covariance matrix
[E,LAM] = eig(Rx);             % eigen-decomposition of Rx
lam = diag(LAM);               % extract eigen-values
lam = sqrt(1./lam);            % square root of inverse
LAM = diag(lam);               % back into matrix form
Z=LAM*E'*X;                    % the whitening transform
scatter(Z(1,:),Z(2,:))   % plot z1 vs. z2
```

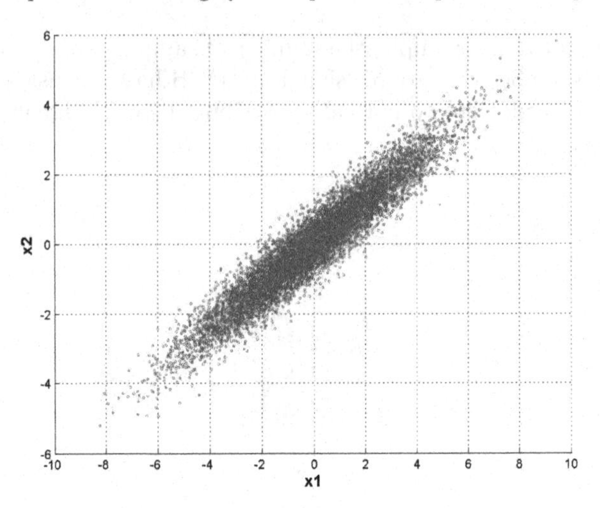

Fig. 6.22. Scatter plot of two mixed signals obtained by linear mixing of two normally distributed source signals.

Now, if we compare the scatter plots of the Gaussian source signals, Fig. 6.21, and signals obtained by PCA (whitening) transform, Fig. 6.23, we see that they look identical. This confirms that PCA is really optimal transform for Gaussian sources. This comparison, however, also shows that there is nothing more to be done by ICA transform if source signals are Gaussian. Thus, when ICA is required it is understood that unknown source signals are statistically independent and at most one of them is Gaussian.

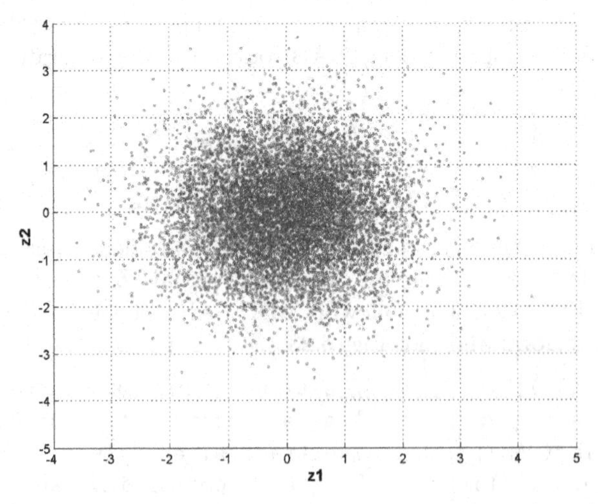

Fig. 6.23. Scatter plot of two signals obtained by whitening (PCA) transform.

6.2 Independent Component Analysis

ICA is a statistical and computational technique for solving BSS problem through recovery of the hidden factors, or the sources, from the measurements. In the model, the measured variables may be linear or nonlinear mixtures of the unknown sources. The mixing system that is active during recording of data from sources is also unknown. As already mentioned, only linear models given by (6.1) are treated here. For previously discussed reasons, the source signals are assumed to be non-Gaussian and statistically independent. The technique of ICA, although not yet by the name, was introduced in early 1980s in [13, 64, 65] and in a journal paper [80]. The introduced neural network, based on neuromimetic architecture, was able to separate independent signals. It was shown later on that the Herault-Jutten network only works when source signals are sub-Gaussian [37]. Cardoso [26] presented an algorithm that minimizes second and fourth order statistical dependence through joint approximate diagonalization of eigenmatrices, which is known as the JADE algorithm. A neural network for blind separation of source signals with the equivariant property (when convergence of the learning process does not depend on the properties of the unknown mixing matrix) was formulated by Cichocki, [36]. Common [38] defined the concept of ICA and proposed a cost function related to approximate minimization of mutual information between the components of the signal matrix $\mathbf{Y}$, (6.3).

In the following years, several algorithms that perform ICA were proposed. They included information maximization (*Infomax*) by Bell and Sejnowski [17], maximum likelihood (ML) [56, 114], negentropy maximization [57], and FastICA [77]. As pointed out by several authors, [91, 23, 112, 132] there is equivalence between *Infomax*, ML, negentropy maximization and even FastICA in a sense that all of them minimize either directly or indirectly mutual information between components of the signal matrix $\mathbf{Y}$. Because of its unifying framework and because it is an exact measure of statistical independence [41], we shall present a mutual information minimization approach to ICA. Before that, however, we want to mention that the BSS problem can be solved using the second order statistics provided that the source signals have colored statistics i.e. that covariances over different time lags are well-defined [139, 102, 18, 161].

In order to derive the minimum mutual information approach to ICA, we need to introduce the concept of entropy and the definition of mutual information. We shall also present some mathematical results without proofs, which will be useful in the subsequent derivations. Entropy of a random variable can be interpreted as the degree of information that the observation of the variable gives. The entropy of a set of variables is called joint entropy. Scatter plots shown in Fig. 6.16 to 6.18 and 6.21 to 6.23 represent visualisation of the joint entropy between the two source signals, mixed signals and separated signals. It is evident from scatter plots that more information about one signal could be inferred from the knowledge of another one in a case of mixed signals than in a case of separated or source signals. Entropy H is formally defined for a discrete-valued random variable X as [41]:

$$H(X) = -\sum_i P(X = a_i) \log P(X = a_i) \tag{6.24}$$

where $P(X = a_i)$ stands for probability of the random variable X having realization a_i. The definition of entropy for discrete-valued random variable can be generalized for continuous-valued variables and vectors, in which case it is called the differential or marginal entropy:

$$H(x) = -\int p_x(\xi) \log p_x(\xi) d\xi \tag{6.25}$$

The joint entropy for a vector random variable is defined as:

$$H(\mathbf{x}) = -\int p_{\mathbf{x}}(\xi) \log p_{\mathbf{x}}(\xi) \mathbf{d}\xi \tag{6.26}$$

The random process with maximum entropy, i.e., maximal randomness among processes with finite support, is a uniformly distributed random process. The random process with maximum entropy among random variables of finite variance and infinite support is the Gaussian or normally distributed random process [110]. Joint entropy of the random vector process $\mathbf{y}$ defined by the linear model (6.3) can be expressed as [76]:

$$H(\mathbf{y}) = H(\mathbf{x}) \ + \log |\det \mathbf{W}| \tag{6.27}$$

From (6.27), we see that differential entropy is increased by a linear transformation. For a square invertible matrix $\mathbf{W}$, the derivative of the determinant of $\mathbf{W}$ with respect to matrix elements w_{ij} is [60]:

$$\frac{\partial}{\partial w_{ij}} \det \mathbf{W} = W_{ij} \tag{6.28}$$

where W_{ij} is a *cofactor* of $\mathbf{W}$ at a position (i,j). (6.28) is written in matrix form as:

$$\frac{\partial}{\partial \mathbf{W}} \det \mathbf{W} = adj\,(\mathbf{W})^T = \det \mathbf{W} \left(\mathbf{W}^T\right)^{-1} \tag{6.29}$$

where $adj(\mathbf{W})$ stands for *adjoint* of $\mathbf{W}$. The previous result also implies:

$$\frac{\partial \log |\det \mathbf{W}|}{\partial w_{ij}} = \frac{1}{\det \mathbf{W}} W_{ij} \tag{6.30}$$

or in matrix form:

$$\frac{\partial \log |\det \mathbf{W}|}{\partial \mathbf{W}} = \left(\mathbf{W}^T\right)^{-1} \tag{6.31}$$

We now define mutual information as the measure of the information that members of a set of random variables have on the other random variables in the set. For that purpose we use an interpretation of the mutual information as a distance, using what is called the Kullback-Leibler divergence [41]:

$$I(y_1, y_2, ..., y_N) = D\left(p(\mathbf{y}) \,||\, \prod_{n=1}^{N} p_n(y_n)\right) = \int p(\mathbf{y}) \log \frac{p(\mathbf{y})}{\prod_{n=1}^{N} p_n(y_n)} \mathbf{dy} \tag{6.32}$$

The Kullback-Leibler divergence can be considered as a distance between the two probability densities. It is always non-negative, and zero if and only if the two distributions are equal [41], which implies:

$$p(\mathbf{y}) = \prod_{n=1}^{N} p_n(y_n) \tag{6.33}$$

so that components of $\mathbf{y}$ are statistically independent (see definition of statistical independence given by (6.5)). Therefore, mutual information can be used as a cost function for ICA algorithms. Using definitions for marginal and joint entropies given by (6.25) and (6.26), mutual information can be written as:

$$I(y_1, y_2, ..., y_N) = \sum_{n=1}^{N} H(y_n) - H(\mathbf{y}) \tag{6.34}$$

Furthermore, using the result for the entropy of the transformation given by (6.27), the mutual information is written as:

$$I(y_1, y_2, ..., y_N) = \sum_{n=1}^{N} H(y_n) - \log|\det \mathbf{W}| - H(\mathbf{x}) \tag{6.35}$$

If we look at the definition for the marginal entropy in (6.25), we see that it can be interpreted as a weighted mean of the term $\log p_x(x)$. Using the definition of mathematical expectation, it can alternatively be written as $H(x) = -E\left[\log p_x(x)\right]$. This allows us to write mutual information as:

$$I(y_1, y_2, ..., y_N) = -\sum_{n=1}^{N} E\left[\log p_n(y_n)\right] - \log|\det \mathbf{W}| - H(\mathbf{x}) \tag{6.36}$$

We next derive a gradient descent algorithm that will enable unsupervised learning of the de-mixing matrix $\mathbf{W}$ through minimization of the mutual information $I(y_1, y_2, ..., y_N)$:

$$\Delta w_{ij} = \frac{\partial I}{\partial w_{ij}} = -\sum_{n=1}^{N} E\left[\frac{1}{p_n(y_n)} \frac{\partial p_n(y_n)}{\partial y_n} \frac{\partial y_n}{\partial w_{ij}}\right] - \frac{\partial}{\partial w_{ij}} \log|\det \mathbf{W}| - \frac{\partial}{\partial w_{ij}} H(\mathbf{x}) \tag{6.37}$$

In (6.37), entropy of the measured signals $H(\mathbf{x})$ does not depend on w_{ij}. Also $\partial y_n / \partial w_{ij} = x_j \delta_{ni}$ where δ_{ni} is Kronecker delta. Now (6.37) reduces to:

$$\Delta w_{ij} = \frac{\partial I}{\partial w_{ij}} = -E\left[\frac{1}{p_i(y_i)}\frac{\partial p_i(y_i)}{\partial y_i}x_j\right] - \frac{1}{\det \mathbf{W}}W_{ij} \tag{6.38}$$

To obtain (6.38), we have also used result given by (6.30). The term:

$$\varphi_i(y_i) = -\frac{1}{p_i(y_i)}\frac{\partial p_i(y_i)}{\partial y_i} \tag{6.39}$$

is also called score or activation function. Combining (6.39) and (6.38), we obtain the de-mixing matrix update as:

$$\Delta \mathbf{W} = E\left[\varphi(\mathbf{y})\mathbf{x}^T\right] - \left(\mathbf{W}^T\right)^{-1} \tag{6.40}$$

The learning rule given by (6.40) involves calculation of the matrix inverse. This is not a very attractive property especially if unsupervised learning has to be performed in an adaptive, i.e., on-line manner. Independent discovery of the *natural* gradient [9] or *relative* gradient [25] helped to overcome this difficulty and obtain computationally scalable learning rules. The key insight behind the natural gradient was that the space of square non-singular matrices is not flat, i.e., Euclidean, but curved, i.e., Riemannian. Therefore, the Euclidean gradient of a scalar function with matrix argument does not point into the steepest direction of the cost function $I(\mathbf{W})$. Instead, it has to be corrected by the metric tensor, which was found to be $\mathbf{W}^T\mathbf{W}$. Applying this correction to the learning rule given by (6.40) we obtain:

$$\Delta \mathbf{W}\mathbf{W}^T\mathbf{W} = \left(E\left[\varphi(\mathbf{y})\mathbf{y}^T\right] - \mathbf{I}\right)\mathbf{W} \tag{6.41}$$

The complete learning equation for de-mixing matrix $\mathbf{W}$ is given by:

$$\mathbf{W}(k+1) = \mathbf{W}(k) + \eta\left(\mathbf{I} - E\left[\varphi(\mathbf{y})\mathbf{y}^T\right]\right)\mathbf{W}(k) \tag{6.42}$$

where k is an iteration index and η is customary learning gain. (6.42) represents a *batch* or *off line* learning rule, meaning that the entire set of T samples is used for learning. At each iteration k, mathematical expectation is evaluated as:

$$E\left[\varphi_i(y_i)y_j\right] \simeq \frac{1}{T}\sum_{t=1}^{T}\varphi_i(y_i(t))y_j(t) \tag{6.43}$$

In some applications, we do not have available the entire data set. Instead, we have to acquire data on the sample-by-sample basis and update de-mixing matrix $\mathbf{W}$ in the same manner. This is called *adaptive* or *on-line* learning which is obtained from (6.42) by simply dropping the expectation operator:

$$\mathbf{W}(t+1) = \mathbf{W}(t) + \eta\left(\mathbf{I} - \varphi(\mathbf{y}(t))\mathbf{y}(t)^T\right)\mathbf{W}(t) \tag{6.44}$$

The de-mixing matrix learning rules given by (6.42) and (6.44) have the equivariant property, which means that convergence of the ICA algorithm does not depend on the properties of the unknown mixing matrix [25, 36]. In order to show that, we combine together (6.3), (6.42) or (6.44) and (6.1) and obtain:

$$\begin{aligned}\mathbf{y}(t) &= \mathbf{W}(t)\mathbf{x}(t) \\ &= \left[\mathbf{W}(t-1) + \eta\left(\mathbf{I} - \varphi(\mathbf{y}(t-1))\mathbf{y}(t-1)^T\right)\mathbf{W}(t-1)\right]\mathbf{A}\mathbf{s}(t) \\ &= \left[\mathbf{Q}(t-1) + \eta\left(\mathbf{I} - \varphi(\mathbf{y}(t-1))\mathbf{y}(t-1)^T\right)\mathbf{Q}(t-1)\right]\mathbf{s}(t)\end{aligned} \tag{6.45}$$

where $\mathbf{Q} = \mathbf{WA}$. Evidently, performance of the ICA algorithm given by (6.42) or (6.44) does not depend on the mixing matrix $\mathbf{A}$. This property is very important for solving BSS problems of the ill-conditioned mixtures, i.e., mixtures for which mixing matrix $\mathbf{A}$ is almost singular. There are two issues which ought to be briefly discussed at this point: the stopping criterion and the value of the learning gain η. In an *on-line* learning given by (6.44) the stopping criterion does not have to be defined because the learning process never stops i.e. it is not iterative. In a *batch* or *off-line* learning the simplest form of the stopping criterion is the relative error criterion defined by:

$$\|\mathbf{W}(k+1) - \mathbf{W}(k)\|_2 < \varepsilon \tag{6.46}$$

where in (6.46) $\|\|_2$ represents L_2 norm and ε is a small parameter that can be set through experiments. Computationally more demanding option is to evaluate mutual information $I(y_1, y_2, ..., y_N)$ at each iteration and to apply relative error criteria on the mutual information itself. Numerical value of the learning gain η depends on the amplitude range of signals $\mathbf{y}(t)$ but also on the value of the to be defined Gaussian exponent that is used to define a nonlinearity $\varphi_n(y_n)$ (see (6.47) and (6.48)). It has to be determined experimentally for each particular scenario.

We see in (6.39) that the score function depends on the knowledge of the pdf of the source signals, which is not available because the problem is *blind.* Despite the fact that exact pdf is not known in general, the ICA works due to the fact that if the model pdf is an approximation to the source signal pdf, then the extracted signals will be the source signals [24, 10]. The function used to approximate super-Gaussian class of distributions is $\varphi(y) = \tanh(y)$, and the function to approximate sub-Gaussian class of distributions is $\varphi(y) = y^3$. Another commonly used activation function for ICA algorithms is derived from (6.39) using the generalized Gaussian distribution as a model for $p_i(y_i)$, [33, 35]. The generalized Gaussian distribution is given by:

$$p_i(y_i, \theta_i) = \frac{\theta_i}{2\sigma_i \Gamma(1/\theta_i)} \exp\left(-\frac{1}{\theta_i}\left|\frac{y_i}{\sigma_i}\right|^{\theta_i}\right) \tag{6.47}$$

where θ_i is called the Gaussian exponent, $\sigma_i^{\theta_i} = E\left[|y_i|^{\theta_i}\right]$ is the dispersion of the distribution and $\Gamma(\theta) = \int_0^\infty t^{\theta-1}\exp(-t)dt$ is a gamma function. By varying the value of the Gaussian exponent θ_i, we can control the shape of distribution. For $\theta_i = 1$ the Laplace distribution is obtained, for $\theta_i = 2$, the Gaussian distribution is obtained and for $\theta_i \to \infty$ a uniform distribution is obtained. Fig. 6.24 shows distributions obtained from (6.47) for values of θ_i equal to 1, 2 and 3 with dispersion parameter $\sigma = 1$. If the generalized Gaussian pdf given by (6.47), is inserted in the score function (6.39), the score function parameterized with the value of Gaussian exponent is obtained as:

$$\varphi_i(y_i) = sign(y_i)\,|y_i|^{\theta_i - 1} \qquad (6.48)$$

The approximation given by (6.48) is valid for super-Gaussian distributions when $1 \leq \theta_i \leq 2$ and for sub-Gaussian distributions when $\theta_i > 2$, [33, 154]. In a majority of applications, we know in advance whether the source signals are sub-Gaussian (most of the communication signals and natural images) or super-Gaussian (speech and music signals). Then, the Guassian exponent can be set to predefined value. Alternative aproaches are to estimate score functions from data using multi-layer-perceptron [136] or using some empirical estimator of the unknown pdf, such as the Gram-Charlier approximation of the pdf [136] or the Parzen window based pdf estimator with Gaussian kernel [118]. In Appendix F, we give a derivation of the batch ICA learning rule (6.42) and adaptive ICA learning rule (6.44) with the score functions (6.39) obtained from an empirical distribution of the unknown pdf. Estimation of the unknown pdf is based on Parzen's window with Gaussian kernel [118]. Corresponding MATLAB code is also given at the end of Appendix F. The last approach provides robust ICA algorithms, but has computational complexity of the order $O(T^2N^2)$, where T represents number of samples and N represents number of sources. Algorithms for estimation of mutual information, score

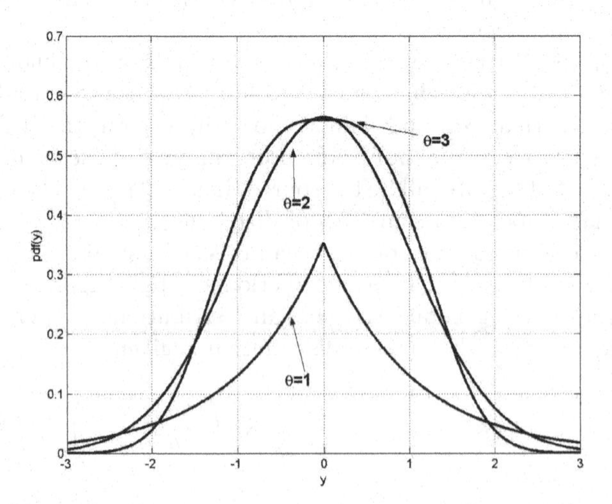

Fig. 6.24. Plot of generalized Gaussian pdf for various values of Gaussian exponent θ_i with $\sigma = 1$.

functions and entropies with smaller computational complexity of the order $O(3^N T + N^2 T)$ are presented in [113]. Further improvement in terms of the computational complexity of the order $O(NT \log T + N^2 T)$ is presented in [124]. We shall not further ellaborate these approaches here.

We now discuss results of the application of the minimum mutual information ICA algorithm given by (6.42) on the blind separation of mixed images and speech source signals, as well as two uniformly distributed source signals. We discuss results for uniformly distributed sources first. We have illustrated by the scatter plot shown in Fig. 6.18 that PCA is not good enough to reduce redundancy present among the mixed signals. We now apply the batch ICA algorithm given by (6.42) on signals obtained by the whitening transform given by (6.11). Because we knew that the source signals are sub-Gaussian, we have chosen value for the Gaussian exponent in the activation function (6.48) to be $\theta_1 = \theta_2 = 3$. The learning gain parameter was $\eta = 0.1$ and result was obtained after 400 iterations.

ICA algorithm given by (6.42) is implemented with the following sequence of MATLAB commands:

```
it=0;
I=eye(N); % identity matrix
W = I; % initial value for de-mixing matrix
while (it<ITMAX)  % iterate until it >ITMAX
it = it+1;
Y = W*Z; % ICA transform (6.3)
for i=1:N
fi(i,:)=sign(Y(i,:)).*power(abs(Y(i,:)),theta(i)-1);% (6.48)
end
        F = I - fi*Y'/T;% (6.42)
        W = W + gain*F*W; % (6.42)
end
```

The symbol N stands for the number of sensors which is assumed to be equal to the number of sources M, and which is equal to 2 in this example, and symbol T stands for the number of samples. The scatter plot of the reconstructed source signals is shown in Fig. 6.25. In comparison with the scatter plot shown in Fig. 6.16, we see that source signals are reconstructed well, which was not case with the PCA reconstructed signals, the scatter plot of which is shown in Fig. 6.18. To further support this statement we give values of the estimated kurtosis parameter for reconstructed source signals as $\kappa(y_1) = -1.194$ and $\kappa(y_2) = -1.194$. Estimated values of the kurtosis of the source signals were $\kappa(s_1) = -1.195$ and $\kappa(s_2) = -1.194$, while estimated values of the kurtosis of the mixed signals were $\kappa(x_1) = -0.798$ and $\kappa(x_2) = -0.566$. Also, we have provided evidence in Figs. 6.19 and 6.20 as well as in Table 6.1 that, unlike PCA, the ICA reduced the level of both the second order and fourth order statistical dependence between the transformed signals y_1 and y_2.

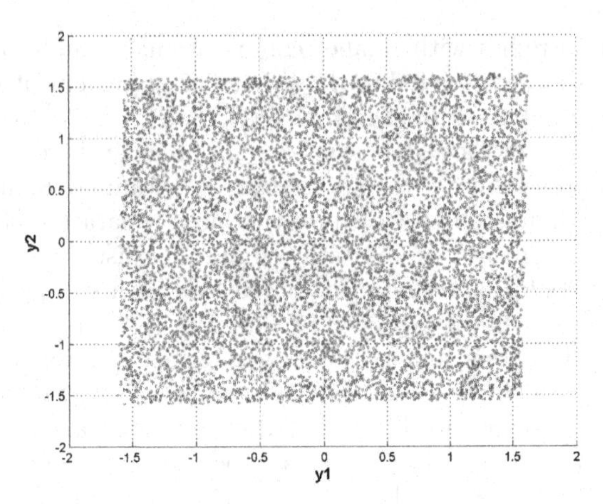

Fig. 6.25. Scatter plot of the signals recovered by the ICA algorithm (6.42).

We now present results of the application of the ICA algorithm given by (6.42) for reconstruction of images from the mixture shown in Fig. 6.2. The images reconstructed by PCA transform are shown in Fig. 6.7 . Because source images have sub-Gaussian statistics, the results obtained by PCA transform were not good. The ICA reconstructed images are shown in Fig. 6.26. Results shown in Fig. 6.26 were obtained after 700 iterations with the learning gain set to $\eta = 4 \times 10^{-8}$ and Gaussian exponent set to $\theta_1 = \theta_2 = 3$. Both parameters, number of iterations and learning gain, were chosen through experiments based on the visual quality of the separated images. Estimated kurtosis values of the reconstructed source images were $\kappa(y_1) = -1.36$ and $\kappa(y_2) = -0.86$. (Recall that for the source signals $\kappa(s_1) = -1.36$ and $\kappa(s_2) = -0.88$ as well as that for the reconstructed source images PCA resulted in $\kappa(s_1) = -0.82$ and $\kappa(s_2) = -0.4$). In comparison with the results obtained by PCA transform, the results obtained by ICA are very good, even though in the background of both reconstructed images remnants of other source image are still slightly visible. This is attributed to the inexact matching between the source pdf's and approximate pdf's obtained using the score function (6.48) with Gaussian exponent $\theta_1 = \theta_2 = 3$.

To further support this statement, we show in Fig. 6.27 results obtained using the JADE algorithm [26], which minimizes only second and fourth order statistical depenedence between the signals y_i. The remnant of the *cameraman* image is visible in the reconstructed *flowers* source image more than in a case of the minimum mutual information ICA algorithm given by (6.42). The MATLAB code for the JADE algorithm can be downloaded from *http://www.tsi.enst.fr/~cardoso/Algo/Jade/jade.m.* Also, for the purpose of comparison, we show in Fig. 6.28 results obtained by the FastICA algorithm [77] with cubic nonlinearity, which is suitable for sub-Gaussian distributions. As in the case of the JADE algorithm, the remnant

Fig. 6.26. Source images reconstructed by the ICA algorithm (6.42)

of the *cameraman* image is more visible in the reconstructed flowers source image than in a case of the minimum mutual information ICA algorithm given by (6.42). The FastICA algorithm is less accurate because of the employed negentropy approximation, although it makes FastICA algorithm very fast. MATLAB code for the FastICA algorithm can be downloaded from *http://www.cis.hut.fi/projects/ica/fastica/*.

Finally, in Fig. 6.29, we show histograms of the images reconstructed by the minimum mutual information ICA algorithm (6.42). Comparing them with the histograms of the source images shown in Fig. 6.8, we see that they are practically the same. This confirms that the principle of statistical independence and the minimization of the mutual information provide a powerful methodology for the solution of the BSS problems. We summarize performance comparisons between PCA and ICA in the blind image reconstruction problem in Table 6.2, where kurtosis values are given for source images, mixed images, PCA reconstructed images and ICA reconstructed images. When presenting

Fig. 6.27. Source images reconstructed by the JADE ICA algorithm.

Fig. 6.28. Source images reconstructed by the FastICA algorithm with cubic non-linearity.

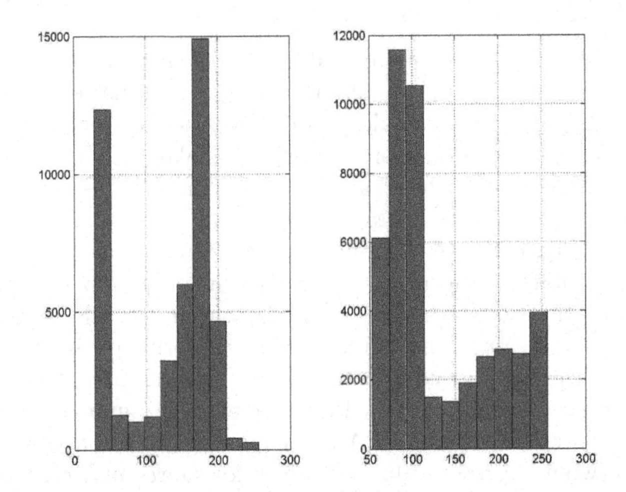

Fig. 6.29. Histograms of the images reconstructed by the ICA algorithm (6.42).

Signal type	κ_1	κ_2
Source images	-1.36	-0.88
Mixed images	-0.93	-0.64
PCA, (6.9), recovered images	-0.82	-0.40
ICA, (6.42), recovered images	-1.36	-0.86

Table 6.2. Kurtosis values of the source, mixed, PCA and ICA recovered images.

PCA results we have taken into account the fact that order of the recovered images was reversed.

We now show in Fig. 6.30 the speech signals reconstructed by the minimum mutual information ICA algorithm. Because speech signals are super-Gaussian, we have selected values for the Gaussian exponent to be $\theta_1 = \theta_2 =$

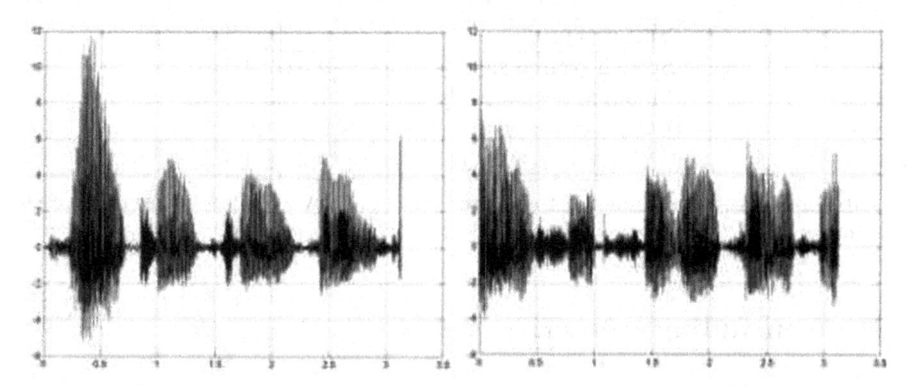

Fig. 6.30. Source speech signals reconstructed by the ICA algorithm (6.42).

1.5. The results shown in Fig. 6.30 were obtained after 400 iterations with the learning gain $\eta = 0.4$. Estimated kurtosis values of the reconstructed speech signals were $\kappa(y_1) = 9.52$ and $\kappa(y_2) = 2.42$. In comparison with the speech signals reconstructed by PCA, which were shown in Fig. 6.12, ICA was evidently very successful in the reconstruction of the source speech signals.

In Fig. 6.31, we show histograms of the recovered speech source signals. In comparison with the histograms of the source speech signals shown in Fig. 6.14, we see that they practically look the same. This again confirms that the principle of statistical independence and the minimization of the mutual information can succeed in solving the BSS problems. We summarize the performance comparisons between PCA and ICA in the blind speech reconstruction problem in Table 6.3, where kurtosis values are given for source speech signals, mixed speech signals, PCA reconstructed speech signals and ICA reconstructed speech signals. When presenting PCA results we have taken into account the fact that order of the recovered signals was reversed.

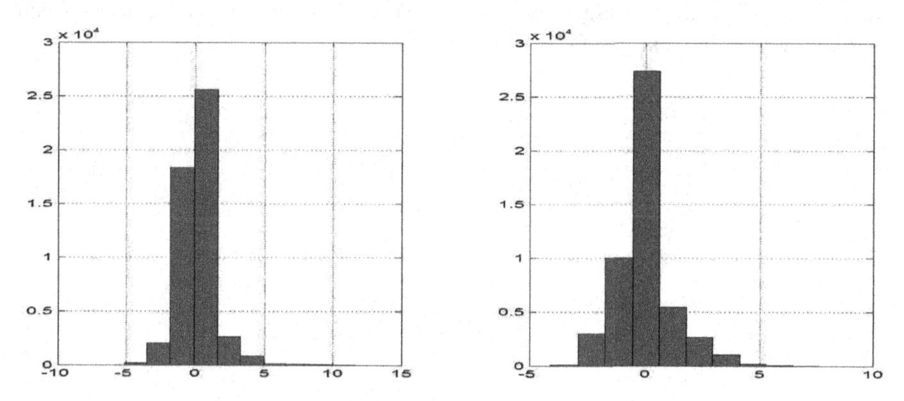

Fig. 6.31. Histograms of the speech signals recovered by the ICA algorithm (6.42).

Signal type	κ_1	κ_2
Speech source signals	9.52	2.42
Mixed speech signals	7.18	3.89
PCA recovered speech signals	6.31	1.86
ICA recovered speech signals	9.52	2.42

Table 6.3. Kurtosis values of the source, mixed, PCA and ICA recovered speech signals.

6.3 Concluding Remarks

ICA and PCA are methods that are used to recover unknown source signals from their linear mixtures in the unsupervised way, i.e., using the observed data only. Both ICA and PCA recover unknown source signals by exploiting the assumption that source signals were generated by independent physical sources and because of that are statistically independent. While PCA makes recovered signals uncorrelated ICA makes them approximately independent. Because of that PCA is optimal transform for Gaussian sources but it is not good enough for non-Gaussian sources. In the later case PCA is usually used as a pre-processing step for the ICA algorithms. The ICA algorithms can be derived in various ways in dependence on the definition of the cost function that measures the statistical (in)dependence between the transformed signals. We have derived batch and adaptive ICA algorithms by minimizing the mutual information between the transformed signals. Mutual information was used for two reasons. First, it represents an exact measure of statistical (in)dependence. Second, it also has unifying character in a sense that ICA algorithms derived from the maximum likelihood standpoint, information maximization standpoint or negentropy maximization standpoint all minimize either directly or indirectly mutual information between the transformed signals. Effectiveness of derived PCA and ICA algorithms is demonstrated on the examples of blind separation of natural images, which have sub-Gaussian statistics, and speech signals, which have super-Gaussian statistics.

A

Support Vector Machines

While introducing the soft SVMs by allowing some unavoidable errors and, at the same time, while trying to minimize the distances of the erroneous data points to the margin, or to the tube in the regression problems, we have augmented the cost $0.5\mathbf{w}^T\mathbf{w}$ by the term $\sum_{i=1}^{n}\left(\xi_i^k + \xi_i^{*k}\right)$ as the measure of these distances. Obviously, by using $k = 2$ we are punishing more strongly the far away points, than by using $k = 1$. There is a natural question then - what choice might be better in application. The experimental results [1] as well as the theoretically oriented papers [16, 131] point to the two interesting characteristics of the L1 and L2 SVMs. At this point, it is hard to say about some particular advantages. By far, L1 is more popular and used model. It seems that this is a consequence of the fact that L1 SVM produces sparser models (less SVs for a given data). Sparseness is but one of the nice properties of kernel machines. The other nice property is a performance on a real data set and a capacity of SVMs to provide good estimation of either unknown decision functions or the regression ones. In classification, we talk about the possibility to estimate the conditional probability of the class label. For this task, it seems that the L2 SVMs may be better. A general facts are that the L1-SVMs can be expected to produce sparse solutions and that L2-SVMs will typically not produce sparse solutions, but may be better in estimating conditional probabilities. Thus, it may be interesting to investigate the relationship between these two properties. Two nice theoretical papers discussing the issues of sparseness and its trade-off for a good prediction performance are mentioned above. We can't go into these subtleties here. Instead, we provide to the reader the derivation of the L2 SVMs model, and we hope the models presented here may help the reader in his/hers own search for better SVMs model.

Below we present the derivation of the L2 soft NL classifier given by (2.29) following by the derivation of the L2 soft NL regressor. Both derivations are performed in the feature space $\mathcal{S}$. Thus the input vector to the SVM is the $\boldsymbol{\Phi}(\mathbf{x})$ vector. All the results are valid for a linear model too (where we work in the original input space) by replacing $\boldsymbol{\Phi}(\mathbf{x})$ by $\mathbf{x}$.

T.-M. Huang et al.: *Kernel Based Algorithms for Mining Huge Data Sets*, Studies in Computational Intelligence (SCI) **17**, 209–215 (2006)
www.springerlink.com

A.1 L2 Soft Margin Classifier

Now, we start from the equation (2.24) but instead of a linear distance ξ_i we work with a quadratic one ξ_i^2 . Thus the task is to

$$\frac{1}{2}\mathbf{w}^T\mathbf{w} + \frac{C}{2}\sum_{i=1}^{n}\xi_i^2, \tag{A.1a}$$

subject to

$$y_i\left[\mathbf{w}^T\boldsymbol{\Phi}(\mathbf{x}_i) + b\right] \geq 1 - \xi_i, \quad i = 1,\ n,\ \xi_i \geq 0, \tag{A.1b}$$

i.e., subject to

$$\mathbf{w}^T\mathbf{x}_i + b \geq +1 - \xi_i, \quad \text{for } y_i = +1,\ \xi_i \geq 0, \tag{A.1c}$$

$$\mathbf{w}^T\mathbf{x}_i + b \leq -1 + \xi_i, \quad \text{for } y_i = -1,\ \xi_i \geq 0, \tag{A.1d}$$

Now, both the $\mathbf{w}$ and the $\boldsymbol{\Phi}(\mathbf{x})$ are the s-dimensional vectors. Note that the dimensionality s can also be infinite and this happens very often (e.g., when the Gaussian kernels are used). Again, the solution to the quadratic programming problem (A.1) is given by the saddle point of the primal Lagrangian $L_p(\mathbf{w}, b, \boldsymbol{\xi}, \boldsymbol{\alpha})$ shown below

$$L_p(\mathbf{w}, b, \boldsymbol{\xi}, \boldsymbol{\alpha}) = \frac{1}{2}\mathbf{w}^T\mathbf{w} + \frac{C}{2}\left(\sum_{i=1}^{n}\xi_i^2\right) - \sum_{i=1}^{n}\alpha_i\{y_i[\mathbf{w}^T\mathbf{x}_i + b] - 1 + \xi_i\} \tag{A.2}$$

Note that the Lagrange multiplier $\boldsymbol{\beta}$ associated with $\boldsymbol{\xi}$ is missing here. It vanishes by combining (2.26e) and (2.26c) which is now equal to . Again, we should find an optimal saddle point $(\mathbf{w}_o, b_o, \boldsymbol{\xi}_o, \boldsymbol{\alpha}_o)$ because the Lagrangian L_p has to be minimized with respect to $\mathbf{w}$, b and $\boldsymbol{\xi}$, and *maximized* with respect to nonnegative α_i. And yet again, we consider a solution in a dual space as given below by using

- standard conditions for an optimum of a constrained function

$$\frac{\partial L}{\partial \mathbf{w}_o} = 0, \qquad \text{i.e., } \mathbf{w}_o = \sum_{i=1}^{n}\alpha_i y_i \boldsymbol{\Phi}(\mathbf{x}_i) \tag{A.3a}$$

$$\frac{\partial L}{\partial b_o} = 0, \qquad \text{i.e., } \sum_{i=1}^{n}\alpha_i y_i = 0 \tag{A.3b}$$

$$\frac{\partial L}{\partial \xi_{io}} = 0, \qquad \text{i.e., } C\xi_i - \alpha_i = 0, \tag{A.3c}$$

- and the KKT complementarity conditions below,

$$\alpha_{io}\{y_i\left[\mathbf{w}^T\boldsymbol{\Phi}(\mathbf{x}_i)+b\right]-1+\xi_i\}=0$$

$$\alpha_{io}\{y_i\left[\sum_{i=1}^{n}\alpha_{jo}y_jk(\mathbf{x}_j,\mathbf{x}_i)+b_o\right]-1+\xi_i\}=0,\quad i=1,\ldots n \tag{A.4}$$

A substitution of (A.3a) and (A.3c) into the L_p leads to the search for a maximum of a *dual Lagrangian*

$$L_d(\alpha)=\sum_{i=1}^{n}\alpha_i-\frac{1}{2}\sum_{i,\,j=1}^{n}y_iy_j\alpha_i\alpha_j\left(k(\mathbf{x}_i,\mathbf{x}_j)+\frac{\delta_{ij}}{C}\right) \tag{A.5a}$$

subject to

$$\alpha_i\geq 0,\quad i=1,n \tag{A.5b}$$

and under the equality constraints

$$\sum_{i=1}^{n}\alpha_iy_i=0 \tag{A.5c}$$

where, $\delta_{ij}=1$ for $i=j$, and it is zero otherwise. There are three tiny differences in respect to the most standard L1 SVM. First, in a Hessian matrix, a term $1/C$ is added to its diagonal elements which ensures positive definiteness of $\mathbf{H}$ and stabilizes the solution. Second, there is no upper bound on α_i and the only requirement is α_i to be non-negative. Third, there are no longer complementarity constraints (2.26e), $(C-\alpha_i)\xi_i=0$.

A.2 L2 Soft Regressor

An entirely similar procedure leads to the soft L2 SVM regressors. We start from the reformulated equation (2.44) as given below

$$R_{\mathbf{w},\,\xi,\,\xi^*}=\left[\frac{1}{2}||\mathbf{w}||^2+C\left(\sum\nolimits_{i=1}^{n}\xi_i^2+\sum\nolimits_{i=1}^{n}\xi_i^{*2}\right)\right] \tag{A.6}$$

and after an introduction of the Lagrange multipliers α_i or α_i^* we change to the unconstrained primal Lagrangian L_p as given below,

$$\begin{aligned}L_p(\mathbf{w},b,\,\xi_i,\,\xi_i^*,\,\alpha_i,\,\alpha_i^*)&=\frac{1}{2}\mathbf{w}^T\mathbf{w}+\frac{C}{2}\sum\nolimits_{i=1}^{n}(\xi_i^2+\xi_i^{*2})\\&-\sum\nolimits_{i=1}^{n}\alpha_i\left[\mathbf{w}^{\mathrm{T}}\mathbf{x}_i+b-y_i+\varepsilon+\xi_i\right]\\&-\sum\nolimits_{i=1}^{n}\alpha_i^*\left[y_i-\mathbf{w}^{\mathrm{T}}\mathbf{x}_i-b+\varepsilon+\xi_i^*\right]\end{aligned} \tag{A.7}$$

Again, the introduction of the dual variables β_i and β_i^* associated with ξ_i and ξ_i^* is not needed for the L2 SVM regression models. At the optimal solution the partial derivatives of L_p in respect to primal variables vanish. Namely,

$$\frac{\partial L_p(\mathbf{w}_o, b_o, \xi_{io}, \xi_{io}^*, \alpha_i, \alpha_i^*)}{\partial \mathbf{w}} = \mathbf{w}_o - \sum_{i=1}^{n} (\alpha_i - \alpha_i^*)\boldsymbol{\Phi}(\mathbf{x}_i) = 0, \quad \text{(A.8a)}$$

$$\frac{\partial L_p(\mathbf{w}_o, b_o, \xi_{io}, \xi_{io}^*, \alpha_i, \alpha_i^*)}{\partial b} = \sum_{i=1}^{n} (\alpha_i - \alpha_i^*) = 0, \quad \text{(A.8b)}$$

$$\frac{\partial L_p(\mathbf{w}_o, b_o, \xi_{io}, \xi_{io}^*, \alpha_i, \alpha_i^*)}{\partial \xi_i} = C\xi_i - \alpha_i = 0, \quad \text{(A.8c)}$$

$$\frac{\partial L_p(\mathbf{w}_o, b_o, \xi_{io}, \xi_{io}^*, \alpha_i, \alpha_i^*)}{\partial \xi_i^*} = C\xi_i^* - \alpha_i^* = 0. \quad \text{(A.8d)}$$

Substituting the KKT above into the primal L_p given in (A.7), we arrive at the problem of the *maximization of a dual variables Lagrangian* $L_d(\alpha, \alpha^*)$ below,

$$\begin{aligned} L_d(\alpha_i, \alpha_i^*) &= -\frac{1}{2} \sum_{i,j=1}^{n} (\alpha_i - \alpha_i^*)(\alpha_j - \alpha_j^*) \left(\boldsymbol{\Phi}^T(\mathbf{x}_i)\boldsymbol{\Phi}(\mathbf{x}_j) + \frac{\delta_{ij}}{C} \right) \\ &\quad - \varepsilon \sum_{i=1}^{n} (\alpha_i + \alpha_i^*) + \sum_{i=1}^{n} (\alpha_i - \alpha_i^*) y_i \\ &= -\frac{1}{2} \sum_{i,j=1}^{n} (\alpha_i - \alpha_i^*)(\alpha_j - \alpha_j^*) \left(k(\mathbf{x}_i, \mathbf{x}_j) + \frac{\delta_{ij}}{C} \right) \\ &\quad - \sum_{i=1}^{n} (\varepsilon - y_i)\alpha_i - \sum_{i=1}^{n} (\varepsilon + y_i)\,\alpha_i^* \end{aligned} \quad \text{(A.9)}$$

subject to constraints

$$\sum_{i=1}^{n} \alpha_i^* = \sum_{i=1}^{n} \alpha_i \text{ or } \sum_{i=1}^{n} (\alpha_i - \alpha_i^*) = 0 \quad \text{(A.10a)}$$

$$0 \le \alpha_i \quad i = 1, \ldots n \quad \text{(A.10b)}$$

$$0 \le \alpha_i^* \quad i = 1, \ldots n \quad \text{(A.10c)}$$

$$\text{(A.10d)}$$

At the optimal solution the following *KKT complementarity conditions* must be fulfilled

$$\alpha_i \left(\mathbf{w}^{\mathrm{T}}\mathbf{x}_i + b - y_i + \varepsilon + \xi_i \right) = 0 \quad \text{(A.11a)}$$

$$\alpha_i^* \left(- \mathbf{w}^{\mathrm{T}}\mathbf{x}_i - b + y_i + \varepsilon + \xi_i^* \right) = 0 \quad \text{(A.11b)}$$

$$\alpha_i \alpha_i^* = 0, \xi_i \xi_i^* = 0 \quad i = 1,\ n. \quad \text{(A.11c)}$$

Note that for the L2 SVM regression models the complementarity conditions (2.53c) and (2.53d) are eliminated here. After the calculation of Lagrange multipliers α_i and α_i^*, and by using (A.8a) we can find an *optimal* (desired) weight vector of the L2 *regression hyperplane* in a feature space as

$$\mathbf{w}_o = \sum_{i=1}^{n} (\alpha_i - \alpha_i^*)\mathbf{\Phi}(\mathbf{x}_i) \tag{A.12}$$

The best L2 regression hyperplane obtained is given by

$$F(\mathbf{x}, \mathbf{w}) = \mathbf{w}_o^T \boldsymbol{\Phi}(\mathbf{x}) + b = \sum_{i=1}^{n} (\alpha_i - \alpha_i^*) k(\mathbf{x}_i, \mathbf{x}) + b \tag{A.13}$$

Same as for the L1 SVM classifiers, there are three tiny differences in respect to the most standard L1 SVM regressors. First, in a Hessian matrix, a term $1/C$ is added to its diagonal elements which ensures positive definiteness of $\mathbf{H}$ and stabilizes the solution. Second, there is no upper bound on α_i and the only requirement is α_i to be nonnegative. Third, there are no longer complementarity constraints (2.53c) and (2.53d), namely the conditions $(C - \alpha_i)\xi_i = 0$ and $(C - \alpha_i^*)\xi_i^* = 0$ are missing in the L2 SVM regressors.

Finally, same as for the L1 SVMs, note that the NL decision functions here depend neither upon $\mathbf{w}$ nor on the true mapping $\boldsymbol{\Phi}(\mathbf{x})$. The last remark is same for all NL SVMs models shown here and it reminds that we *neither have to express, nor to know* the weight vector $\mathbf{w}$ and the true mapping $\boldsymbol{\Phi}(\mathbf{x})$ at all. The complete data modeling job will be done by finding the dual variables $\alpha_i^{(*)}$ and the kernel values $k(\mathbf{x}_i, \mathbf{x}_j)$ only.

A.3 Geometry and the Margin

In order to introduce the new concepts of a margin as given in (2.9) and an optimal canonical hyperplane, we present some basics of analytical geometry here. The notion of a distance D between the point and a hyperplane is both very useful and important. Its definition follows. In $\Re^m$ let there be a given point $P(x_{1p}, x_{2p}, \ldots, x_{mp})$ and a hyperplane $d(\mathbf{x}, \mathbf{w}, b) = 0$ defined by $w_1x_1 + w_2x_2 + \ldots + w_nx_n \pm b = 0$. The distance D from point P to a hyperplane d is given as

$$D = \frac{|(\mathbf{w}\mathbf{x}_p) \pm b|}{\|\mathbf{w}\|} = \frac{|w_1x_{1p} + w_2x_{2p} + \cdots + w_mx_{mp} \pm b|}{\sqrt{w_1^2 + w_2^2 + \cdots + w_m^2}} \tag{A.14}$$

Thus, for example, the distance between the point (1, 1, 1, 1) and a hyperplane $x_1 + 2x_2 + 3x_3 + 4x_4 - 2 = 0$ is $|\,[1\,2\,3\,4]\,[1\,1\,1\,1]^T - 2|/\sqrt{30} = 8/\sqrt{30}$.

At this point we are equipped enough to present an optimal canonical hyperplane, i.e., a canonical hyperplane having the maximal margin. Among all separating canonical hyperplanes there is the unique one having a maximal margin. The geometry needed for this presentation is shown in Fig. A.1.

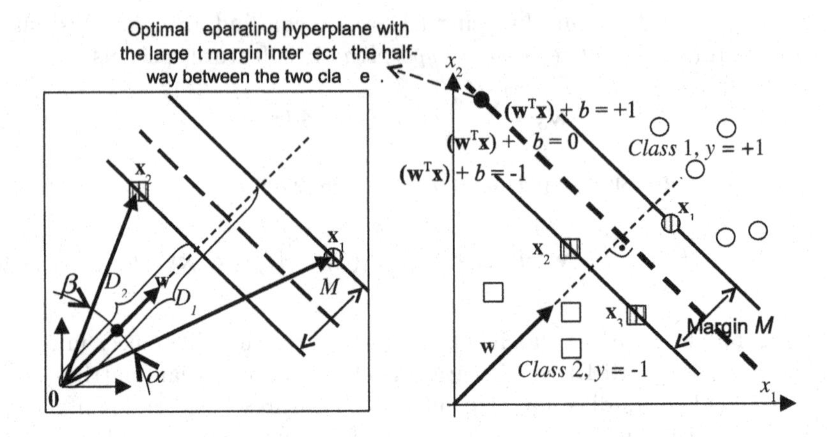

Fig. A.1. Optimal canonical separating hyperplane (OCSH) with the largest margin intersects halfway between the two classes. The points closest to it (satisfying $y_j \left|\mathbf{w}^T\mathbf{x}_j + b\right| = \mathbf{1}, j = 1, N_{SV}$) are *support vectors* and the OCSH satisfies $y_i(\mathbf{w}^T\mathbf{x}_i + b) \geq 1 i = 1,\ n$ (where n denotes the number of training data and N_{SV} stands for the number of SV). Three support vectors (x_1 from class 1, and x_2 and x_3 from class 2) are training data shown textured by vertical bars. Sketch for the margin M calculation -framed.

The margin M to be maximized during the training stage is a projection, onto the separating hyperplane's normal (weight) vector direction, of a distance between any two support vectors belonging to different classes. In the example shown in the framed picture in Fig. A.1 this margin M equals

$$M = (\mathbf{x}_1 - \mathbf{x}_2)_{\mathbf{w}} = (\mathbf{x}_1 - \mathbf{x}_3)_{\mathbf{w}} \tag{A.15}$$

where the subscript $_w$ denotes the projection onto the weight vector $\mathbf{w}$ direction. The margin M can now be found by using support vectors $\mathbf{x}_1$ and $\mathbf{x}_2$ as follows

$$D_1 = \|\mathbf{x}_1\| \cos(\alpha),\ D_2 = \|\mathbf{x}_2\| \cos(\beta) \text{ and } M = D_1 - D_2 \tag{A.16}$$

where $\boldsymbol{\alpha}$ and $\boldsymbol{\beta}$ are the angles between $\mathbf{w}$ and $\mathbf{x}_1$ and between $\mathbf{w}$ and $\mathbf{x}_2$ respectively as given by,

$$\cos(\alpha) = \frac{\mathbf{x}_1^T\mathbf{w}}{\|\mathbf{x}_1\|\ \|\mathbf{w}\|}, \text{ and } \cos(\beta) = \frac{\mathbf{x}_2^T\mathbf{w}}{\|\mathbf{x}_2\|\ \|\mathbf{w}\|} \tag{A.17}$$

Substituting (A.17) into (A.16) results in

$$M = \frac{\mathbf{x}_1^T\mathbf{w} - \mathbf{x}_2^T\mathbf{w}}{\|\mathbf{w}\|} \tag{A.18}$$

and by using the fact that $\mathbf{x}_1$ and $\mathbf{x}_2$ are support vectors satisfying $y_j \left|\mathbf{w}^T\mathbf{x}_j + b\right| = 1, j = 1, 2$, i.e.,$\mathbf{w}^T\mathbf{x}_1 + b = 1$ and $\mathbf{w}^T x_2 + b = -1$ we finally obtain equation (2.9) in the main part of the chapter

$$M = \frac{2}{\|\mathbf{w}\|}. \tag{A.19}$$

In deriving this important result, we used a geometric approach with the graphical assistance of the framed sketch in Fig. A.1 Alternatively, we could have used a shorter, algebraic, approach to show the relationship between a weight vector norm $\|\mathbf{w}\|$ and a margin M. We could have simply used (A.14) to express distance D between any support vector and a canonical separating plane. Thus, for example, for two-dimensional inputs shown in Fig. A.1 the distance D between a support vector x_2 and a canonical separating line equals half of the margin M, and from (A.14) it follows that $D = M/2 = |\mathbf{w}^T\mathbf{x}_2 + b|/\|\mathbf{w}\| = 1/\|\mathbf{w}\|$. This again gives $M = 2/\|\mathbf{w}\|$ where we used the fact that x_2 is a support vector. In this case the numerator in the above expression for D equals 1.

With these remarks we left the SVMs models developed and presented in the chapter to both the mind and the hand of a curious reader. However, we are aware that the most promising situation would be if the kernel models reside in the heart of the reader. We wish and hope that this chapter paved, at least, a part of the way to this veiled place.

B

Matlab Code for ISDA Classification

```
%*********************************************************************
%
% ISDA Classification with Gaussian RBF kernel
% ISDA_C.m-The main function for ISDA
%
% Input
% n - Number of training data
% x - Inputs of the training data
% Y - Labels for the training data
% s - Sigma parameter of the Gaussian RBF kernel
% k - Constant 1/k added to the kernel matrix
% C - Penalty parameter
% stopping= Stopping criterion of the algorithm
%
% Output
% alpha - Lagrange Multiplier α
% bias - bias term b
%
%*********************************************************************
function [alpha,bias] = ISDA_C(n,x,Y,s,C,k,stopping)
YE=ones(n,1)*-1; %YE= Y_i E_i, n=number of data, s=σ
dim = size(x,2);% Dimensionality of the inputs
max_YE=inf;
q=n; % n is size of training data.
iter=0; iter_pre=0; % Counter for number of iteration
h=1000 % Number of steps before shrinking
set_A=[1:n]; %Set A includes entire data set at beginning
alpha=zeros(n,1); tol=1e-12;
omega=1;% Learning rate
xsquare=sum(x.^2,2);% Precomputing x_i x_i term in (3.52)
Kij=zeros(n,1);%Variables for storing one row of Kernel matrix
```

T.-M. Huang et al.: *Kernel Based Algorithms for Mining Huge Data Sets*, Studies in Computational Intelligence (SCI) **17**, 217–221 (2006)
www.springerlink.com

```
buf_dealp=zeros(n,1);%Caching changes in α
YEm=YE;%Caching YiEi in the previous iterations.
while (1)
 iter = iter +1; alphaold=alpha;
 max_i=1; max_YE=0;
 if(h==0) % Performing shrinking after h steps
 h=1000;
 [set_A,max_i]=shrinking(alpha,YE,tol,C,set_A); % shrinking
 %Check if the reduced optimization problem (3.45) is solved
 if(abs(YE(set_A(max_i)))<stopping)
 if(length(find(buf_dealp(sv_ind)~=0))==length(sv_ind))
 % Compute YiEi from scratch
 [YE]=recompute_gradient(YE,alpha,set_A,n,x,Y,k,dim,s);
 YEm=YE; buf_dealp=zeros(n,1);iter_pre=iter;
 else
 %Updating YiEi uses cached YiEi and Δα
 [YE]=update_grad(YE, YEm,alpha,set_A,n,x,Y,k,dim,s...
 ,buf_dealp);

 YEm=YE; buf_dealp=zeros(n,1);
 end
 set_A=[1:n]; % Reset A
 end%if inn
 % Shrinking after updating of YiEi
 [set_A,max_i]=shrinking(alpha,YE,tol,C,set_A);
 % The algorithm stops if KKT conditions (3.6) is fulfilled with τ.
 if(abs(YE(set_A(max_i)))<stopping) break; end%if
 end
 %Check KKT conditions (3.6) to find KKT violator
 ind_vio= find((alpha(set_A)<C)&(YE(set_A)<0));
 ind_vio2=find((alpha(set_A)>0)&(YE(set_A)>tol));
 to_vio=[ind_vio;ind_vio2];
 %Find the worst KKT violator
  [max_YE max_i]=max(abs(YE(set_A([ind_vio;ind_vio2]))));
 max_i=to_vio(max_i);max_i=set_A(max_i);
 xy=x(max_i,:)*x';
 disti=ones(length(set_A),1)*xsquare(max_i)-2*xy(set_A)'...
 +xsquare(set_A);
 % Work out i-th row of the reduced RBF kernel matrix K_A
 Kij(set_A)=exp(-0.5*disti/s^2)+1/k;
 alphaold=alpha(max_i);
 %Updating α with (3.8)
 alpha(max_i) = alpha(max_i) -omega* YE(max_i)/(Kij(max_i));
 %Clipping using (3.9)
 alpha(max_i) = min(max(alpha(max_i)-tol,0), C);
```

```
%Find out change in α
de_alpha = (alpha(max_i)-alphaold)*Y(max_i);
%Updating the cache of Δα
buf_dealp(max_i)=buf_dealp(max_i)+de_alpha;
%Updating the cache of Y_iE_i with (3.10)
YE(set_A)=YE(set_A)+de_alpha*Y(set_A).*Kij(set_A);
h=h-1;
end% of while
bias= 1/k*alpha'*Y;%Work out the optimal bias term b_o

%*************************************************************
%
% shrinking.m- Perform Step 7 of Algorithm 3.1
% Input
% alpha - Lagrange multiplier
% YE    - KKT conditions
% tol   - Precision of the KKT conditions to be fulfilled.
% set_A - Set A in Algorithm 3.1
% C - Penalty parameter
%
% Output
% set_A - New set A after shrinking
% max_i - The index of the worst KKT violator in set A
%
%*************************************************************

function [set_A,max_i]=shrinking(alpha,YE,tol,C,set_A)
%Find KKT violators
ind_vio= find((alpha(set_A)<C)&(YE(set_A)<tol));
ind_vio2=find((alpha(set_A)>0)&(YE(set_A)>tol));
ind_sv=find(alpha(set_A)>0);
set_A= union(set_A(ind_vio),set_A(ind_vio2));%Shrink set A
set_A= union(set_A(ind_sv),set_A);
ind_vio= find((alpha(set_A)<C)&(YE(set_A)<tol));
ind_vio2=find((alpha(set_A)>0)&(YE(set_A)>tol));
%Find the worst KKT violator
[max_YE max_i]=max(abs(YE(set_A([ind_vio;ind_vio2]))));
to_vio=[ind_vio;ind_vio2]; max_i=to_vio(max_i);%
%Recomputing Y_iE_i that is not in the set A from scratch

%*************************************************************
%
% recompute_gradient.m- Routine for recomputing the
% KKT conditions from scratch using (3.49).
%
```

```
% Input
% YE    - KKT conditions
% alpha - Lagrange multiplier
% set_A - Set A in Algorithm 3.1
% n - Number of training data
% x - Input of the training data
% y - Desired Labels of the training data
% k - Constant 1/k added to the kernel matrix
% dim - Dimensionality of the input
% s - Sigma of the Gaussian RBF kernel
%
% Output
% YE - Recomputed KKT conditions
%
%************************************************************

function [YE]=recompute_gradient(YE,alpha,set_A,...
n,x,y,k,dim,s)
sv_set=find(alpha>0);
non_set_A=setdiff([1:n],set_A);
sv_size=length(sv_set);
non_q=length(non_set_A);
YE(non_set_A)=0;
xsquare=sum(x.^2,2);
Kij=zeros(n,1);
for(i=1:sv_size)
ind_i=sv_set(i);
xy=x(ind_i,:)*x';
disti=ones(non_q,1)*xsquare(ind_i)-2*xy(non_set_A)'...
+xsquare(non_set_A);
Kij(non_set_A)=exp(-0.5*disti/s^2)+1/k;
YE(non_set_A)=YE(non_set_A)+y(ind_i)*alpha(ind_i)...
*Kij(non_set_A);
end;
YE(non_set_A)=YE(non_set_A).*y(non_set_A);
YE(non_set_A)=YE(non_set_A)-1;
%Updating Y_iE_i that is not in the set A

%************************************************************
%
% update_grad.m - Routine for updating the KKT conditions
% according to (3.51).
% Input
% YE    - Most recent KKT conditions in the cache
% YEm - KKT conditions stored in the cache after
```

```
% the pervious execution of the recompute_gradient.m.
% alpha - Lagrange multiplier
% set_A - Set A in Algorithm 3.1
% n - Number of training data
% x - Input of the training data
% y - Desired Labels of the training data
% k - Constant 1/k added to the kernel
% dim - Dimensionality of the input
% s - Sigma of the Gaussian RBF kernel
% delalpha - change in alpha after the previous execution
% of recompute_gradient.m.
%
% Output
% YE - Updated KKT conditions.
%
%**********************************************************

function [YE]=update_grad(YE, YEm,alpha,set_A,n,x,y,k,...
dim,s,delalpha)
set_alp=find(delalpha~=0); non_set_A=setdiff([1:n],set_A);
non_q=length(non_set_A); xsquare=sum(x.^2,2); Kij=zeros(n,1);
for(i=1:length(set_alp))
ind_i=set_alp(i);
xy=x(ind_i,:)*x';
disti=ones(non_q,1)*xsquare(ind_i)-2*xy(non_set_A)'...
+xsquare(non_set_A);
Kij(non_set_A)=exp(-0.5*disti/s^2)+1/k;
YEm(non_set_A)=YEm(non_set_A)+delalpha(ind_i)*...
y(non_set_A).*Kij(non_set_A);
end
YE(non_set_A)=YEm(non_set_A);
```

C

Matlab Code for ISDA Regression

```
%*************************************************************
%
% ISDA Regression with Gaussian RBF Kernel
%
% ISDA_R.m-main function for ISDA regression
%
% Input
% n - Number of training data
% x - Inputs of the training data
% Y - Desired outputs for the training data
% s - Sigma parameter of the Gaussian RBF kernel
% k - Constant 1/k added to the kernel matrix
% C - Penalty parameter
% ep - Width of the insensitivity zone
% stopping - Stopping criterion of the algorithm
%
% Output
% alpha - (alphaup - alphadown)
% bias - Bias term b
%*************************************************************
function [alpha, bias] = ISDA_R(n,x,Y,s,C,ep,k,stopping)
dim = size(x,2)
tol = 1e-12
iter = 0;
alphaup = zeros(n,1);
alphadown = zeros(n,1);
Eu=-Y+ep; Eum=Eu;
El=ep+Y;  Elm=El;
bufdel_alph=zeros(n,1);
set_A=[1:n];
omega=1;
```

T.-M. Huang et al.: *Kernel Based Algorithms for Mining Huge Data Sets*, Studies in Computational Intelligence (SCI) **17**, 223–228 (2006)
www.springerlink.com

```
h=100;
xsquare=sum(x.^2,2);
Kij=zeros(n,1);
while(1)
if(h==0)
h=100;
[set_A, max_Eu]=shrinking_reg(alphaup,alphadown,Eu,El...
,tol,C,set_A);
if(max_Eu<stopping)
sv_ind=find((alphaup-alphadown)~=0);
if(length(find(bufdel_alph(sv_ind)~=0))==length(sv_ind))
[Eu, El]=recompute_gradientreg(Eu,El,alphaup,alphadown,...
set_A,n,x,Y,k,dim,s,ep);
Eum=Eu; Elm=El; bufdel_alph=zeros(n,1);
else
[Eu,El]=update_grad_reg(Eu, El, Eum,Elm,set_A,n,x,Y,k,...
dim,s,bufdel_alph);
Eum=Eu;Elm=El;bufdel_alph=zeros(n,1);
end
set_A=[1:n];
end
[set_A, max_Eu]=shrinking_reg(alphaup,alphadown,...
Eu,El,tol,C,set_A);
if(max_Eu<stopping) break; end
end
h=h-1;
u_vio=find((alphaup(set_A)>0)&(Eu(set_A)>tol));
u_vio=union(u_vio,find((alphaup(set_A)<C)&(Eu(set_A)<tol)));
l_vio=find((alphadown(set_A)>0)&(El(set_A)>tol));
l_vio=union(l_vio,find((alphadown(set_A)<C)&(El(set_A)<tol)));
[max_Eu max_iu]=max(abs(Eu(set_A(u_vio))));
[max_El max_il]=max(abs(El(set_A(l_vio))));
if(isempty(max_Eu)) max_Eu=0; end; if(isempty(max_El))...
max_El=0; end;
if(max_El>max_Eu)
max_il=l_vio(max_il); max_il=set_A(max_il);
max_Eu=max_El
xy=x(max_il,:)*x';
disti=ones(length(set_A),1)*xsquare(max_il)-2*xy(set_A)'...
+xsquare(set_A);
disti=exp(-0.5*disti/s^2);
Kij(set_A)=disti;
alpha_old =alphadown(max_il);
alphadown(max_il)=alphadown(max_il)-omega*El(max_il)...
/(Kij(max_il));
```

```
alphadown(max_il) = min(max(alphadown(max_il),0), C);
del_alp=-(alphadown(max_il)-alpha_old);
bufdel_alph(max_il)=bufdel_alph(max_il)+del_alp;
Eu(set_A)=Eu(set_A)+del_alp*Kij(set_A);
El(set_A)=El(set_A)-del_alp*Kij(set_A);
else
max_Eu
max_iu=u_vio(max_iu); max_iu=set_A(max_iu);
xy=x(max_iu,:)*x';
disti=ones(length(set_A),1)*xsquare(max_iu)-2*xy(set_A)'...
+xsquare(set_A);
disti=exp(-0.5*disti/s^2);
Kij(set_A)=disti;
alpha_old =alphaup(max_iu);
alphaup(max_iu)=alphaup(max_iu)-omega*Eu(max_iu)...
/(Kij(max_iu));
alphaup(max_iu) = min(max(alphaup(max_iu),0), C);
del_alp=(alphaup(max_iu)-alpha_old);
bufdel_alph(max_iu)=bufdel_alph(max_iu)+del_alp;
Eu(set_A)=Eu(set_A)+del_alp*Kij(set_A);
El(set_A)=El(set_A)-del_alp*Kij(set_A);
end
end%while
bias= sum((alphaup-alphadown))*1/k;
%[(alphaup-alphadown) Eu El f' Y]
Number_of_Iteration_cycles = iter
alpha =[alphaup;alphadown];
%*************************************************************
%
% recompute_gradientreg.m
% - using (3.54) and (3.55) to recompute the KKT conditions
%  from scratch.
%
% Input
% Eu    - KKT conditions for alphaup
% El - KKT conditions for alphadown
% alphaup - Lagrange multiplier α
% alphadown - Lagrange multiplier α*_i
% set_A - Set A in Algorithm 3.2
% n - Number of training data
% x - Inputs of the training data
% y - Desired outputs of the training data
% k - Constant 1/k added to the kernel
% dim - Dimensionality of the input
% s - Sigma of the Gaussian RBF kernel
```

```
% ep - Width of the insensitivity zone
%
% Output
% Eu - Recomputed KKT conditions for alphaup
% El - Recomputed KKT conditions for alphadown
%*********************************************************
function [Eu,El]=recompute_gradientreg(Eu,El,alphaup...
,alphadown,set_A,n,x,y,k,dim,s,ep)
sv_set=find((alphaup-alphadown)~=0);
non_set_A=setdiff([1:n],set_A);
sv_size=length(sv_set);
non_q=length(non_set_A);
Eu(non_set_A)=0;
El(non_set_A)=0;
xsquare=sum(x.^2,2);
Kij=zeros(n,1);
for(i=1:sv_size)
ind_i=sv_set(i);
xy=x(ind_i,:)*x';
disti=ones(length(non_set_A),1)*xsquare(ind_i)...
-2*xy(non_set_A)'+xsquare(non_set_A);
disti=exp(-0.5*disti/s^2);
Kij(non_set_A)=disti;
Eu(non_set_A)=Eu(non_set_A)+(alphaup(ind_i)...
-alphadown(ind_i))*Kij(non_set_A);
end;
El(non_set_A)=Eu(non_set_A);
for(j=1:non_q)
ind_j=non_set_A(j);
Eu(ind_j)= Eu(ind_j)-y(ind_j)+ep;
El(ind_j)=ep+y(ind_j)-El(ind_j)
end;
%*********************************************************
%
% update_grad_reg.m
% -Update KKT conditions using (3.57).
%
% Eu - Most recent KKT conditions for alphaup in the cache
% Eum - KKT conditions stored in the cache for alphaup after
% the pervious execution of the recompute_gradientreg.m.
% Elm - KKT conditions for alphadown in the cache
% after the pervious execution of the
% recompute_gradientreg.m.
% El - Most recent KKT conditions for alphadown in the cache
% set_A - Set A in Algorithm 3.2
```

```
% n - Number of training data
% x - Input of the training data
% y - Desired Labels of the training data
% k - Constant 1/k added to the kernel
% dim - Dimensionality of the input
% s - Sigma of the Gaussian RBF kernel
% delalpha - change in (alphaup-alphadown) after previous
% execution of recompute_gradientreg.m.
%
% Output
% Eu - Updated KKT conditions for alphaup
% El - Updated KKT conditions for alphadown
%*********************************************************
function [Eu,El]=update_grad_reg(Eu, El, Eum,Elm,...
set_A,n,x,y,k,dim,s,delalpha)
set_alp=find(delalpha~=0);
non_set_A=setdiff([1:n],set_A);
non_q=length(non_set_A);
xsquare=sum(x.^2,2);
Kij=zeros(n,1);
for(i=1:length(set_alp))
ind_i=set_alp(i);
xy=x(ind_i,:)*x';
disti=ones(length(non_set_A),1)*xsquare(ind_i)-...
2*xy(non_set_A)'+xsquare(non_set_A);
disti=exp(-0.5*disti/s^2)+1/k;
Kij(non_set_A)=disti;
Eum(non_set_A)=Eum(non_set_A)+delalpha(ind_i)*Kij(non_set_A);
Elm(non_set_A)=Elm(non_set_A)-delalpha(ind_i)*Kij(non_set_A);
end
Eu(non_set_A)=Eum(non_set_A);
El(non_set_A)=Elm(non_set_A);

%*********************************************************
%
% shrinking_reg.m - perform Step 8 of Algorithm 3.2
%
% Input
% alphaup - Lagrange multiplier α
% alphadown- Lagrange multiplier α_i^*
% Eu - KKT conditions for alphaup
% El - KKT conditions for alphadown
% tol- Precision of the KKT conditions to be fulfilled.
% set_A - set A in Algorithm 3.1
% C - Penalty parameter
```

```
% Output
% set_A - New set A after shrinking
% max_Eu - The KKT condition for the worst KKT violator
%
%**********************************************************


function [set_A, max_Eu]=shrinking_reg(alphaup,alphadown...
,Eu,El,tol,C,set_A)
u_vio=find((alphaup(set_A)>0)&(Eu(set_A)>tol));
u_vio=union(u_vio,find((alphaup(set_A)<C)&(Eu(set_A)<tol)));
l_vio=find((alphadown(set_A)>0)&(El(set_A)>tol));
l_vio=union(l_vio,find((alphadown(set_A)<C)&...
(El(set_A)<tol)));
[max_Eu max_iu]=max(abs(Eu(set_A(u_vio))));
[max_El max_il]=max(abs(El(set_A(l_vio))));
if(isempty(max_Eu)) max_Eu=0; end
if(isempty(max_El)) max_El=0; end
if(max_El>max_Eu) max_Eu=max_El; end
set_A=union(set_A(u_vio),union(set_A(l_vio),set_A(sv_ind)));
```

D

Matlab Code for Conjugate Gradient Method with Box Constraints

```
%*******************************************************
%
% cggsemi_sup1.m - This program implements Algorithm 5.4
% i.e. solving min 0.5*x'Ax-hx
%  s.t. low_b <= x <=up_b
%
% Input
% A - Matrix A
% h - Vector h
% up_b - Upper bound for x
% low_b - Lower bound for x
% stopping - Stopping criterion of the algorithm
% Recommanded value stopping <= 10e-5
% tol - small number to take care of numerical error
% Recommanded value tol = 10e-12.
%
% Output
% x - Vector x, solution of the optimization problem
%*******************************************************

function [x] = cggsemi_sup1(A,h,up_b,low_b,tol,stopping)
[m,n]=size(A);
x =zeros(m,1);
%r= current rk = next step -*gradient at current direction
rk = h-A*x;
r=rk;
r_dot =rk;
act_i = find(r~=0);
counter =1;
r_dot(act_i) =0;
p=r_dot;
```

T.-M. Huang et al.: *Kernel Based Algorithms for Mining Huge Data Sets*, Studies in Computational Intelligence (SCI) **17**, 229–231 (2006)
www.springerlink.com

```
in_comp=1:m;
in_comp=in_comp';
iter=0;
%tic;
while (1)
%Step 2 and 3 of the Algorithm 5.4
    x_zero= find(abs(up_b-x)<=tol);
    % upper bound in semi-learning = 1
    x_upbound = find(abs(low_b-x)<=tol);
    % lower bound in semi- learning is -1
    act_i_temp = find(r(x_zero)>=0);
    act_up = find(r(x_upbound)<=0);
    act_i = [x_zero(act_i_temp);x_upbound(act_up)];
    if(length(act_i)~=0) non_act_i =setdiff(in_comp,act_i);
    else
    non_act_i = in_comp;
    end
    if(abs(r(non_act_i))<=stopping) break; end
     %Step 4 of the Algorithm 5.4
    r_dot =r;
    r_dot(act_i)=0;
    p=r_dot;
    while(1)
        %Step 5 of the Algorithm 5.4
        s = A*p;
        c= p'*r;
        d=p'*s;
        a=c/d;
        xk=x+a*p;
         iter=iter+1;
        rk = r-a*s;
        %checking
        x_lower = find(xk-up_b>0);
        x_upper = find(low_b-xk>0);
        s_out = [x_lower;x_upper];
        if(length(s_out)>0)
        %Step 6 of the Algorithm 5.4
            rato_low = -(x(x_lower)-up_b)./p(x_lower);
            rato_up =(low_b-x(x_upper))./p(x_upper);
            rato = [rato_low;rato_up];
            %Step 7 of the Algorithm 5.4
            [minra,I] = min(rato);
            xk= x+minra*p;
          if abs(low_b-xk(s_out(I)))<tol xk(s_out(I))=low_b;
            else xk(s_out(I))= up_b;
```

```
        end
          r = r-minra*s;
    act_i= [find(abs(xk-up_b)<tol);find(abs(low_b-xk)<tol)];
            non_act_i =setdiff(in_comp,act_i);
      % upper bound in semi-learning = 1
     x_zero= find(abs(up_b-xk)<=tol);
     % lower bound in semi- learning is -1
    x_upbound = find(abs(low_b-xk)<=tol);
    act_i_temp = find(r(x_zero)>=0);
    act_up = find(r(x_upbound)<=0);
    %x_zero = cat(1,x_zero,x_upbound);
    act_i = [x_zero(act_i_temp);x_upbound(act_up)];
    if(length(act_i)~=0) non_act_i =setdiff(in_comp,act_i);
    else
    non_act_i = in_comp;
    end
            if(abs(r(non_act_i))<=stopping)
             x=xk;
            break;
            end
           %Step 4 of the Algorithm 5.4
            r_dot =r;
            r_dot(act_i)=0;
            p=r_dot;
            x=xk;
        elseif length(find(abs(rk(non_act_i))>=stopping))==0
            r=rk;
            x=xk;
            break;
        else
            r_dot=rk;
            r_dot(act_i) = 0;
            b =-(s'*r_dot)/d;
            pk = r_dot+b*p;
            p=pk;
            r=rk;
            x=xk;
        end
    end%while inner
end %while outer
iter;
x;
```

E

Uncorrelatedness and Independence

We have emphasized several times that ICA assumed that source signals are statistically independent and also that PCA only decorrelates mixed signals. We have also said that for normal or Gaussian processes uncorrelatedness and independence are equivalent statements. In addition, we have said that for some random process $\mathbf{x}$ having statistically independent components the following applies:

$$p(\mathbf{x}) = \prod_n p_n(x_n) \tag{E.1}$$

Let us assume that x_n are normal processes with $N(\mu_n, \sigma_n)$ where μ_n denotes mean and σ_n denotes standard deviation. Pdf of each individual process x_n is given by:

$$p_n(x_n) = \frac{1}{(2\pi)^{1/2}\sigma_n} \exp\left(-\frac{(x_n - \mu_n)^2}{2\sigma^2}\right) \tag{E.2}$$

Joint pdf of $\mathbf{x}$ is given as:

$$p_{\mathbf{x}}(\mathbf{x}) = \frac{1}{(2\pi)^{N/2} (\det \mathbf{R_x})^{1/2}} \exp\left(-\frac{1}{2}(\mathbf{x} - \mu_{\mathbf{x}})^T \mathbf{R}_{\mathbf{x}}^{-1} (\mathbf{x} - \mu_{\mathbf{x}})\right) \tag{E.3}$$

where N represents dimensionality of $\mathbf{x}$, $\mathbf{R_x}$ represents data covariance matrix and $\mu_{\mathbf{x}}$ represents vector of the mean values. Elements of the data covariance matrix at position (i, j) are correlations among the components of the vector $\mathbf{x}$ given by:

$$\mathbf{R_x}(i, j) = E[x_i x_j] \tag{E.4}$$

where E denotes mathematical expectation. If processes x_n are uncorrelated then:

$$\mathbf{R_x}(i, j) = \delta_{ij} \tag{E.5}$$

where symbol δ_{ij} denotes Kronecker delta equal to one for $i = j$ and zero otherwise. Under this condition $\mathbf{R_x}$ becomes diagonal matrix with variance

T.-M. Huang et al.: *Kernel Based Algorithms for Mining Huge Data Sets*, Studies in Computational Intelligence (SCI) **17**, 233–236 (2006)
www.springerlink.com

on the main diagonal i.e. $\mathbf{R}_\mathbf{x}(i,i) = \sigma_i^2$. It is easy to show that joint pdf $p_\mathbf{x}(\mathbf{x})$ can be written as:

$$p_\mathbf{x}(\mathbf{x}) = \prod_n \frac{1}{(2\pi)^{1/2}\sigma_n} \exp\left(-\frac{(x_n-\mu_n)^2}{2\sigma^2}\right) = \prod_n p_n(x_n) \tag{E.6}$$

which proves that for Gaussian processes uncorrelatedness implies statistical independence. This is why PCA is optimal transform for Gaussian processes. If processes however are non-Gaussian (E.5) will not lead to (E.1) i.e. uncorrelatedness will not imply statistical independence. In order to quantify statistical independence between the stochastic processes we have to introduce cumulants [22, 97]. Before that we have to introduce first and second characteristic functions [126, 22]. The first characteristic function of the random process x is defined as the Fourier transform of the pdf $p(x)$:

$$\Psi_x(\omega) = E\left[\exp(j\omega x)\right] = \int_{-\infty}^{+\infty} \exp(j\omega x)\, p(x) dx \tag{E.7}$$

where j in (E.7) represents imaginary unit i.e. $j = \sqrt{-1}$. If term $\exp(j\omega x)$ is expanded into power series first characteristic function can be written with the moments as the coefficients in the expansion [126, 22]:

$$\begin{aligned}\Psi_x(\omega) &= E\left[1 + j\omega x + \frac{(j\omega)^2}{2!}x^2 + .. + \frac{(j\omega)^n}{n!}x^n + ...\right] \\ &= 1 + E[x]\, j\omega + E\left[x^2\right]\frac{(j\omega)^2}{2!} + ... + E\left[x^n\right]\frac{(j\omega)^n}{n!} + ...\end{aligned} \tag{E.8}$$

Second characteristic function $K_x(\omega)$ is obtained by applying natural logarithm on the first characteristic function:

$$K_x(\omega) = \ln \Psi_x(\omega) \tag{E.9}$$

If expanded version of $\Psi_x(\omega)$ given by (E.8) is inserted into (E.9) the power series version of the second characteristic function is obtained with cumulants as the coefficients in the expansion:

$$K_x(\omega) = C_1(x)(j\omega) + C_2(x)\frac{(j\omega)^2}{2!} + ... + C_n(x)\frac{(j\omega)^n}{n!} + ... \tag{E.10}$$

Cumulants have several properties that make them very useful in various computations with stochastic processes. We refer to [22, 100] for details. In the context of ICA important property of the cumulants is that, unlike moments, cumulants of the order higher than 2 are equal to zero if random process is Gaussian [126, 22]. Therefore they are natural measure for non-Gaussianity.

It will be shown that kurtosis introduced by (6.14) is actually normalized fourth order cumulant. Because odd order statistics (cumulants) are zero for random processes which are symmetrically distributed the first statistics of the order higher than two which is in use is the fourth order statistics. In practice cumulants have to be estimated from data and for that purpose it is useful their relations with moments [97, 100]. Relations between moments and cumulants are derived from the following identity:

$$\exp\{K_x(\omega)\} = \Psi_x(\omega) \tag{E.11}$$

For zero mean random process x we give here relation between cumulants and moments up to the order four:

$$\begin{aligned} C_2(x) &= E[x^2] \\ C_3(x) &= E[x^3] \\ C_4(x) &= E[x^4] - 3E^2[x^2] \end{aligned} \tag{E.12}$$

It is easy to verify that kurtosis given by (6.14) follows from definition:

$$\kappa(x) = \frac{C_4(x)}{\left(C_2(x)\right)^2} \tag{E.13}$$

In order to measure statistical independence between stochastic processes cross-cumulants have to be defined. The cross-cumulant of the order two is defined with:

$$C_{11}(x_i, x_j) = E[x_i(t)x_j(t+\tau)] \tag{E.14}$$

where τ represents the relative shift between the two sequences. We shall assume here that $\tau = 0$. We observe that C_{11} is nothing else but cross-correlation between x_i and x_j. It measures the second order statistical dependence between the two processes. In order to make cross-correlation invariant with respect to the fluctuations of the signal amplitude it is normalized as:

$$\overline{C_{11}}(x_i, x_j) = \frac{E[x_i(t)x_j(t)]}{\left(E[x_i^2(t)]\right)^{\frac{1}{2}} \left(E[x_j^2(t)]\right)^{\frac{1}{2}}} \tag{E.15}$$

which implies $-1 \leq \overline{C_{11}}(x_i, x_j) \leq 1$. In order to measure the fourth order statistical dependence between the stochastic processes we need to define the fourth order (FO) cross-cumulant [97, 100]. If we assume that shifts between sequences are equal to zero then three FO cross-cumulants between zero mean random processes x_i and x_j are defined as:

$$C_{13}(x_i, x_j) = E[x_i(t)x_j^3(t)] - 3E[x_i(t)x_j(t)]E[x_j^2(t)] \tag{E.16}$$

$$C_{22}(x_i, x_j) = E[x_i^2(t)x_j^2(t)] - E[x_i^2(t)]E[x_j^2(t)] - 2\left(E[x_i(t)x_j(t)]\right)^2 \tag{E.17}$$

$$C_{31}(x_i, x_j) = E[x_i^3(t)x_j(t)] - 3E[x_i(t)x_j(t)]E[x_i^2(t)] \tag{E.18}$$

We could proceed with the higher order cross-cumulants but relations between them and moments become very complex and therefore computationally very expensive [97]. In order to prevent dependence of the sample estimate of FO cross-cumulants on the signal amplitude we normalize them according to:

$$\overline{C_{pr}}(x_i, x_j) = \frac{C_{pr}(x_i, x_j)}{\left(E[x_i^2(t)]\right)^{\frac{p}{2}} \left(E[x_j^2(t)]\right)^{\frac{r}{2}}} \tag{E.19}$$

F

Independent Component Analysis by Empirical Estimation of Score Functions i.e., Probability Density Functions

In Chap. 6 we have derived batch and adaptive ICA learning rules through the minimization of the mutual information between the output signals $\mathbf{y}$. Batch form of the learning algorithm was given by (6.42) as:

$$\mathbf{W}(k+1) = \mathbf{W}(k) + \eta \left(\mathbf{I} - E\left[\varphi\left(\mathbf{y}\right) \mathbf{y}^T \right] \right) \mathbf{W}(k) \tag{F.1}$$

while adaptive learning algorithm was obtained by dropping mathematical expectation operator $E[o]$ from (F.1) and replacing iteration index k by "time" index t:

$$\mathbf{W}(t+1) = \mathbf{W}(t) + \eta \left(\mathbf{I} - \varphi\left(\mathbf{y}(t)\right) \mathbf{y}(t)^T \right) \mathbf{W}(t) \tag{F.2}$$

Optimal form of the nonlinear vector function $\varphi(\mathbf{y})$, called score or activation function, was shown in (6.39) to be:

$$\varphi_i(y_i) = -\frac{1}{p_i(y_i)} \frac{\partial p(y_i)}{\partial y_i} \tag{F.3}$$

Unknown probability density function $p_i(y_i)$ was modelled by generalized Gaussian distribution [33, 35] given by (6.47) as:

$$p_i(y_i) = \frac{\theta_i}{2\sigma_i \Gamma\left(1/\theta_i\right)} \exp\left(-\frac{1}{\theta_i} \left| \frac{y_i}{\sigma_i} \right|^{\theta_i} \right) \tag{F.4}$$

where with the value of the parameter θ_i, called Gaussian exponent, $1 \leq \theta_I \leq 2$ (F.4) can model sparse or super-Gaussian distributions and with value of the parameter $\theta_I > 2$ (F.4) can model bimodal or sub-Gaussian distributions. In (F.4) $\Gamma(o)$ represents the gamma function. If (F.4) is inserted in (F.3) the parametric representation of the score function is obtained as in (6.48):

$$\varphi_i(y_i) = sign(y_i) \left| y_i \right|^{\theta_i - 1} \tag{F.5}$$

If it is not possible to know in advance to which statistical class the signals belong the parameterized score function (F.5) will lead to suboptimal

T.-M. Huang et al.: *Kernel Based Algorithms for Mining Huge Data Sets*, Studies in Computational Intelligence (SCI) **17**, 237–240 (2006)
www.springerlink.com

or even divergent learning algorithms (6.1) and (6.2). In such a case one of the solutions is to calculate activation function based on the Parzen's window based empirical estimate of the unknown probability density function with the Gaussian kernel such as proposed in [118]:

$$\hat{p}_i(y_i(t), \mathbf{y}_i) = \frac{1}{T}\sum_{tt=1}^{T} G\left(y_i(t) - y_i(tt), \sigma^2\mathbf{I}\right) \tag{F.6}$$

where $\mathbf{y}_i$ denotes the whole sample, T denotes sample size and $G(o)$ is Gaussian probability density function:

$$G\left(y_i(t), \sigma^2\mathbf{I}\right) = \frac{1}{\sqrt{2\pi}\sigma}\exp\left(-\frac{y_i^2(t)}{2\sigma^2}\right) \tag{F.7}$$

σ^2 represents variance of the empirical estimator which is usually in the interval $0.01 \leq \sigma^2 \leq 0.1$. From (F.6) and (F.7) an empirical estimator for the derivative of the probability density function follows as:

$$\frac{d\hat{p}_i(y_n)}{dy_i} = -\frac{1}{T}\sum_{tt=1}^{T}\frac{y_i(t) - y_i(tt)}{\sigma^2}G\left(y_i(t) - y_i(tt), \sigma^2\mathbf{I}\right) \tag{F.8}$$

Now using (F.3), (F.6), (F.7) and (F.8) batch and adaptive learning rules for the de-mixing matrix $\mathbf{W}$ could be obtained as in (F.1) and (F.2) respectively, with the nonlinear activation function directly estimated from data. The potential drawback of this approach could be high computational complexity in comparison with approach based on the parametric model of the nonlinear activation functions. The computational complexity of the approach based on empirical estimate of the score functions is $O(T^2N^2)$ where N represents number of sources or classes.

Algorithms for estimation of mutual information, score functions and entropies with smaller computational complexity of the order $O(3^NT+N^2T)$ are presented in [113]. Further improvement in terms of the computational complexity of the order $O(NT\log T + N^2T)$ are presented in [124]. We shall not further elaborate these approaches here. MATLAB implementation of batch ICA learning rule (F.1) (also given by (6.42)) with the empirical estimator of the score functions (given by (F.3), (F.6)-(F.8)) is given below.

```
%***********************************************************
%
% Minimum mutual information batch ICA algorithm with
% the score function estimated from data
%
% input arguments:
%    X - vector of zero mean measured signals with the
%  dimension NSOURCExT.
%    gain - learning gain
```

```
%    ITMAX - maximal number of iterations in batch
%  learning procedure.
%    NSOURCE - number of source signals
%    T - sample size
%
% output arguments:
%   Y - vector of separated signals of dimensions NSOURCExT
%
%*****************************************************

function [Y]= mut_inf_ica_score_estimate(X,gain,ITMAX,NSOURCE,T)

I = eye(NSOURCE);          % initial value for demixing matrix
W = I;
it = 1;

for n=1:NSOURCE
    X(n,:) = X(n,:) - mean(X(n,:));       % make data zero mean
end;

while (it < ITMAX)
      it = it+1;
      FE = zeros(NSOURCE,NSOURCE);
      for k=1:T
          Y(:,k) = W*X(:,k);
      end

      [fi,pdfest]= pdf_score_est(Y,0.01);  % eq. (F.6)-(F.8)

      for k=1:T
          FE = FE + fi(:,k)*Y(:,k)';
      end
      FE = FE/T;
      F = I - FE;
      W = W + gain*F*W;      % eq. (F.1)
end
%**********************************************
% pdf and score function estimator
%
% input arguments:
%     Y - signal matrix NxT (N - number of signals,
%   T - number of samples)
%     sigm - variance of the pdf estimator
%
% output arguments:
```

```
%    fi - score function estimate NxT
%    pdfest - estimate of the pdf
%*********************************************

function [fi,pdfest]= pdf_score_est(Y,sigm)

[N T] = size(Y);
cf = 1/(sqrt(2*pi)*sigm);
sigm2=sigm*sigm;

for t=1:T
    for n=1:N
        pdfest(n,t)=0;
        dpdfest(n,t)=0;
        for tt=1:T
            gk = cf*exp(-power(Y(n,t)-Y(n,tt),2)/2/sigm/sigm);
            pdfest(n,t) = pdfest(n,t) + gk;
            dpdfest(n,t) = dpdfest(n,t) - gk*(Y(n,t)-Y(n,tt))...
            /sigm2;
        end
        pdfest(n,t)=pdfest(n,t)/T; % Eq. (F.6)
        dpdfest(n,t)=dpdfest(n,t)/T; % Eq. (F.8)
        fi(n,t)=-dpdfest(n,t)/pdfest(n,t); % Eq. (F.3)
    end
end
```

G

SemiL User Guide

SemiL (Copyright (C) 2004 Te-Ming Huang and Vojislav Kecman) is efficient software for solving large-scale semi-supervised learning problem using graph based approaches that is presented in Chap. 5. It is available at

www.learning-from-data.com

SemiL is designed to solve semi-supervised learning problems using all the models listed in Tables 5.3 and 5.4. In summary, it implements the approach listed below:

1. Hard label approach with the maximization of smoothness, and
2. Soft label approach with the maximization of smoothness,

for all three types of models (i.e., Basic model, Norm Constrained Model and Bound Constrained Model presented in Sect. 5.6) by using either Standard or Normalized Laplacian, e.g., with option *-l 1 -h 1 -mu 0.0101 -lambda 0.0101*, the following formulation is used.

$$\mathcal{Q}(F) = \frac{1}{2}\left(\sum_{i,j=1}^{n} W_{ij}\left\|\frac{1}{\sqrt{D_{ii}}}F_i - \frac{1}{\sqrt{D_{jj}}}F_j\right\|^2 + \lambda\sum_{i=1}^{n}\|F_i - Y_i\|^2 + \mu\sum_{j=1}^{n}\|F_j\|^2\right) \tag{G.1}$$

It is important to note that *lambda* controls the amount of penalty on the empirical error of the labeled point and *mu* controls the norm of the output of the unlabeled points. It is the same as the consistency method (CM) proposed in [155] when α in Algorithm 5.2 is equal to 0.99.

G.1 Installation

- **For Windows User**: Unzip the file *semil.zip* and it will self-extract itself into the folder *SemiL*. The executable file *SemiL* will be in the folder *Windows*. Rename the file *SemiL* to *SemiL.exe*[1] to execute. Once extracted,

[1] The windows version of SemiL is developed in Visual C++ 6.0.

T.-M. Huang et al.: *Kernel Based Algorithms for Mining Huge Data Sets*, Studies in Computational Intelligence (SCI) **17**, 241–245 (2006)
www.springerlink.com

and after running MS-DOS, the working directory in Command Prompt (MS-DOS) may be *c:/SemiL/Windows*. That means, once you are in Command Prompt type in:

```
cd c:/semil/Windows
```

- **For Linux User**: The executable file for Linux (SemiL) will be in *.../SemiL/Linux* folder.

Before using please go through the options. By typing

```
semil
```

all the SemiL routine's options will be displayed. They are as follows:

```
-t Distance type
        1 = Euclidean distance
        2 = Cosine distance
-d Degree of the graph (this is a design parameter and
 for each problem the final model may have different degree)
-m Cachesize : set cache memory size in MB (default 0)
        (m is needed when working with dense raw data only,
        to speed up the calculations)
-l Standard or normalised Laplacian
        0 = standard Laplacian
        1 = normalized Laplacian
-h  Hard or soft label
        0 = hard label
        1 = soft label
-k Kernel_type
        0 = RBF function exp(-(|u-v|^2)/gamma)\n
        1 = Polynomial function (not an option at the moment)
-p Degree of polynomial (at the moment not implemented)
-mu Penalty parameter valid for the
Norm Constrained Model only.
-lambda Penalty parameter for the empirical error, valid for
        the Soft Label approach only.
-r Number of random experimental runs for a given setting
-g Gamma value for the RBF kernel
(shape parameter of the n-dimensional Gaussian)
-pl Percentage of the labeled points in the experiment
-stp Precision for the solver (default = 1e-5)
-up_b Upper bound for the bounded constraint (default = inf)
-low_b Lower bound for the bounded constraint (default = -inf)
-nr Normalization of the output i.e., F* matrix
        0 = Without a Normalization
        1 = With a Normalization
-ocl One class labeling (Default 1)
```

```
0 = Two Class labeling  (-1 and +1)
1 = One Class labeling(+1 only)
```

G.2 Input Data Format

SemiL can take two different types of data as the input. For first time solving a given problem with SemiL, you need to convert your data set into the raw data file format as given in the following section.

G.2.1 Raw Data Format:

<label1> <index1>:<value1> <index2>:<value2>...
<label2> <index1>:<value1> <index2>:<value2>...

<label1> is the desired label of the first data point and <label2> is the desired label of the second data point. The <index1> is an integer value starting from one (1) and it tells to the program which dimension <value> belongs to. SemiL can take raw data in sparse form or in dense form. For data point i with unknown label, set the value of <label$i$> to zero.

Example G.1. we have 7 4-dimensional measurements belonging to three classes and only one measurement per class is labeled. The data are given in Table G.1.

	Dimension of the input			
Label	value 1	value 2	value 3	value 4
1	0	1.1	0.3	-1.1
0	-2	0	1.1	0.7
0	1.1	-3.1	0	1.1
2	0	0	0	2
3	5	-0.5	1	2.3
0	2	0	-4.1	0
0	0	1.1	0	3.7

Table G.1. Example input data.

Data in DENSE format are to be given as follows:

```
1 1:0 2:1.1 3:0.3 4:-1.1
0 1:-2 2:0 3:1.1 4:0.7
0 1:1.1 2:-3.1 3:0 4:1.1
2 1:0 2:0 3:0 4:2
3 1:5 2:-0.5 3:1 4:2.3
0 1:2 2:0 3:-4.1 4:0
0 1:0 2:1.1 3:0 4:3.7
```

and the data in SPARSE format are to be given as:

```
1 2:1.1 3:0.3 4:-1.1
0 1:-2 3:1.1 4:0.7
0 1:1.1 2:-3.1 4:1.1
2 4:2
3 1:5 2:-0.5 3:1 4:2.3
0 1:2 3:-4.1
0 2:1.1 4:3.7
```

After solving the problem for the first time, SemiL will generate a distance matrix file (you should specify the name at the prompt) and a label file having the same name augmented by the label extension. You can use these two files during the design runs playing with various design parameters without an evaluation of a distance matrix each time. In Windows version of SemiL, Intel BLAS is incorporated to improve the performance on evaluating the distance matrix when data is dense. You can specify the amount of cache by defining an option "-m". The program can run in the following two modes,

1. **Experiment Mode (ExM)**: ExM tests different types of semi-supervised learning algorithms by inputting data set with all the data labeled. In this mode, it will randomly select a fixed number of data points as labeled points, and then it will try to predict the label for the rest of the points. By comparing the predicted labels and the true labels, the user can examine the performance of different settings for a semi-supervised learning. The number of data points to be selected is specified by option "-pl", which stands for percentage of data point to be labeled from all data. The user can specified how many experiments should be run by the option "-r ". To activate this mode, the user only needs to supply the routine with ALL the data labeled.
2. **Predicting mode (PM)**: The routine will run in PM as long as there is at least one label equal to zero. In the predicting mode, the program will predict the label of ALL the unlabeled data. To activate this mode, the user simply set the label of unlabeled points equal to 0 in the data file.

G.3 Getting Started

1. Prepare your data in the format readable by the program. If your data is in Matlab, use the *convtosp.m* or *convtode.m* to convert them into the format readable by SemiL. To use these routines, you need to put the label of your data points as *the first column* of your Matlab variable in Matlab. *Convtosp.m* will convert your full Matlab variable into the proper format as a sparse input data. *Convtoden.m* will convert your full Matlab variable into a dense input data for the program.

2. Once the data is prepared, you can use the command line to run the program. Below, we first run the problem *20 News Group Recreation* (the same one used in Sect. 5.4) for which the data are extracted (by using the Rainbow software [96]) and stored in the file *rec.txt* (in a sparse format).
3. To perform the run, type in the following line in the directory of the exe file

```
Semil -t 2 -d 10 -m 0 -l 0 -h 0 -k 0 -u 0 -g 10 -r 50
-pl 0.003 -lambda 0 -mu 0.5 rec.txt
```

4. Thus, the user starts with the raw data input to the program which will compute the distance matrix (used for the RBF model's only) and save it separately from the labels. It will produce a file named by us. Here we named it *rec2_10d.dat* for the output of the solver which will be saved as the file. Additionally, two more files will be created, namely *rec2_10d.dat.output* and *rec2_10d.dat.label.* At the same time the error rate for each run will be recorded in the file *error_rate.dat.*

G.3.1 Design Stage

5. After the distance matrix is calculated and associated with the corresponding labels (which are stored in separate files) a design by changing various model parameters (settings e.g., l, h, k, g, r, pl lambda, and mu) can start by typing in the following line.

```
Semil -l 0 -h4 0 -k 0 -u 0 -g 10 -r 50 -pl 0.003
-lambda 0 -mu 0 rec2_10d.dat rec2_10d.dat.label
```

Note that the filenames will be different if you name the two files with different names. The above line will implement GRFM [160]. To use CM model [155] use the following line.

```
Semil -l 1 -h 1 -k 0 -u 0 -g 10 -r 50 -pl 0.003
-lambda 0.0101 -mu 0.0101 rec2_10d.dat rec2_10d.dat.label
```

The two examples above are the original CM and GRFM models given in Tables 5.3 and 5.4 (These models are marked by star in the corresponding tables.). We did not test for other ten models given in the Tables 5.3 and 5.4, they are left to the interested readers to explore.
In this setting, the computation of distances will be skipped and the program will read the distance matrix from file and use it for the simulation.
6. Same as in the run with raw data the results will be saved in three files: *rec2_10d.dat.output*, *rec2_10d.dat.label* and in *error_rate.dat.* Also, the errors in each run will be printed on the screen.

References

1. S. Abe. *Support Vector Machines for Pattern Classification.* Springer-Verlag, London, 2004.
2. J. B. Adams and M. O. Smith. Spectral mixture modeling: a new analysis of rock and soil types at the Viking lander 1 suite. *J. Geophysical Res.*, 91(B8):8098–8112, 1986.
3. J. B. Adams, M. O. Smith, and A. R. Gillespie. *Image spectroscopy: interpretation based on spectral mixture analysis*, pages 145–166. Mass: Cambridge University Press, 1993.
4. M.A. Aizerman, E.M. Braverman, and L.I. Rozonoer. Theoretical foundations of the potential function method in pattern recognition learning. *Automation and Remote Control*, 25:821–837, 1964.
5. H. Akaike. A new look at the statistical model identification. *IEEE Trans. on Automatic Control*, 19(12):716–723, 1974.
6. J. M. Aldous and R. J. Wilson. *Graphs and applications : an introductory approach.* Springer, London, New York, 2000.
7. A.A. Alizadeh, R. E. Davis, and Lossos MA, C. Distinct types of diffuse large B-cell lymphoma identified by gene expression profiling. *Nature*, (403):503–511, 2000.
8. U. Alon, N. Barkai, D. A. Notterman, K. Gish, S. Ybarra, D. Mack, and A. J. Levine. Broad patterns of gene expression revealed by clustering analysis of tumor and normal colon cancer tissues probed by oligonucleotide arrays. In *Proc. of the Natl. Acad. Sci. USA*, pages 6745–6750, USA, 1999.
9. S. Amari. Natural gradient works efficiently in learning. *Neural Computation*, 10(2):251–276, 1998.
10. S. Amari. Superefficiency in blind source separation. *IEEE Transactions on Signal Processing*, 47:936–944, 1999.
11. C. Ambroise and G.J. McLachlan. Selection bias in gene extraction on the basis of microarray gene-expression data. In *Proc. of the Natl. Acad. Sci. USA*, volume 99, pages 6562–6566, 2002.
12. J. K. Anlauf and M. Biehl. The Adatron- An adaptive perceptron algorithm. *Europhysics Letters*, 10(7):687–692, 1989.
13. B. Ans, J. Hérault, and C. Jutten. Adaptive neural architectures: detection of primitives. In *Proc. of COGNITIVA'85*, pages 593–597, Paris, France, 1985.

14. H. Barlow. Possible principles underlying the transformation of sensory messages. *Sensory Communication*, pages 214–234, 1961.
15. R. Barrett, M. Berry, T.F. Chan, and J. Demmel. *Templates for the Solution of Linear Systems: Building Blocks for Iterative Methods.* Society for Industrial and Applied Mathematics, Philadelphia, 1994.
16. P. L. Bartlett and A. Tewari. Sparseness vs estimating conditional probabilities: Some asymptotic results., 2004. submitted for a publication and taken from the P. L. Bartlett's site.
17. A. J. Bell and T. J. Sejnowski. An information-maximization approach to blind separation and blind deconvolution. *Neural Computation*, 7(6):1129–1159, 1995.
18. A. Belouchrami, K.A. Meraim, J.F. Cardoso, and E. Moulines. A blind source separation technique based on second order statistics. *Transactions on Signal Processing*, 45(2):434–444, 1997.
19. K. Bennett and A. Demiriz. Semi-supervised support vector machines. In *Advances in Neural Information Processing Systems*, volume 19. The MIT Press, 1998.
20. S. Berber and M. Temerinac. *Fundamentals of Algorithms and Structures for DSP.* FTN Izdavastvo, Novi Sad, 2004.
21. M. Black. Lecture notes of statistical analysis of gene expression microarray data, 2004.
22. D. R. Brillinger. *Time Series Data Analysis and Theory.* McGraw-Hill, 1981.
23. J. F. Cardoso. Infomax and maximum likelihood for blind source. *IEEE Signal Processing Letters*, 4:112–114, 1997.
24. J. F. Cardoso. On the stability of source separation algorithms. *Journal of VLSI Signal Processing Systems*, 26(1/2):7–14, 2000.
25. J. F. Cardoso and B. Laheld. Equivariant adaptive source separation. *IEEE Trans. Signal Processing*, 44(12):3017–3030, 1996.
26. J. F. Cardoso and A. Soulomniac. Blind beamforming for non-gaussian signals. In *Proc. IEE-Part F*, volume 140, pages 362–370, 1993.
27. C.C. Chang and C.J. Lin. LIBSVM: A library for support vector machines, 2002.
28. O. Chapelle and A. Zien. Homepage of low density separation, 2005.
29. O. Chapelle and A. Zien. Semi-supervised classification by low density separation. In *Proc. of the* 10th *International Workshop on Artificial Intelligence and Statistics, AI STATS 2005*, Barbados, 2005.
30. J. Chen and X. Z. Wang. A new approach to near-infrared spectal data analysis using independent component analysis. *J. Chem. Inf.*, 41:992–1001, 2001.
31. Y. Chen, G. Want, and S. Dong. Learning with progressive transductive support vector machines. *Pattern Recognition Letters*, 24:1845–1855, 2003.
32. V. Cherkassky and F. Mulier. *Learning From Data: Concepts, Theory and Methods.* John Wiley & Sons, Inc., New York, 1998.
33. S. Choi, A. Cichocki, and S. Amari. Flexible independent component analysis. *Journal of VLSI Signal Processing*, 20:25–38, 2000.
34. F. Chu and L. Wang. Gene expression data analysis using support vector machines. In C. Donald, editor, *Proc. of the 2003 IEEE International Joint Conference on Neural Networks*, pages 2268–2271, New York, 2003. IEEE Press.
35. A. Cichocki and S. Amari. *Adaptive Blind Signal and Image Processing-Learning Algorithms and Applications.* John Wiley, 2002.

36. A. Cichocki, R. Unbehaunen, and E. Rummert. Robust learning algorithm for blind separation of signals. *Electronic Letters*, 28(21):1986–1987, 1994.
37. M. Cohen and A. Andreou. Current-mode subthreshold MOS implementation of the Herault-Jutten autoadaptive network. *IEEE Journal of Solid-State Circuits*, 27(5):714–727, 1992.
38. P. Common. Independent component analysis- a new concept? *Signal Processing*, 36(3):287–314, 1994.
39. C. Cortes. *Prediction of Generalization Ability in Learning Machines.* PhD thesis, Department of Computer Science, University of Rochester, 1995.
40. C. Cortes and V. Vapnik. Support-vector networks. *Machine Learning*, 20(3):273–297, 1995.
41. T. M. Cover and J. A. Tomas. *Elements of Information Theory.* John Wiley, 1991.
42. Nello Cristianini and John Shawe-Taylor. *An introduction to support vector machines and other kernel-based learning methods.* Cambridge University Press, Cambridge, 2000.
43. D. Decoste and B. Schölkopf. Training invariant support vector machines. *Journal of Machine Learning*, 46:161–190, 2002.
44. Jian-Xion Dong, A. Krzyzak, and C. Y. Suen. A fast SVM training algorithm. In *Proc. of the International workshop on Pattern Recognition with Support Vector Machines*, 2002.
45. H. Drucker, C.J.C. Burges, L. Kaufman, A. Smola, and V. Vapnik. Support vector regression machines. In *Advances in Neural Information Processing Systems 9*, pages 155–161, Cambridge, MA, 1997. MIT Press.
46. Q. Du, I. Kopriva, and H. Szu. Independent component analysis for classifying multispectral images with dimensionality limitation. *International Journal of Information Acquisition*, 1(3):201–216, 2004.
47. Q. Du, I. Kopriva, and H. Szu. Independent component analysis for hyperspectral remote sensing imagery classification. In *Optical Engineering*, 2005.
48. C. Eisenhart. Roger Joseph Boscovich and the combination of observationes. In *Actes International Symposium on R. J. Boskovic*, pages 19–25, Belgrade - Zagreb - Ljubljana, YU, 1962.
49. T. Evgeniou, M. Pontil, and T. Poggio. Regularization networks and support vector machines. *Advances in Computational Mathematics*, 13:1–50, 2000.
50. Sebastiani Fabrizio. Machine learning in automated text categorization. *ACM Computing Surveys*, 34(1):1–47, 2002.
51. B. Fischer and J. M. Buhmann. Path-based clustering for grouping of smooth curves and texture segmentation. *IEEE transactions on pattern analysis and machine intelligence*, 25:513–518, April 2003.
52. B. Fischer, V. Roth, and J. M. Buhmann. Clustering with the connectivity kernel. In S. Thrun, L. Saul, and B. Schölkopf, editors, *Proc. of the Advances in Neural Information Processing Systems 2004*, volume 16. MIT Press, 2004.
53. T. Fries and R. F. Harrison. Linear programming support vectors machines for pattern classification and regression estimation and the set reduction algorithm. Technical report, University of Sheffield,, Sheffield, UK, 1998.
54. T.-T. Friess, N. Cristianini, and I.C.G. Campbell. The kernel adatron: A fast and simple learning procedure for support vector machines. In J. Shavlik, editor, *Proc. of the 15th International Conference of Machine Learning*, pages 188–196, San Francisco, 1998. Morgan Kaufmann.

55. B. Gabrys and L. Petrakieva. Combining labelled and unlabelled data in the design of pattern classification systems. *International Journal of Approximate Reasoning*, 35(3):251–273, 2004.
56. M. Gaeta and J.-L. Lacoume. Source separation without prior knowledge: the maximum likehood solution. In *Proc of EUSIPCO*, pages 621–624, 1990.
57. M. Girolami and C. Fyfe. Generalised independent component analysis through unsupervised learning with emergent busgang properties. In *Proc. of ICNN*, pages 1788–1891, 1997.
58. F. Girosi. An equivalence between sparse approximation and support vector machines. Technical report, AI Memo 1606, MIT, 1997.
59. T. Graepel, R. Herbrich, B. Schölkopf, A. Smola, P. Bartlett, K.-R. Müller, K. Obermayer, and R. Williamson. Classification on proximity data with lp-machines. In *Proc. of the 9th Intl. Conf. on Artificial NN, ICANN 99*, Edinburgh, Sept 1999.
60. S. I. Grossman. *Elementary Linear Algebra*. Wadsworth, 1984.
61. I. Guyon, J. Weston, S. Barnhill, and V. Vapnik. Gene selection for cancer classification using support vector machines. *Machine Learning*, 46:389–422, 2002.
62. I. Hadzic and V. Kecman. Learning from data by linear programming. In *NZ Postgraduate Conference Proceedings*, Auckland, Dec. 1999.
63. T. Hastie, R. Tibshirani, B. Narasimhan, and G. Chu. Pamr: Prediction analysis for microarrays in R. Website, 2004.
64. Hérault, J. C. Jutten, and B. Ans. Détection de grandeurs primitives dans un message composite par une architechure de calcul neuromimétique en aprentissage non supervis. In *Actes du Xéme colloque GRETSI*, pages 1017–1022, Nice, France, 1985.
65. J. Herault and C. Jutten. Space or time adaptive signal processing by neural network models. In *Neural networks for computing: AIP conference proceedings 151*, volume 151, New York, 1986. American Institute for physics.
66. M. R. Hestenes. *Conjugate Direction Method in Optimization*. Springer-Verlag New York Inc., New York, 1 edition, 1980.
67. G. Hinton and T. J. Sejnowski. *Unsupervised Learning*. The MIT Press, 1999.
68. C.W. Hsu and C.J. Lin. A simple decomposition method for support vector machines. *Machine learning*, (46):291–314, 2002.
69. T.-M. Huang and V. Kecman. Bias b in SVMs again. In *Proc. of the* 12th *European Symposium on Artificial Neural Networks, ESANN 2004*, pages 441–448, Bruges, Belgium, 2004.
70. T.-M. Huang and V. Kecman. Semi-supervised learning from unbalanced labelled data: An improvement. In M. G. Negoita, R.J. Howlett, and L. C. Jain, editors, *Proc. of the Knowledge-Based Intelligent Information and Engineering Systems, KES 2004*, volume 3 of *Lecture Notes in Artificial Intelligence*, pages 802–808, Wellington, New Zealand, 2004. Springer.
71. T.-M. Huang and V. Kecman. Gene extraction for cancer diagnosis by support vector machines. In Duch W., J. Kacprzyk, and E. Oja, editors, *Lecture Notes in Computer Science*, volume 3696, pages 617–624. Springer -Verlag, 2005.
72. T.-M. Huang and V. Kecman. Gene extraction for cancer diagnosis using support vector machines. *International Journal of Artificial Intelligent in Medicine -Special Issue on Computational Intelligence Techniques in Bioinformatics*, 35:185–194, 2005.

73. T.-M. Huang and V Kecman. Performance comparisons of semi-supervised learning algorithms. In *Proceedings of the Workshop on Learning with Partially Classified Training Data, at the* 22nd *International Conference on Machine Learning, ICML 2005*, pages 45–49, Bonn, Germany, 2005.
74. T.-M. Huang and V. Kecman. Semi-supervised learning from unbalanced labeled data: An improvement. *International Journal of Knowledge-Based and Intelligent Engineering Systems*, 2005.
75. T.-M. Huang, V. Kecman, and C. K. Park. SemiL: An efficient software for solving large-scale semi-supervised learning problem using graph based approaches, 2004.
76. A. Hyvärinen, J. Karhunen, and E. Oja. *Independent Component Analysis.* John Wiley, 2001.
77. A. Hyvärinen and E. Oja. A fast fixed-point algorithm for independent component analysis. *Neural Computation*, 9(7):1483–1492, 1997.
78. T. Joachims. Making large-scale SVM learning practical. In B. Schölkopf, C. Burges, and A. Smola, editors, *Advances in Kernel Methods- Suppot Vector Learning.* MIT-Press, 1999.
79. T. Joachims. Transductive inference for text classification using support vector machines. In *ICML*, pages 200–209, 1999.
80. C. Jutten and J. Herault. Blind separation of sources, Part I: an adaptive algorithm based on neuromimetic architecture. *IEEE Trans. on Signal Processing*, 24(1):1–10, 1991.
81. V. Kecman. *Learning and soft computing : support vector machines, neural networks, and fuzzy logic models.* Complex Adaptive Systems. The MIT Press, Cambridge, Mass., 2001.
82. V. Kecman, T. Arthanari, and I. Hadzic. LP and QP based learning from empirical data. In *IEEE Proceedings of IJCNN 2001*, volume 4, pages 2451–2455, Washington, DC., 2001.
83. V. Kecman and I. Hadzic. Support vectors selection by linear programming. In *Proceedings of the International Joint Conference on Neural Networks (IJCNN 2000)*, volume 5, pages 193–198, 2000.
84. V. Kecman, T.-M. Huang, and M. Vogt. *Iterative Single Data Algorithm for Training Kernel Machines From Huge Data Sets: Theory and Performance*, volume 177 of *Studies in Fuzziness and Soft Computing*, pages 255–274. Springer Verlag, 2005.
85. V. Kecman, M. Vogt, and T.M. Huang. On the equality of kernel adatron and sequential minimal optimization in classification and regression tasks and alike algorithms for kernel machines. In *Proc. of the* 11th *European Symposium on Artificial Neural Networks*, pages 215–222, Bruges, Belgium, 2003.
86. I. Kopriva, Q. Du, H. Szu, and W. Wasylkiwskyj. Independent component analysis approach to image sharpening in the presence of atmospheric turbulence. *Coptics Communications*, 233(1-3):7–14, 2004.
87. B. Krishnapuram. *Adaptive Classifier Design Using Labeled and Unlabeled Data.* Phd, Duke University, 2004.
88. R. H. Lambert and C. L. Nikias. *Blind deconvolution of multipath mixtures*, volume 1, chapter 9. John Wiley, 2000.
89. K. Lang. 20 newsgroup data set, 1995.
90. C. I. Lawson and R. J. Hanson. Solving least squares problems. Prentice-Hall, Englewood Cliffs, 1974.

91. T-W. Lee. *Independent Component Analysis- Theory and Applications.* 1998.
92. Y. Li, A. Cichocki, and S. Amari. Analysis of sparse representation and blind source separation. *Neural Computation*, 16(6):1193–1234, 2004.
93. S. Makeig, A. J. Bell, T. Jung, and T. J. Sejnowski. Independent component analysis of electroencephalographic data. In *Advances in Neural Information Processing Systems 8*, pages 145–151, 1996.
94. O. L. Mangasarian. Linear and nonlinear separation of patterns by linear programming. *Operations Research*, 13:444–452, 1965.
95. O.L. Mangasarian and D.R Musicant. Successive overrelaxation for support vector machines. *IEEE, Trans. Neural Networks*, 11(4):1003–1008, 1999.
96. A. K. McCallum. Bow: A toolkit for statistical language modeling, next retrieval, classification and clustering, 1996.
97. P. McCullagh. *Tensor Methods in Statistics.* Chapman and Hall, 1987.
98. M. Mckeown, S. Makeig, G. Brown, T. P. Jung, S. Kinderman, T.-W. Lee, and T. J. Sejnowski. Spatially independent activity patterns in functional magnetic resonance imaging data during the stroop color-naming task. In *Proceedings of the National Academy of Sciences*, volume 95, pages 803–810, 1998.
99. G. J. McLachlan, K.-A. Do, and C. Ambrois. *Analyzing Microarray Gene Expression Data.* Wiley-Interscience, 2004.
100. J. Mendel. Tutorial on higher-order statistics (spectra) in signal processing and system theory: Theoretical results and some applications. In *Proc. IEEE*, volume 79, pages 278–305, 1991.
101. J. Mercer. Functions of positive and negative type and their connection with the theory of integral equations. *Philos. Trans. Roy. Soc.*, 209(415), 1909.
102. L. Molgedey and H. G. Schuster. Separation of mixture of independent signals using time delayed correlations. *Physical Review Letters*, 72:3634–3636, 1994.
103. S. A. Nene, S. K. Nayar, and H. Murase. Columbia object image library (coil-20). Technical Report CUCS-005-96, Columbia Unversity, 1996.
104. D. Nuzillard and A. Bijaoui. Blind source separation and analysis of multispectral astronomical images. *Astronomy and Astrophysics Suppl. Ser.*, 147:129–138, 2000.
105. D. Nuzillard, S. Bourg, and J. Nuzillard. Model-free analysis of mixtures by NMR using blind source separation. *Journal of Magn. Reson.*, 133:358–363, 1998.
106. E. Oja. Blind source separation: neural net principles and applications. In *Proc. of SPIE*, volume 5439, pages 1–14, Orlando, FL, April 2004.
107. A. M. Ostrowski. *Solutions of Equations and Systems of Equations.* Academic Press, New York, 1966.
108. E. Osuna, R. Freund, and F. Girosi. An improved training algorithm for support vector machines., 1997.
109. E. Osuna, R. Freund, and F. Girosi. Support vector machines: Training and applications. Ai memo 1602, Massachusetts Institute of Technology, Cambridge, MA, 1997.
110. A. Papoulis. *Probability, Random Variables, and Stochastic Processes.* McGraw-Hill, 3 edition, 1991.
111. C. K. Park. Various models of semi-supervised learning, 2004. Personal Communication.

112. B. Pearlmutter and L. Parra. A context-sensitive generalization of ica. In *Advances in Neural Information processing Systems*, volume 9, pages 613–619, 1996.
113. D. T. Pham. Fast algorithm for estimating mutual information, entropies and score functions. In S. Amari, A. Cichocki, S. Makino, and N. Murata, editors, *Proc. of the Fourth International Conference on Independent Component Analysis and Blind Signal Separation (ICA'2003)*, pages 17–22, Nara, Japan, 2003.
114. D.T. Pham. Blind separation of mixtures of independent sources through a guasimaximum likelihood approach. *IEEE Trans. on Signal Processing*, 45(7):1712–1725, 1997.
115. J.C. Platt. Fast training of support vector machines using sequential minimal optimization. In B. Schölkopf, C. Burges, and A. Smola, editors, *Advances In Kernel Methods- Support Vector Learning*. The MIT Press, Cambridge, 1999.
116. T. Poggio, S. Mukherjee, R. Rifkin, A. Rakhlin, and A. Verri. b. Technical report, Massachusetts Institute of Technology, 2001.
117. M.F. Porter. An algorithm for suffix stripping. *Program*, 3:130–137, 1980.
118. J. C. Principe, D. Xu, and J. W. Fisher III. *Information-Theoretic Learning*, volume 1, chapter 7. John-Wiley, 2000.
119. A. Rakotomamonjy. Variable selection using SVM-based criteria. *Journal of Machine Learning*, (3):1357–1370, 2003.
120. J. Rissanen. Modeling by shortest data description. *Automatica*, 14:465–471, 1978.
121. T. Ristaniemi and J. Joutensalo. Advanced ICA-based recivers for block fading DS-CDMA channels. *Signal Processing*, 85:417–431, 2002.
122. J. R. Schewchuk. An introduction to the conjugate gradient method without the agonizing pain, 1994.
123. Bernhard Schölkopf and Alexander J. Smola. *Learning with kernels : Support vector machines, regularization, optimization, and beyond.* Adaptive computation and machine learning. The MIT Press, Cambridge, Mass., 2002.
124. S. Schwartz, M. Zibulevsky, and Y. Y. Schechner. ICA using kernel entropy estimation with NlogN complexity. In *Lecture Notes in Computer Science*, volume 3195, pages 422–429, 2004.
125. J. J. Settle and N. A. Drake. Linear mixing and estimation of ground cover proportions. *Int. J. Remote Sensing*, 14:1159–1177, 1993.
126. K. S. Shamugan and A. M. Breiphol. *Random Signals- Detection, Estimation and Data Analysis*. John-Wiley, 1988.
127. R. B. Singer and T. B. McCord. Mars: large scale mixing of bright and dark surface materials and implications for analysis of spectral reflectance. In *Proc.10th Lunar Planet. Sci. Conf.*, pages 1835–1848, 1979.
128. A. Smola, T.T. Friess, and B. Schölkopf. Semiparametric support vector and linear programming machines. In *Advances in Neural Information Processing Systems 11*, 1998.
129. A. Smola and B. Schölkopf. On a kernel-based method for pattern recognition, regression, approximation and operator inversion. Technical report, GMD Technical Report, Berlin, 1997.
130. A. J. Smola and R. Kondor. Kernels and regularization on graphs. In *COLT/Kernel Workshop*, 2003.

131. I. Steinwart. Sparseness of support vector machines. *Journal of Machine Learning Research*, 4:1071–1105, 2003.
132. J.V. Stone. *Independent Component Analysis-A Tutorial Introduction.* The MIT Press, 2004.
133. Y. Su, T.M. Murali, V. Pavlovic, M. Schaffer, and S. Kasif. Rankgene: A program to rank genes from expression data, 2002.
134. Johan A. K. Suykens. *Least squares support vector machines.* World Scientific, River Edge, NJ, 2002.
135. M. Szummer and T. Jaakkola. Partially labelled classification with markov random walks. In *Proc. of the Advance in Neural Information Processing Systems*, volume 14, 2001.
136. A. Taleb and C. Jutten. Source separation in post-nonlinear mixtures. *IEEE Transactions on Signal Processing*, 47(10):2807–2820, 1999.
137. R. Tibshirani, T. Hastie, B. Narasimhan, and G. Chu. Diagnosis of multiple cancer types by shrunken centroids of gene expression. In *Proc. of the National Academy of Sciences of the United States of America*, volume 99, pages 6567–6572, USA, 2002.
138. R. Tibshirani, T. Hastie, B. Narasimhan, and G. Chu. Class prediction by nearest shrunken centroids, with applications to DNA microarrays. *Statistical Science*, 18(1):104–117, 2003.
139. L. Tong, R.W. Liu, V.C. Soon, and Y. F. Huang. Indeterminacy and identifiability of blind identification. *IEEE Trans. on Circuits and Systems*, 38:499–509, 1991.
140. K. Torkkola. *Blind Separation of Delayed and convolved sources*, chapter 8. John Wiley, 2000.
141. V. Vapnik. *Estimation of Dependences Based on Empirical Data [in Russian].* Nauka, Moscow., 1979. English translation: 1982, Springer Verlag, New York.
142. V. Vapnik, S. Golowich, and A. Smola. Support vector method for function approximation, regression estimation, and signal processing. In *In Advances in Neural Information Processing Systems 9*, Cambridge, MA, 1997. MIT Press.
143. V. N. Vapnik. *Statistical Learning Theory.* J.Wiley & Sons, Inc., New York, NY, 1998.
144. V.N. Vapnik. *The Nature of Statistical Learning Theory.* Springer Verlag Inc, New York, 1995.
145. V.N. Vapnik and A.Y. Chervonenkis. On the uniform convergence of relative frequencies of events to their probabilities. *Doklady Akademii Nauk USSR*, 181, 1968.
146. V.N. Vapnik and A.Y. Chervonenkis. The necessary and sufficient condititons for the consistency of the method of empirical minimization [in Russian]. *Yearbook of the Academy of Sciences of the USSR on Recognition,Classification, and Forecasting*, 2:217–249, 1989.
147. K. Veropoulos. *Machine Learning Approaches to Medical Decision Making.* PhD thesis, The University of Bristol, 2001.
148. M. Vogt. SMO algorithms for support vector machines without bias. Technical report, Institute of Automatic Control Systems and Process Automation, Technische Universitat Darmstadt, 2002.
149. M. Vogt and V. Kecman. An active-set algorithm for support vector machines in nonlinear system identification. In *Proc. of the* 6th *IFAC Symposium on*

Nonlinear Control Systems (NOLCOS 2004),, pages 495–500, Stuttgart, Germany, 2004.

150. M. Vogt and V. Kecman. *Active-Set Method for Support Vector Machines*, volume 177 of *Studies in Fuzziness and Soft Computing*, pages 133–178. Springer-Verlag, 2005.
151. Wikipedia. Machine learning — wikipedia, the free encyclopedia, 2005. [Online; accessed 4-June-2005].
152. M. E. Winter. N-FINDER: an algorithm for fast autonomous spectral endmember determination in hyperpsectral data. In *Proceeding of SPIE*, volume 3753, pages 266–275, 1999.
153. Matt Wright. SVM application list, 1998.
154. L. Zhang, A. Cichocki, and S. Amari. Self-adaptive blind source separation based on activation function adaptation. *IEEE Trans. on Neural Networks*, 15:233–244, 2004.
155. D. Zhou, O. Bousquet, T. N. Lal, J. Weston, and B. Schölkopf. Learning with local and global consistency. In S. Thrun, L. Saul, and B. Schölkopf, editors, *Advances in Neural Information Processing Systems*, volume 16, pages 321–328, Cambridge, Mass., 2004. MIT Press.
156. D. Zhou and B. Schölkopf. Learning from labeled and unlabeled data using random walks. In *DAMG'04: Proc. of the 26th Pattern Recognition Symposium*, 2004.
157. D. Zhou and B. Schölkopf. A regularization framework for learning from graph data. In *Proc. of the Workshop on Statistical Relational Learning at International Conference on Machine Learning*, Banff, Canada, 2004.
158. D. Zhou, J. Weston, and A. Gretton. Ranking on data manifolds. In S. Thrun, L. Saul, and B. Schölkopf, editors, *Advances in Neural Information Processing Systems*, volume 16, pages 169–176, Cambridge, 2004. MIT press.
159. X. Zhu. *Semi-Supervised Learning with Graphs*. PhD thesis, Carnegie Mellon University, 2005.
160. X. Zhu, Z. Ghahramani, and J. Lafferty. Semi-supervised learning using Gaussian fields and harmonic functions. In *Proc. of the* 20th *International Conference on Machine Learning (ICML-2003)*, Washington DC, 2003.
161. A. Ziehe, K.R. Müller, G. Nolte, B.M. Mackert, and G. Curio. TDSEP- an efficient algorithm for blind separation using time structure. In *Proc. of International Conference on Artifical Neural Network (ICANN'98)*, volume 15, pages 675–680, Skovde, Sweden, 1998.

Index